East Africa's Human Environment Interactions

Rob Marchant

East Africa's Human Environment Interactions

Historical Perspectives for a Sustainable Future

palgrave
macmillan

Rob Marchant
Department of Environment
and Geography
University of York
York, UK

ISBN 978-3-030-88986-9 ISBN 978-3-030-88987-6 (eBook)
https://doi.org/10.1007/978-3-030-88987-6

Cover image: Rob Marchant

This Palgrave Macmillan imprint is published by the registered company Springer Nature Switzerland AG
The registered company address is: Gewerbestrasse 11, 6330 Cham, Switzerland

Acknowledgements

My own approach to working across the East African region and being able to connect to its' peoples, places, and history has been profoundly influenced by the voices I have encountered over the past thirty years working in Uganda, Kenya, and Tanzania. Numerous people have offered friendship, support, and comment on the work that has underpinned this book at various stages; without these contributions, this work would not have been possible and indeed I am greatly indebted to the people I have been fortunate to meet along the way! The open generosity, particularly from those with the 'least', has been quite humbling and I thank the countless people that have helped roadside breakdowns, provided plenty of tea and food, and provided great conversation, insight, and depth over the years. Thanks must go to the people who have helped me during some fantastic experiences—first and foremost is Stephen Rucina Mathai from the National Museums of Kenya—shiakamoo Mzee for so many chapters! Individual researchers who I have worked with are far too many numerous to list here, and of course, there is a fear of missing people out—you know who you are, and I am indebted to working with you! Particularly I thank Dave Taylor for starting me on this road and Henry Hooghiesmtra for continued friendship and mentorship. Friends, colleagues, and family have shared these adventures; in particular Antje Ahrends, Oli Boles, Mike Bunguard, Claudia Capitani, Colin Courtney Mustaphi, Aida Cuni Sanchez, Nicholas Dere, Jemma Finch, Esther Githumbi, Robert Halcrow, Nwabueze, Igu, Tabatha Kabora, Rebecca Kariuki, Rahab Kinyanjui, Paul Lane, Carol Lang, Edem Mahu, Esther Makinde, Amos Majule, Andrew Marshall, Colin McClean, Cassian Mumbi, Veronica Muiruri, Rebecca Newmann, Tobias Nyumba, Dickens Odeny, Anthony Onyekuru, Dan Olago, Peter Omeny, Olivia Norfolk, Paramita Punwong, Marion Pfeifer, Phil Platts, Suzi Richer, Hamidu Seki, Evans, Sitati, Daryl Stump, Jessica Thorn, Lucy Waruingi, Jonah Western, Jeff Worden, and Pius Yanda. Rebecca Kariuki deserves special mention for helping produce some of the figures presented in the book. I was helped greatly in edits and research for this book by Chloë, Caroline, Laura and indeed none of what I am would have been possible without you.

Institutions such as The National Museums of Kenya, the British Institute in Eastern Africa, the Africa Conservation Centre, World Conservation Society, Institute for Tropical Forest Conservation, WWF-Tanzania and WWF-Kenya, Universities of Nairobi, Nelson Mandela Institute of Science and Technology, Makerere, Dar es Salaam, and Sokinie, where some of these ideas started to be developed.

Numerous funders have supported the work over the years particularly the European Commission, Past Global Changes (PAGES), Vetenskapsrådet, Formas, SIDA, The Royal Society, the Natural and Environmental Research Council, Economic and Social Research Council, the Arts and Humanities Research Council, the Swiss National Science Foundation, The Mountain Research Initiative, WorldWild Fund, World Conservation Society, Liz Claiborne Foundation, and the Percy Sladen Memorial Trust.

Contents

1 Foundations: The Environment, Ecosystems, and Cultures of East Africa 1

1.1 Introduction 1

1.2 East African Climate: Foundation for Diversity of Life and Livelihoods 3

1.3 East African Geology, Topography, and Drainage 9

1.4 Soils and Edaphic Factors Influencing Ecosystem Composition in East Africa 12

1.5 East African Ecosystem Composition and Distribution 15

1.5.1 The Coastal and Lowland Vegetation 17

1.5.2 Highland Vegetation 20

1.6 Disturbance Factors Influencing Ecosystem Composition 23

1.7 Cultural and Land Use Diversity 24

References 27

2 Origins and Migration: Environmental and Cultural Change Over the Last 300,000 Years in East Africa 39
2.1 Introduction 39
2.2 Detection of Environmental Change Across East Africa 41
2.3 Detection of Cultural Changes in East African Geoarchives 44
2.4 Chronological and Methodological Considerations 51
2.5 Environmental and Human and Cultural History 52
2.5.1 Foundations of Modern Humans 53
2.5.2 Last Glacial Period: Forest Refugia and Imprints on Today's Landscape 56
2.5.3 The Holocene and the Shaping of Human Impacted Ecosystems 60
2.6 Environmental-Human Interconnections: Linking Environmental and Cultural Change 80
References 86

3 Trading Language, New Crops, New Relationships: Digging Anthropocene Foundations 115
3.1 The Last 1000 Years: Continued Environmental Variability 115
3.2 Developing and Managing Mixed Agricultural Complexes 121
3.3 Managing Water and the Growth of Industrial Agriculture 130
3.4 The Development of Modern Pastoralist Societies 134
3.5 Globalised and Commodified World: The Rise of the Swahili Coast 138
References 148

4 Elephants, Maize, and Pervasive Societal Environmental Transformations 163
4.1 Introduction 163
4.2 The Arterial Caravan Routes—Founding the Transport Network 164

4.3 Impact of Ivory Trade on African Elephant Populations 176
4.4 The Slave Trade and the Caravan Routes 179
4.5 The Arrival of Maize, Potatoes, and Tobacco 184
4.6 Social, Land, and Legacies of Ivory and Slave Trade 188
4.7 Ecological Impacts and Legacies of Caravan Trade 191
4.8 Social Impacts Late Nineteenth Century Drought 195
4.9 Shifting World Views; Establishing Rather than Hunting Controls and National Park Foundations 197
References 199

5 Colonial Transitions 205
5.1 Colonial Foundations—An Atypical Moment in Time 205
5.2 The Colonial Explorers and Missionaries 207
5.3 The Colonial Partitioning of East Africa 209
5.4 Land Use Transformations 218
5.4.1 Rangelands 220
5.4.2 Agricultural Lands 223
5.4.3 Forest Transformations 225
5.5 Protected Area Foundations 235
References 238

6 Postcolonial Transitions and Recent Political History 245
6.1 Postcolonial Pathways: Early Transitions 245
6.1.1 Kenya 246
6.2 Colonial Legacies and New Forms of Land Management 255
6.2.1 Forest Management 258
6.2.2 Agricultural Transformations 268
6.3 Postcolonial Approach to Wildlife and Protected Areas 277
6.3.1 Marine Protected Areas 286
6.4 The Rise and Fall of Fortress Conservation Through the Poaching Crises 288

6.5 Development of New Partnerships: The Belt and Road 291
References 297

7 **Using the Past to Chart Future Pathways?** 311
7.1 Introduction 311
7.2 Developing Data and Methodologies to Better Understand Human Environmental Interactions 313
7.2.1 Data Gaps and How to Fill Them 313
7.2.2 Harmonising Datasets and Interlinking Databases and Disciplines 317
7.3 Linking the Past to the Future Through Modelling Frameworks 321
7.3.1 Climate Models 322
7.3.2 Ecosystem Models 326
7.3.3 Land Use Scenarios 328
7.3.4 Remotely Sensed Land Use Transformations 332
7.4 Future Challenges for East Africa 333
7.4.1 Agricultural Expansion and Consequences of Land Conversion 334
7.4.2 Climate Change 340
7.4.3 Population Growth in East Africa and Expansion of Urban Centres 342
7.5 Past to the Present and Towards the Future: Underpinning Sustainability Science with the Long Term 345
7.6 Commodification of Nature and the Rise of Ecosystem Services: Nature-Based Solutions to Challenges 347
7.6.1 Carbon: Reforest Africa—The New Green Revolution? 349
7.6.2 Water: Adapting to the Climate Variability Challenge 352
7.6.3 Soil: The Foundation for Agricultural Production 354

7.7 Current Conservation Challenge: Future-Proofing Protected Areas 356
7.8 Next Steps 365
References 366

References 385

Index 389

List of Figures

Fig. 1.1	East Africa comprises the countries of Kenya, Tanzania, and Uganda (dark green in insert). There are a variety of other countries in different definitions of East or Eastern Africa (light green in insert) that are not included in this volume	2
Fig. 1.2	Mean annual distribution of rainfall and mean annual temperate across East Africa. The highly variable climate regime and spatial variation is also characterised by high intra and inter annual variability, particularly on the timing and duration of rainfall based on WorldClim (Hijmans et al., 2005) as derived by AfriClim 3.0 (Platts et al., 2015)	3
Fig. 1.3	Schematic documenting the main drivers of the rainy season in East Africa. Due to the complex interaction between regional and global circulation systems, local geographic factors, coastal and land influences along with remote forcing such as the Indian Ocean Dipole and ENSO there is massive variability in intra and inter annual timing and duration of rainfall	4

Fig. 1.4 Highland areas across East Africa are also known as 'Water Towers' as they collect moisture from rainfall event, clouds, and non-precipitating mists. Rainfall and collected water is then stored in the soil systems and vegetation that regulate the flow of water into rivers. Mountain Water Towers provide critical connections from the mesic highlands to the more arid lowlands. Examples here include the Aberdares (**a**,**d**,**e**), Kilimanjaro views from the Amboseli basin (**e**), Mount Kenya (**b**), and the Udzungwa Mountains (**c**) 6

Fig. 1.5 Eastern Africa and study site locations and geographic features discussed in text (**a**). Lakes and the varied topography dominate the region that is topographically very diverse (**b**) elevation derived from the SRTM-250 m Digital Elevation data (http://srtm.csi.cgiar.org) 10

Fig. 1.6 Soils vary from the highly productive supporting industrial farming in Loitoktok (**a**) to highly degraded soils at the foothills of the Pare Mountains (**b**) The diversity and distribution of soil types in East Africa derived from the Soil Atlas of Africa, World Reference Base classification (**c**). All photographs: Rob Marchan 13

Fig. 1.7 Lowland ecosystems of East Africa (**a**). Coastal mangrove forest on intertidal mudflats, Unguja, Zanzibar (**b**). Mangrove grades into Coral rag vegetation—Chumbe Island (**c**). Extensive coconut palm groves cover much of the coastal belt (**d**). Coastal Forest, previously much more extensive, can still be found in protected areas such as the Shimba Hills, Kenya (**e**). Desert extend across much of northern Kenya and Uganda such as the Chalbi Desert in northern Kenya (**f**). Continuous savannah with intermittent trees or different composition, dominated by different species of *Acacia* extend through much of East Africa (**f**, **g**, **i**, **k**). Edaphic influences can locally be very important in controlling the presence of local woody species as demonstrated by these views of gallery forest Engaruka (**h**) and Tsavo (**l**). Savannah pastoral landscape, close to Ngorongoro (**j**) (all photographs: Rob Marchant) 15

Fig. 1.8 Mountainous biomes of East Africa (**a**) view from Engaruka towards surrounding highlands (**b**). High alpine vegetation on the rim of Ngorongoro (**c**). Montane vegetation within the crater of Mount Lognonot, Kenya (**d**). Extensive Montane forest in the Aberdares National Park (**e**). *Olea* spp. on the rim of Mt Marsabit, northern Kenya (**f**). Most of the montane ecosystems across East Africa have been heavily impacted in the past and are in a regenerating state, often with extensive stands of montane bamboo *Arundinaria alpina* such as on Mount Elgon (**g**). Montane forest species competing for light on Mt Elgon (**h**). Many of the montane forests are covered in *Usnea* that add to their efficiency to strip moisture out of non-precipitating clouds—Uluguru Mountains. At higher altitudes (**k**) or where the topography leads to locally dry conditions (**i**) continuous alpine grasslands form. At higher altitudes, such as on Mount Kilimanjaro *Scenecio* spp. can be locally dominant (**k**). Montane forests can be dominated by large species such as this stand of *Podocarpus* on Mt Elgon (**j**). Many of the mountains across East Africa are extinct volcanoes and often have wetlands in their caldera such as Lake Paradise on Mt Marsabit (**l**). All photographs: Rob Marchant 16

Fig. 2.1 Sources of palaeoenvironmental information. (**a**) Glacial ice archive on Kilimanjaro (photo: Rob Marchant). (**b**) Maua mire, at nearly 4000 m asl south-eastern Kilimanjaro (photo: Rob Marchant). (**c**) At over 2100 m asl, Rumuiku Swamp sits within a volcanic crater on eastern Mount Kenya (photo: Rob Marchant). (**d**) Palustrine swamp, a source of organic-rich sediments in the relatively arid areas of Amboseli (photo: Rob Marchant). (**e**) Cyperaceae-dominated wetland northern Kenya (photo: Rob Marchant). (**f**) Mangrove sediments, Zanzibar (photo: Rob Marchant). (**g**) Fluvially incised exposed soil facies northern Tanzania (photo: Rob Marchant). (**h**) Cut timbers at Mpingo showing tropical

tree growth rings (photo: Neil Burgess). (**i**) A stalagmite collected from Cold Air Cave, South Africa can provide a palaeoclimatic record (photo: Karin Holmgren) 42
Fig. 2.2 Sources of palaeoenvironmental geoarchives can be quite varied. (**a**) Mubwindi Swamp, Bwindi Forest, Uganda. (**b**) Lake Momella, Arusha National Park. (**c**) Lake Victoria, views from Speke Bay Tanzania. (**d**) Mangrove sediments, Zanzibar. (**e**) Amboseli Swamp. (**f**) Lake Amboseli during a period of flooding. (**g**) Lake Naivasha during a relatively high stand (as demonstrated by the drowned forest). (**h**) A salt lake Serengeti National Park). (**i**) Paradise Lake, an extinct volcanic crater in Mount Marsabit National Park (**j**) Sediments can accumulate within small depressions in Kopi—Serengeti National Park. (**k**) Deva Deva Swamp, Uluguru Mountains, Tanzania. (**l**) Lake Magadi, Ngorongoro crater, Tanzania (All photographs: Rob Marchant) 43
Fig. 2.3 Location of key Palaeoecological and Archaeological sites across East Africa (**a**). There is a clear concentration of Swahili sites along the coast that have yielded many trade goods (**b**) such as Gedis (**c**) and sites on Zanzibar (**d**). These are increasing complemented by a growing number of palaeoecological sites such as Lake Paradise, Marsabit (**e**), Deva Deva Swamp, Uluguru, (**f**) and Maua Swamp, Kilimanjaro (**g**) (All photographs: Rob Marchant) 47
Fig. 2.4 Location of past forest refugia across Africa depicting the potential extent of the core (dark green) and peripheral area (light green). Present-day forests that have been located within refugia over geological timescales are ancient forests. Forest refugia occupy areas of relatively stable past climatic regimes that could provide the best opportunity to conserve maximum biodiversity during a period of future uncertain climatic change. Although of crucial importance to future biodiversity conservation the nature of predicted future climate change is very different to the climate variability experienced in the past 59

Fig. 2.5 Examples of archaeological sites. (**a**) Kuumbi cave on Zanzibar (photo: Ceri Shipton). (**b**) Rock art from Kondoa rock art World Heritage Site, Central Tanzania. Rock art associated with Holocene hunter-gatherers rituals of the Sandawe and the Hadzabe that date to 5000–20,000 years ago (photo: Emmanuel Bwasiri). (**c**) Early Holocene shell beds at Lothagam in northern Kenya (photo: Larry Robbins). (**d**) The Great Mosque of Kilwa Kisiwani, along the Swahili coast of Tanzania (photo: Stephanie Wynne-Jones). (**e**) The 'Bwogero' earthworks at Ntuusi, Uganda (photo: Andrew Reid). (**f**) Sirikwa holes, Kenya, dating from 850 to 550 yr BP (photo: John Sutton). (**g**) The abandoned irrigated agricultural site of Engaruka, Tanzania, occupied from c. 600 to 150 BP with former habitation platforms in foreground and former field system in the plain beyond (photo: Daryl Stump). (**h**) Traditional irrigation reservoir (Ndiva), North Pare Mountains, Tanzania (photo: Daryl Stump). (**i**) Contemporary boma (pastoral enclosure) site in northern Tanzania (photo: Suzi Richer) 61
Fig. 2.6 Examples of archaeological evidence. (**a**) Kanysore Ware, reconstruction by Ceri Ashley (photo: Paul Lane). (**b**) Barbed bone point (harpoon) in situ in early–mid Holocene shell beds at Lothagam, Kenya (photo: Larry Robbins). (**c**) A worked bone implement from the site of Luxmanda, Tanzania, interpreted by Langley et al. (2019) as a matting needle, since it exhibits signs of wear associated with working hides or plant material (photo: Mary Prendergast). (**d**) Urewe Ware from Lolui Island, Uganda (photo: Oliver Boles). (**e**) Excavated iron smelting furnace, Mwanga, Tanzania, this example dated to 650 to 510 cal BP, scales in 10 cm increments (photo: Daryl Stump, 2010). (**f**) Wound glass beads from 14–fifteenth centuries CE, from excavations at Songo Mnara

(photo: Steph Wynne-Jones). (**g**) Hair Cell Phytolith from Songo Mnara, Tanzania in association with a wattle and daub structure, scale is 20 microns (photo: Hayley McParland). (**h**) Micrograph of soils from within an historically irrigated field at Engaruka, Tanzania (photo: Carol Lang). (**i**) Non-charred (left) and charred (right) *Portulaca oleraceae* seeds. This species is a beneficial agricultural weed in East Africa with shallow roots that help retain soil moisture in dry environments (photo: Senna Thornton-Barnett) 65

Fig. 2.7 Location of key Palaeoecological and Archaeological sites across East Africa (left) used for the synthesis. At 6000 years BP there were relatively few palaeoecological and archaeological sites due to the antiquity, the relatively sparsely populated nature of the region and many sedimentary sites having a mid-Holocene hiatus in sediment accumulation. Those sites that do date to this period (right) record a largely subsistence economy based on the exploitation of seasonal resources and fish (Maps produced by Oli Boles and modified by the author) 66

Fig. 2.8 The first major transition in food production, away from subsistence livelihoods, was the expansion of pastoralism and the keeping of domestic livestock, particularly focused on cattle (**a**), sheep and goats (**b**). Although the timing and direction of this pastoral spread is continually being refined the spread appears to originate from the north into East Africa funnelled along the rift valley (All photographs: Rob Marchant) 69

Fig. 2.9 History of forest clearance across the Rukiga Highlands. Initial forest clearance appears to be part of the spread of agriculture associated with Bantu arrival into Uganda and dates from around 2200 years BP (**a**). The focus of this early forest clearance appears to be at the highest altitudes before these were subsequently abandoned and forest clearance spread rapidly across the landscape (**c**). Regenerating forest marks the location of previously cleared forest (**b**). The present day boundary of Bwindi

Impenetrable Forest National Park (**c**, **d**) is a strongly demarcated where the current day forested 'island' is surrounded by a 'sea' of agriculture (**d**) (Photograph: Rob Marchant) 85

Fig. 3.1 Selected late Holocene palaeoenvironmental records from eastern Africa (**a**). The graphs are described here from top to bottom of the page: Modeled reconstructions of historical population and cropland coverage estimates (KK10; Kaplan et al., 2011; Kaplan & Krumhardt, 2011). Lake level reconstructions for interpreting hydroclimatic variability from Lake Turkana (Garcin et al., 2012) and Lake Edward (Thompson et al., 2002). Kilimanjaro northern ice field (**b**) dust concentration reconstruction as a proxy for vegetation cover in the lowland dust-entrainment source areas (Thompson et al., 2002). Afromontane pollen sums (% total pollen) as a proxy of forest cover from Ahakagyezi Swamp, Rukiga Highlands, Uganda (Taylor, 1993). Lake Challa (**c**) hydroclimate reconstruction using BIT (Verschuren et al., 2009). Re-digitised arboreal pollen relative abundance (%) as a proxy for tree cover from Lake Masoko sediment record (Thevenon et al., 2003) (All photographs: Rob Marchant) 116

Fig. 3.2 A palaeoenvironmental record generated from a swamp on Mount Shengena, Pare Mountains northern Tanzania that records the past 1300 years of environmental and human history within the catchment. There are increasing signs of human modification of the Montane forest with recent pervasive change detected by an increase in fire activity from 1300 CE and again with the arrival of *Zea mays* at 1810 CE. The arrival of colonial forest officers who were interested East African camphor (*Ocotea usambarensis*) and the planting of fast-growing exotics such as *Eucalyptus,Pinus* and *Acacia* (Modified from Finch et al. [2017]) 120

Fig. 3.3 The extent of agricultural systems at 1000 CE across East Africa. It is likely a mix of crops, particular sorghum, millet, yams, cassava, beans and more recently rice (**c**) would have been cultivated. In fertile mountain areas such as the Rukiga Highlands (**a**) intensive intercropping was likely (Map produced by Oliver Boles and modified by the author) (All photographs: Rob Marchant) 122
Fig. 3.4 Spread of pastoralism at 1000 CE in East Africa. This was particularly focused on cattle and was augmented by the arrival of new pastoral groups and expansion through the 'Rangelands'. This land use allowed for the seasonal migration and transhumance to access grazing in highland areas (Map produced by Oliver Boles and modified by the author) (All photographs: Rob Marchant) 128
Fig. 3.5 Management of water has always been key to allowing populations to grow crops in a highly seasonal environment as seen at three large irrigation clusters. Elaborate terraces were established to allow water from adjacent highlands (**a**), that flowed down a series of incised valleys (**b**), to be channelled across a lowland plain via a complex series of walls and revetments (**c**). This photographic sequence are taken from Engaruka that was part of the Sonjo cluster where an elaborate and highly organised population farmed an area of some 20 km^2 until their rapid demise around 1500 CE, possibly due to rapid erosion of the headwater sediments (**a**) combine with a phase of climate change (All photographs: Rob Marchant) 131

Fig. 3.6 Spread of pastoralism at 1500 CE in East Africa. This was particularly focused on cattle with new pastoral groups, such as the Maasai arriving into the north of the region and expanding rapidly through the 'Rangelands'. This migration into East Africa was likely driven, in part, by droughts and transhumance would continue to be important to buffer seasonal variation in grazing resource via access to highland areas (Map produced by Oliver Boles and modified by the author) (All photographs: Rob Marchant) 136
Fig. 3.7 Swahili Towns along coast East Africa and islands communities—these would have formed trading nodes the for the annual trading missions along the coast that were associated with the monsoonal winds. Given the coastal nature there was a particular impact and extensive exploitation of mangrove timber for boat construction (**a**), buildings (**b**) and poles for export to the Middle East (**c**) 139
Fig. 3.8 The extent of agricultural systems at 1500 CE in East Africa. New crops started to arrive into East Africa, particularly from the Americas after 1608 CE when Maize (**a**) was first recorded on Pemba—this was quickly followed by potatoes, tomatoes, and squash. Coconut groves (**c**) would have been important on the coast to supply the rapidly developing Swahili trade (Maps produced by Oliver Boles and modified by the author) (All photographs: Rob Marchant) 145
Fig. 4.1 East African elephants. The numbers of elephants across the East African landscape is massively reduced compared to historical estimates. In addition to overall depopulation, the number of large elephants is relatively low as the large 'tusker' elephants would have been selectively killed—impacting on the gene pool and current population structure (All photographs: Rob Marchant) 165

Fig. 4.2 Nineteenth century Caravan routes across East Africa redrawn from Coutu et al. (2016), Beachey (1967), Rockel (2006), Lane (2010), and Cummings (1973). The caravans varied in size some industrial ventures comprised some 2000 porters! Photograph (**a**) shows a small caravan crossing a river in the Congo with (**c**) a collection of ivory at Zanzibar ready for export (edited from the ED Moore Collection, Ivoryton Library Association and Treasure of Connecticut Libraries) 168
Fig. 4.3 Coastal and mangrove forests were at the forefront of increased deforestation and exploitation, particularly for boat building and export. Large mangrove trees are virtually absent across East Africa although their large stumps attest to their durability as a boat-building timber (All photographs: Rob Marchant) 174
Fig. 4.4 Elephant population shifts occurred as a moving front—as populations became locally extinct from the coastal zone caravan trades developed to access the large herds inland. Elephant population range and change in this at 1840 and 1890 across East Africa. Redrawn from Alpers (1977), Coutu et al. (2016), Milner-Gulland and Beddington (1993), Thorbahn (1979) and Håkansson (2004) 175
Fig. 4.5 Plot of historical ivory exports from Zanzibar from 1800 to 1860. This is just from one export point at Zanzibar and clearly shows the rise in ivory exports to the UK while those to China were stagnant and decreasing. Redrawn from Sheriff (1987) 177
Fig. 4.6 Elephants as ecosystem engineers and the rapid reduction in numbers would have profound effects on structure, composition, and distribution of ecosystems over a very short period of time. Demonstrated clearly by this animal exclosure from Amboseli National Park (Photograph: Rob Marchant) 180

Fig. 4.7 Slavery and the slave trade can be traced back more than two millennia in East Africa. The slave trade grew rapidly in the late eighteenth century to provide labour in clove plantations, and growing trade in sisal (**a**, **b**). In addition to forced labour within the region there was also slave export from coastal towns such as Zanzibar Town (**c**) (Photograph: Rob Marchant) 181
Fig. 4.8 Transformed open landscapes of the 1890s in Chome, Pare Mountains Tanzania that was visited by German missionaries. There is clear increase in tree cover with both picture storied documenting a clear increase in tree cover over the past century. The previously highly transformed landscape is likely associated with supply of caravan trades traversing the Pangani Basin. All photographs from Paul Lane and Pauline van Hellermann 192
Fig. 4.9 Spread of pastoralism at 1850 CE in East Africa. This was particularly focused on cattle with pastoral groups, such as the Maasai, well established and becoming increasing sedentary. There would likely have been a diversification of livelihoods both in trading and providing services to the caravans. Transhumance would continue to be important to buffer seasonal variation in grazing resource via access to highland areas. Map produced by Oliver Boles and modified by the author (All photographs: Rob Marchant) 193
Fig. 4.10 Extent of agriculture at 1850 CE in East Africa. This expanded and diversified to take advantage of new market-based crops such as avocado (**b**) and coffee (**c**) that would be supplied to the caravan trade. Map produced by Oliver Boles and modified by the author (All photographs: Rob Marchant) 194
Fig. 5.1 British and German explorer routes; these started from the coastal staging posts of Unguja, Bagamoyo or Mombasa. The numerous people who ventured to 'discover' East Africa were largely driven by exploration, missionary activities, to claim territories or secure natural resources for the developing colonial influence 208

Fig. 5.2 Large tracts of land were assigned over to extensive plantations that took different forms, Cotton (**b**) and Sisal (**c**) plantations in the lowland for cloth and fibre, respectively. Much of the highland areas were cleared of forest and used for extensive Tea plantations (**a**, **d**) 212

Fig. 5.3 British railway route. One of the key infrastructure developments to take place was the establishment of a rail network that would have replaced the need for large caravans to transport goods and services in and out of East Africa. Dar es Salaam station (**a**), constructing the Uganda Railway through the savannah (**b**) and arrival at Kisumu (**c**) 213

Fig. 5.4 Uganda was divided into a series of administrative kingdoms (**a**), some of them such as Buganda were highly centralised with a king (Kabaka), council and large well trained and extensive army (**c**) with reports of 250 large outrigger canoes that functioned as a navy (**b**) 216

Fig. 5.5 As land was subdivided fences became an increasingly common feature of the landscape such as these across the Laikipia Plateau (All photographs: Rob Marchant) 220

Fig. 5.6 With the colonial administration came further forest clearance to open land for plantation and provide timber for the contribution and fuel for mechanisms transport. This growing demand for forest product let to extensive tree nurseries being established and (**b**) clear boundaries between forest reserves and agricultural land being formed (**c**) 227

Fig. 5.7 Forest clearance was particularly acute in coastal areas where there was low lying ground suitable for large scale plantations and the timber resource was highly accessible. Some 95% of the coastal forests has been degraded or impacted on by forestry operations (**a**, **b**) and the ensuing conversion of wood to charcoal (**c**, **d**, **e**) for increasing energy demand (Photographs **a** and **b**: Antje Ahrends, Photographs **c**, **d** and **e**: Rob Marchant) 230

Fig. 5.8 The number of National Parks and Protected Areas increased dramatically throughout the colonial period and continue to do so through to the present day resulting in large of East Africa being gazetted. Some National Parks are small and shrinking, such as Nairobi National Park, while others were vast; resulting in large amounts of the area of East Africa being gazetted, particularly in Tanzania (All photographs: Rob Marchant) 237

Fig. 5.9 The Ngorongoro Conservation Area consisting of the Ngorongoro Crater (**a**) and communal grazing land (**b**) extending to the Serengeti where the Maasai pastoralists have secure tenure to live alongside wildlife within a multiple land-use protected area (All photographs: Rob Marchant) 238

Fig. 6.1 Ujuma village were constructed around a common blueprint. There was a nucleus of public space and central services, political offices and a town hall. Homes tended to be a common size and design with set allotments of land for agricultural production 253

Fig. 6.2 The Chinese led construction boom has included a large program of road construction with a new main road that now connects Nairobi to northern Kenya and Kenyan-Ethiopian border town of Moyale. There are also the development of new cities such as the Konzo technology city (**c**) and new road bridges in cities such as Dar es Salam (**d**) (All photographs: Rob Marchant) 256

Fig. 6.3 Around Bwindi Impenetrable Forest National Park in Uganda a series of Multiple Use zones were established around the National Park through a Development through Conservation project; people could enter the National Park and collect non-timber forest products such as basketry and medicinal plant resources and other income generating schemes such as honey production (All photographs: Rob Marchant) 262

Fig. 6.4 To reduce the impact on forest resource there have been a wide series of income generating schemes such as developing butterfly farms to sell pupae (**a**) round the world. There is a long history of bee-keeping with honey a good alternative income source from forests (**b**) (All photographs: Rob Marchant) 263
Fig. 6.5 Historical events, such as the gazetting of land or shifts in government policy, impact on community decisions around how land will be used and have ramifications to the present day. Timeline of key events that have shaped land use land cover change around the Ngorongoro Highlands (Kariuki et al., 2021) 265
Fig. 6.6 Agriculture continues to spread rapidly in the past decades as depicted in a comparison of MODIS land cover from 2001 and 2013. Most notable is the increase in agriculture (yellow); particularly along the coastal strip and around Lake Victoria. Image analysis P. Platts 270
Fig. 6.7 Increased sedentarisation and agricultural expansion across southern Kenya around the Amboseli Basin. The concentration of animal impacts into single locations, rather than being spread across the wider transhumance landscape, can lead to degradation. New areas of cultivation tend to be increasingly marginal and thus prone to environmental shocks, particularly those associated with drought and land degradation (All photographs: Rob Marchant) 272

Fig. 6.8 Invasive alien plant species. A particular challenging one has been the spread of *Opuntia* (**a**, **d**) prickly pear that was imported as fence species and now spread across large area. Ingested by animals, the thorns can kill and blind species that eat this, particularly during drought periods. *Pontederia crasipes*—water hyacinth—is an aquatic plant native to the Amazon basin, and is often a highly problematic invasive species in a number of large lakes such as Lake Naivasha (**b**). *Opuntia*. Although not an invasive *Acacia reficiens* (**c**)—false umbrella thorn—spreads extensively in savanna areas and shades out the grazing resource leading significant local hardship to pastoral populations (All photographs: Rob Marchant) 273

Fig. 6.9 The Kenya-Tanzania Borderland area is characterised by a wide range of land uses and a large proportion of the land dedication to conservation. The area comprises 14 protected areas including the Serengeti, Ngorongoro and Masai Mara 278

Fig. 6.10 Conservancies, such as the Siana, Nashulai are located around the Masai Mara where there are some 14 conservancies. Kalama, located close to Lewa in northern Kenya, clearly shows the link between pastoral communities and conservation with the milking of an elephant! Community conservancies around National Parks, such as the expanding number around Amboseli, are increasingly common as pastoral communities get land title deeds (All photographs: Rob Marchant) 281

Fig. 6.11 The large number and area covered by National Parks and Protected Areas provide opportunities for employment and providing good and services to support the tourism industry (All photographs: Rob Marchant) 285

Fig. 6.12 One of the flagship projects from the Chinese led construction boom through the Belt and Road Initiative has been the Standard Gauge Railway that is running from Mombasa through Kenya and towards Kisumu and onwards through Uganda and ultimately connecting East to West Africa. This bisects some of the National Parks (**a**), carrying people and freight (**b**) (All photographs: Rob Marchant) 292
Fig. 6.13 New linear infrastructure, such as roads and railways, are not without its challenges. Although there are mitigations measures in the design of bridges, noise screens and underpasses to reduce negative impact on wildlife, how effective these mitigation measures is yet to be determined both directly (**a**) and through the growth of cities (**d**) and facilitating the spread of invasive plants (**c**) (All photographs: Rob Marchant) 295
Fig. 7.1 One of the key rationales in exploring the past linkages between land cover change and human interaction is to examine the potential applications of past insights into contemporary issues, and how these past insights can be used to guide future pathways. By combining data from a wide range of evidence it is possible to explore social–environmental interactions past, present and future 312
Fig. 7.2 Although the number of archaeological and palaeoecological sites, and the spectrum of techniques applied, have greatly increased over the last couple of decades, many spatial and/or temporal gaps remain and biases in past human-environmental interaction (All photographs: Rob Marchant) 314
Fig. 7.3 There is a need to combine different disciplines such as palynology, ecology and remote sensing to truly understand the challenges facing ecosystems and how these have evolved so that future scenario tools can be based on enhanced understanding of ecosystem–human interactions 316

Fig. 7.4 Understanding of past environmental change and how people have been enmeshed within this to shape our contemporary ecosystem can be shown in a number of ways. This stylised carton of landscape and ecosystem evolution for the Amboseli basin can help make palaeoecological insights accessible to a wider user community 318
Fig. 7.5 A number of disciplines and insights combine to understand the role of societies in shaping landscape evolution. 'Biocultural Heritage' can be used to inform contemporary thinking and decision-making and rooted in the understanding of the past this will be very different in different landscapes such as Olduvai Gorge, vs. newly inhabited landscapes such as the Pare Mountains (All photographs: Rob Marchant) 320
Fig. 7.6 Regionally downscaled climate change futures where Global Climate Models are used to depict climate futures at 1 km^2 for continental Africa using the AFRICLIM product. The figure shows changes in precipitation at 2055 and 2085. While these are produced for a wide range of climate parameters, as in the past, future timing and quantity of rainfall are going to key and have the biggest impact on ecosystems—as shown by pictures of Tarangire in the dry (**a**) and wet (**b**) season. Climate predictions suggest some areas getting drier and others getting wetter; trends are likely to be enhanced in the future (Platts et al., 2015) (All photographs: Rob Marchant) 323
Fig. 7.7 Ecosystems are shaped by the interaction of a changing climate, raised levels of atmospheric CO_2, burning regime and changing densities of ecosystem engineers such as elephants, sheep, goats, and cows. Reduced levels of herbivory and fires combined with increased levels of carbon dioxide lead to greater water-use efficiency of drought-adapted trees that is resulting in the spread of woody biomass in savannas as shown by this pair of images from Laikipia taken in 1935 and 2005 325

Fig. 7.8 Niche modelling schematic from collating of data to running the models and describing species distribution across the landscape—run here for *Newtonia bucannanii*. In addition to being a good tool for assessing impacts of climate change on species distribution, niche models can be used to guide botanical fieldwork (**a**) and combined over multiple distributions to show concentrations of diversity. **b** shows the globally important Eastern Arc Biodiversity Hotspot where 570 niche models are combined to assess diversity patterns (Platts, 2012) 327
Fig. 7.9 Communication is key. Due to the diversity and often highly context-specific nature of challenges we need to start the research process with a conversation that brings the different communities and perspectives together. Conversations are at the heart of understanding the scale of the apparent challenges, the key questions that need addressing and what are the potential solutions (All photographs: Rob Marchant) 329
Fig. 7.10 KESHO participatory scenarios has emerged as a key tool to investigate potential futures through the lens of potential land cover change. KESHO has been applied from national to local levels, looking at coastal development, pastoral-agricultural-conservation interactions, coffee futures, and impacts of new railways. Participants are tasked with imagining future land cover and the drivers behind land use transformation, be it natural, policy-induced, or through human agency. In the application shown future forest/carbon cover is shown at 2030 under a Green Economy and a Business as Usual scenario—the difference are massive and could equate to US$10 billion in carbon payments 330

Fig. 7.11 Land use ultimately underpins key ecosystem services and hosts biodiversity. Future land use transitions can be used to explore the Interactions between biodiversity, water and carbon. Red–Green–Blue plot showing the combined impacts on carbon stocks (black to bright green), biodiversity (black to bright red) and water balance (black to bright blue) 332
Fig. 7.12 Population distribution in East Africa derived from WorldPop. It is clear the highest population densities are centred along the coastal strip, around fertile mountains (**c**), the low-lying areas around the large lakes and expanding urban centres (**a, b**) (All photographs: Rob Marchant) 334
Fig. 7.13 Alternative livelihoods for many communities are being developed such as these Masai pastoralists working in the tourist industry or selling bone products (All photographs: Rob Marchant) 339
Fig. 7.14 One of the key climate change challenges ahead is going to be associated with the timing and quantity of rainfall. Although the climate is predicted to be wetter, in common with the result of the world, increased extremes are going to have the biggest impact. These flash floods near Suswa (Kenya) have been exacerbated by upland forest clearance (All photographs: Rob Marchant) 341
Fig. 7.15 With the main climate change challenge linked with rainfall, key to mitigating the impact will be expansion of catchment-based approaches where water towers (**a**, **b**) are managed to capture water, deliver this is a managed way to avoid unintended downstream impacts (**c**). East African mountains are key ecosystems that capture water from low-level clouds drifting through the forest to allow these water towers to regulate supply to down-stream users (**c**) (All photographs: Rob Marchant) 342

Fig. 7.16 East Africa is characterised by inherently strong interactions between nature, human populations, and the environment. The drivers behind land use transformation ultimately impact on the foundation for all life—that of plant life 345
Fig. 7.17 Ecosystem services, or Natures contributions to people, are increasingly being recognised as vital to managing some of the Sustainable Development challenges as well as underpinning future economic development. Be it carbon storage or water cycling, tourism or pollination, key to mainstreaming ecosystem services is being able to adequately recognize, demonstrate and capture the value of the myriad of Ecosystem services 348
Fig. 7.18 Conservation vs agriculture in the balance. Around many Protected Areas increasingly intensive and irrigated farms, such as this one near Amboseli, is reducing space and functionality of the landscape to support wildlife populations and increasing human–wildlife conflict that is shifting the balance towards agriculture and reducing space for wildlife, particularly migratory species like elephants. Key habitats, such as these swamps (**a**) are drained for agricultural production (**b**) further increasing human-wildlife conflict, particularly during periods of drought (All photographs: Rob Marchant) 360

1

Foundations: The Environment, Ecosystems, and Cultures of East Africa

1.1 Introduction

East Africa covers an area of about 6.2 million km^2 and, for the purposes of this book, comprises the countries of Kenya, Tanzania, and Uganda that surround Lake Victoria (Fig. 1.1). The region is characterised by great social and environmental contrasts and environmental variability comprises the great East African Rift System, volcanic mountains, extensive fringing highlands, and dramatic transitions from highlands to lowland. Between the Eastern and Western Rift valleys the area is characterised by large lakes including Lake Victoria, one of the major sources of the Nile River and the second biggest lake in the world.

The region includes several iconic mountains, such as Kilimanjaro, Mount Kenya, and the Rwenzori—the 'Mountains of the Moon'—that form the boundary with the Congo Basin to the west. In contrast to these highlands, there are arid low-lying depressions such as the Chalbi Desert of Northern Kenya. Thus, East Africa varies from wet and cold highlands with remnants of permanent ice through to arid lowlands, from hyper-humid coastal areas through to extensive seasonal savannas. Although

R. Marchant, *East Africa's Human Environment interactions*, https://doi.org/10.1007/978-3-030-88987-6_1

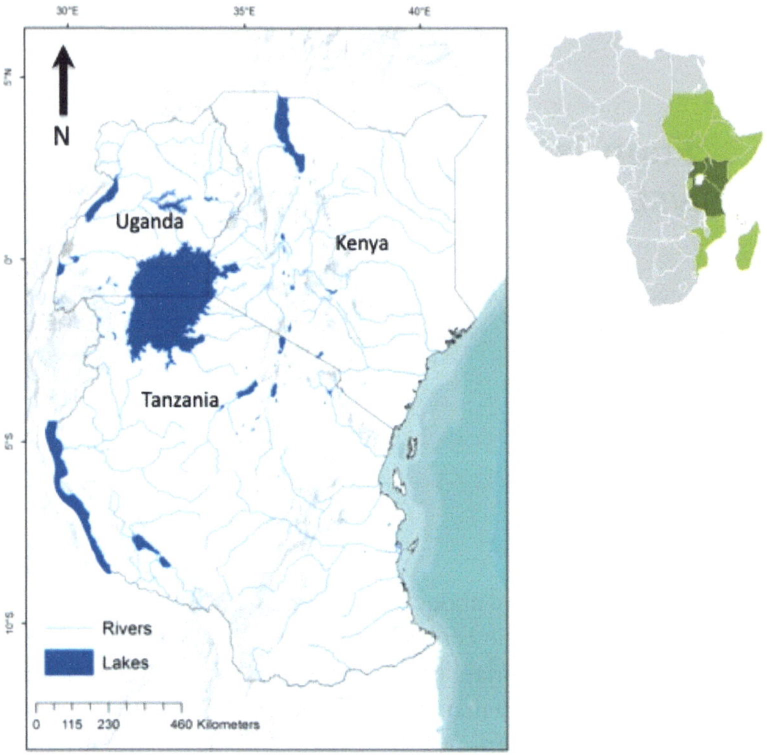

Fig. 1.1 East Africa comprises the countries of Kenya, Tanzania, and Uganda (dark green in insert). There are a variety of other countries in different definitions of East or Eastern Africa (light green in insert) that are not included in this volume

Kenya, Tanzania, and Uganda are a somewhat arbitrary collection of countries, they do reflect the current political combination of the East African Union and a shared colonial history, but more crucially, a deep-rooted history of agricultural transitions, similar land use systems and trading links. This commonality is documented by the use of Kiswahili, a common and uniting language that has been adopted for this region over the past 1000 years or so.

1.2 East African Climate: Foundation for Diversity of Life and Livelihoods

The primary environmental control across East Africa is the climate, and how this varies across the region (Fig. 1.2). The climatological patterns in East Africa are complex due to the interaction of major global convergence zones with more regional climate. These are then mediated by the land feedbacks; particularly associated with the lakes and the varied topography that dominate the region.

A good example of the influence of topography on the climate can be seen from the highlands associated with the Western Rift valley; these block the flow of moist air from the Congo Basin and create rain shadows on the lee slopes of mountains towards the east. Conversely, highlands that receive rainfall derived from the Indian Ocean, such as Kilimanjaro,

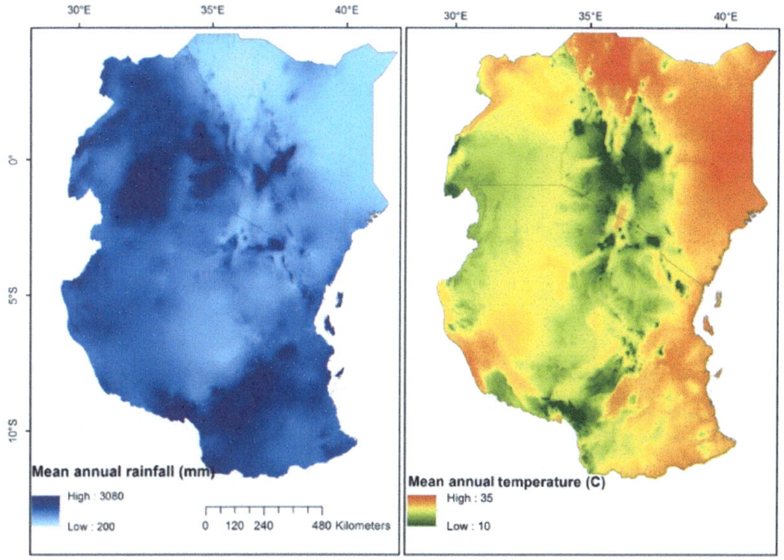

Fig. 1.2 Mean annual distribution of rainfall and mean annual temperate across East Africa. The highly variable climate regime and spatial variation is also characterised by high intra and inter annual variability, particularly on the timing and duration of rainfall based on WorldClim (Hijmans et al., 2005) as derived by AfriClim 3.0 (Platts et al., 2015)

Mount Kenya, and Mount Marsabit have western slopes that are considerably drier than the eastern flanks. The large lakes are also sufficiently extensive to generate their own climate character. Topographical variation associated with the Great Rift Valley and large inland water bodies (e.g. Lake Victoria) greatly influence the spatial distribution of seasonal rainfall in the region (Oettli & Camberlin, 2005) (Fig. 1.2). Tropical cyclones over the Indian Ocean indirectly contribute to intra-seasonal events by creating a west–east pressure gradient resulting in enhanced rainfall over much of the Western and central equatorial sub-region of East Africa alongside contrasting dry conditions over the Eastern equatorial sub-region (Shanko & Camberlin, 1998) (Fig. 1.2). Hence, the climate regime of East Africa (Fig. 1.2). is a product of a combination of global drivers, regional moderation, and local edaphic and topographic feedbacks (Fig. 1.3).

The equatorial location of East Africa dictates that the large-scale climate regime is dominated by the Inter-Tropical Convergence Zone (ITCZ) that is characterised by a mainly north–south-north annual

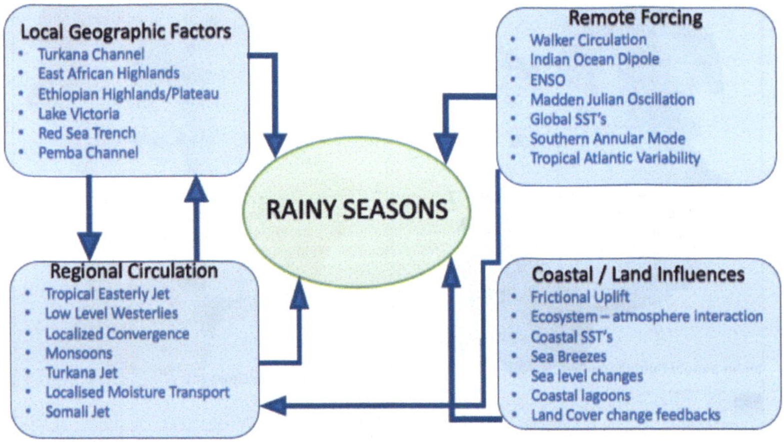

Fig. 1.3 Schematic documenting the main drivers of the rainy season in East Africa. Due to the complex interaction between regional and global circulation systems, local geographic factors, coastal and land influences along with remote forcing such as the Indian Ocean Dipole and ENSO there is massive variability in intra and inter annual timing and duration of rainfall

migration, along with the related rainfall front. Other regional climate systems include the Congo Air Boundary (CAB) (with airflows from the west and southwest), the northeast trade winds, and the southeast trade winds with their associated monsoonal rainfall (Costa et al., 2014). The ITCZ is directly linked to the meeting of the trade winds from both hemispheres, however, the subtropical high-pressure cells, that are located 20–30° north and south of the equator, shift with the seasons (Nicholson, 1994, 1996, 2000, 2017; Ogallo et al., 1988). As the passage of the ITCZ migrates it creates a strongly bimodal pattern of seasonal rainfall, although this exhibits considerable variation across the region (Bergonzini et al., 2004). East Africa's bimodal rainfall regime is characterised by rain falling from June to August and from December to February. For much of equatorial East Africa, monsoonal winds from the southeast in March to May bring long periods of rain (Yang et al., 2015) that are generally termed the 'long rains' or 'Masika' and are characterised by heavier and longer rainfall. In comparison, the 'short rains', locally termed 'Vuli', extend from October to December, brought in by the north-easterly monsoon. The combination of these two rainy periods accounts for the majority of equatorial East Africa's annual rainfall. However, there can be large interannual variability to the point where one or more of the rainy seasons can fail (Nelson et al., 2012), and the amount of short-season precipitation in any given year yields little to no predictive power over the following long rains (Lyon, 2014). Over most of East Africa annual rainfall is between 600 and 1500 mm yr^{-1} (Cuní-Sanchez et al., 2016), but this can be much higher over the highlands and much lower over north-eastern Kenya and Uganda. During the short rains the most direct and most robust influence is that of the low-level flow in the equatorial Indian Ocean, but even its link to rainfall variability wanes from time to time. The Lake Victoria Basin and Eastern Rift Valley areas can often experience a third rainfall regime during the northern hemisphere summer (June–August) when the zonal and the meridional arms of the ITCZ are displaced to the north and east resulting in enhanced convergence of moist air mass derived from the Congo Basin (Nicholson, 2001). Combined effects of the Lake Victoria trough and the highlands west of the Rift Valley also contribute to the third rainfall peak (Omeny et al., 2008).

Atmosphere, insolation, topography, and vegetation all interact to influence the environmental conditions of the isolated massifs and mountain chains of East Africa (Fig. 1.3). These mountain chains act as 'water towers', collecting moisture from advecting clouds and non-precipitating mists, then storing it in root and soil systems thereby regulating the flow of water into rivers (Fig. 1.4): demonstrating how mountains provide critical connections from the mesic highlands to the more arid lowlands (Cuní-Sanchez et al., 2016). Climatic variations with altitude are complex because of various factors superimposed on the general climatic regime as described above. This factor can be summarised as an adiabatic lapse rate.

Fig. 1.4 Highland areas across East Africa are also known as 'Water Towers' as they collect moisture from rainfall event, clouds, and non-precipitating mists. Rainfall and collected water is then stored in the soil systems and vegetation that regulate the flow of water into rivers. Mountain Water Towers provide critical connections from the mesic highlands to the more arid lowlands. Examples here include the Aberdares (**a**, **d**, **e**), Kilimanjaro views from the Amboseli basin (**e**), Mount Kenya (**b**), and the Udzungwa Mountains (**c**)

Although lapse rates have often been 'determined' from altitudinal changes in temperature (Marchant & Hooghiemstra, 2004), results vary considerably and are known to fluctuate through time (Loomis et al., 2017). One standard that is constant is the dry adiabatic lapse rate; this can be defined as the cooling of an absolute dry parcel of air as it passes to higher altitudes due to changes in pressure (Barry & Chorley, 2009): laws of thermodynamics fix this rate of cooling at 9.8 °C of 1000 m^{-1}. However, cooling air invariably produces condensation and the liberalisation of latent heat, which reduces the rate of temperature decrease (Barry & Chorley, 2009). The saturated adiabatic lapse rate varies considerably with ambient temperature; at high temperatures the rate can be as low as −4 °C 1000 m^{-1}, and increases with decreasing ambient temperature to approaching −9 °C 1000 m^{-1} at −40 °C (Barry & Chorley, 2009). Thus, the lapse rate varies according to levels of ambient moisture and temperature. The environmental lapse rate commonly used for Eastern and Central Africa (−6.5 °C 1000 m^{-1}; Kenworthy, 1966) will vary according to aspect, topography, and radiation balance, and has been different under past climatic regimes (Loomis et al., 2017; Nicholson et al., 2013). Thus, applying lapse rates to time points in the past, or indeed different areas from where the rate was computed, must be done with caution.

The complex climate variability of East Africa can be exemplified by changes on the numerous mountains (Fig. 1.4). For example, Mount Marsabit receives moist air and low-level clouds derived from the Indian Ocean via the Turkana Channel Jet (Nicholson, 2001); the latter is associated with dominant, almost continuous, south-easterly winds at Mt Marsabit throughout the year (Fig. 1.2). As is common with many mountains, the water supply to Mt Marsabit has diminished over the last few decades; rainfall decreased from 1961 to 2010 by about 10 mm yr^{-1} and the number of days with fog has declined by about 50% over the past three decades (Los et al., 2019). There is a strong feedback between the character of the ecosystem, and the use of this with related impacts on local water availability. For example, the cutting of trees, largely for cattle fodder, has reduced the ability of the forest to capture occult precipitation (Cuní-Sanchez et al., 2016; Dinku et al., 2011).

Increased demand for groundwater has also resulted in increased pressures on the forest harvesting and further curtailed its ability to capture occult precipitation thus exacerbating the decline in water resource. A similarly spatially complex rainfall pattern can be seen on Kilimanjaro following extensive monitoring and logging of key climate parameters from 60 climate stations over the past decade shows that rainfall spatial pattern has changed in the early twenty-first century (Nicholson, 2001). Thus, the amount of rainfall can fluctuate immensely over short distances and there is considerable seasonality however there is a relatively strong coherence in the patterns of interannual variability across the region due to the macro and mesoscale drivers of the climate system (Figs. 1.2 and 1.3).

It is challenging to document rainfall variability for East Africa where, as we have seen, rainfall is influenced by a range of global, regional, and local-scale drivers and feedbacks (Fig. 1.2). However, the major characteristic climate change across East Africa is increased rainfall variability manifesting as changes in the intensity, duration, or timing of the precipitation regime (Adler et al., 2021). The long rains are more reliable with most of the interannual rainfall variability being associated with the short rains that often 'fail' (Thorn et al., 2021). Much of the rainfall variability that occurs during the short rainy season is linked to large-scale climate systems such as the El Niño Southern Oscillation (ENSO) (Indeje et al., 2000; Mutai & Ward, 2000; Ogallo et al., 1988) and the Indian Ocean Dipole (IOD) (Behera et al., 2003, 2005; Cai et al., 2014; Owiti et al., 2008; Saji et al., 1999; Marchant et al., 2007). While ENSO originates in the Pacific Ocean, other tropical ocean basins also exert influence on climate variability over land. For example, variations in the horizontal structure of tropical Atlantic sea surface temperatures (SSTs) influence the position of the ITCZ and consequently the local climate over East Africa. Conditions over the Indian Ocean are clearly part of the mechanism by which the short rains are modulated via the Indian Ocean Dipole system. Warm SSTs in the west and cold SSTs in the east (positive phase of the IOD) are associated with a weakened Walker circulation over the Indian Ocean, while the reverse SST pattern of the IOD's negative phase acts to enhance the Walker circulation over the Indian Ocean. The other main factor that appears to play a role in creating the seasonal cycle and determining the strength of the link between the Indian Ocean and continental land is the low-level Turkana Jet Stream, which appears to suppress rainfall from June to October (Hamilton, 1982). Clearly there are wide series of interacting systems that will shift in location, duration, strength, and ability to interact over time. With so many moving parts it is hardly surprising that the climate of East Africa is complex! Although a series of recent studies have made progress in understanding the seasonal cycle in this region, further work is required to truly understand the system and allow us to use this understanding to make informed decisions about living with environmental change.

1.3 East African Geology, Topography, and Drainage

East Africa is characterised by rapid transitions from extensive lowlands to cathedral-like mountains: some of these are geological young, isolated volcanos (e.g. Kilimanjaro) and others ancient uplifted massive mountain chains (e.g. the Ruwenzori). East Africa has undergone tectonic dislocation from the Mesozoic (Braile et al., 2006; Macgregor, 2015; Summerfield, 2005), which resulted in the formation of the Western and Eastern Rift Valleys with uplift ranging from 700 m asl (metres above sea level) to over 5800 m asl (Calais et al., 2006; Hamilton, 1982) (Fig. 1.5).

The most active tectonic phase took place from mid-Tertiary times, centred around 22, 6 and 2.5 million years ago (Maslin et al., 2014), and was accompanied by extensive volcanism which produced many of the singularities of the East African topography such as Kilimanjaro and Mount Kenya. In contrast to this volcanic activity, the Ruwenzori and Eastern Arc mountains were formed by upthrust of Precambrian crystalline rocks with their origins aligning around 30 million years ago (Schlüter, 2008). Tectonic activity also produced the high plateau that characterises much of East Africa (Fig. 1.5). Most of the tectonic uplift had occurred before the start of the Quaternary glaciations (Karlén et al., 1999). With each phase of uplift, the associated erosion and deposition of the uplifted geology resulted in the topographically complex landscape that we know today as the East African Rift System with

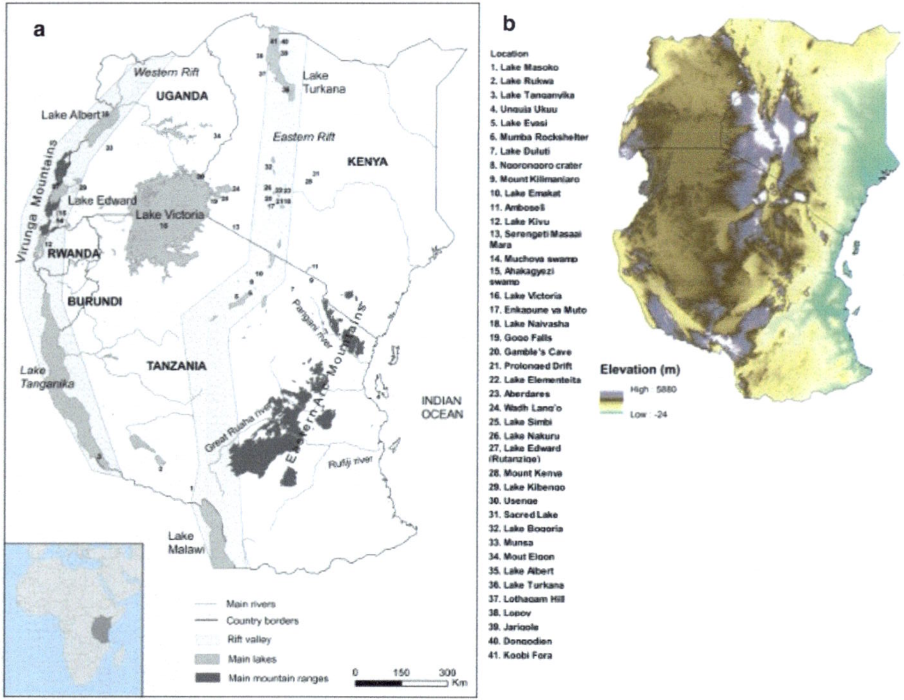

Fig. 1.5 Eastern Africa and study site locations and geographic features discussed in text (**a**). Lakes and the varied topography dominate the region that is topographically very diverse (**b**) elevation derived from the SRTM-250 m Digital Elevation data (http://srtm.csi.cgiar.org)

numerous lakes located along its span (Trauth et al., 2007, 2014). Towards the south of the region, Lake Tanganyika dominates a chain of large, deep lakes (Kivu, Edward, and Albert) along the floor of the western (Albertine) branch of the East African Rift System from south to north (Fig. 1.5). The many lakes that occupy the Eastern Rift Valley are typically smaller and shallower. The rifting process has also had considerable effect on central African watersheds outside of the main centres of rifting (Walling, 1999) causing a reversal of previously west-flowing rivers, thereby creating Lake Victoria between the Western (Albertine) and Eastern (Gregory) Rift Valleys (Street & Grove, 1975) (Fig. 1.5). As a result of the ongoing tectonic activity, the direction and flow of rivers is changing. For example, The Mau Highlands of central Kenya host the headwaters of many major rivers, one, the Mara, flows into Lake Victoria whereas the Tana flows eastwards into the Indian Ocean. The Mara River's flow source is around 3000 m in the Mau Forest in Kenya and flows across different landscapes before draining into Lake Victoria at Musoma Bay in Tanzania. The Mara River, like all river systems across East Africa, is impacted by widespread human activities such as deforestation and subsequent cultivation of land beginning at the headwaters in the Mau Forest complex and extending all the way through the lowlands. The Tana River Basin covers an area of 100,000 km^2 and is the largest in Kenya flowing over 1000 km starting from Mount Kenya, Mau and the Aberdare Mountains and ending at the Indian Ocean. Like many of East Africa's rivers the Tana River winds through a densely forested ecosystem at its headwaters before transitioning to agricultural areas and rangelands downstream where the river flows for 700 km through semi-arid flood plains and terminates in a large delta at Ungwana Bay in the Indian Ocean. The upper and middle sections have several hydroelectric dams while the lower basin hosts large irrigation schemes; a common characteristic with many of the region's rivers.

1.4 Soils and Edaphic Factors Influencing Ecosystem Composition in East Africa

Soils are the interface between the climate system and the earth and are essentially a function of geology, climate, topography, ecosystems, and management. As well as being a product of growth, soils form a crucial edaphic control on what can grow where, how productive an area can be, and the density of population that can be sustained. The diversity and distribution of soil types in East Africa result from the range of geological and climatic conditions of the area acting upon a tectonically modified topography (Fig. 1.5). According to Bamutaze (2015), 25 main soil types that can be divided into three broad units characterise the area: the Rift Valley area, the Tanzanian Craton, and Eastern Kenya and Tanzania (Fig. 1.6). The Rift Valley formation is dominated by the formation of Fluvisols (e.g. Sakané et al., 2011), young soils developed in alluvial deposits that are usually fertile and productive (IUSS, 2015). In locations with high evapotranspiration, saline soils such as Solonchacks and Solonetz soils appear. For example, in the semi-arid basin of Lake Turkana some 15–20% of the rangelands are affected by soil salinity, large extensions of Calcisols, sodic Solonetz and weakly developed Regosols (Mace et al., 2012; Mbaluka & Brown, 2016) (Fig. 1.6). On either side of the Rift Valley weathering of volcanic materials (basalts, ash, and lava deposits) produces fertile and sandy-loam Andosols with thick, dark, and organic-rich topsoil as a result of the accumulation of organic matter (Funakawa et al., 2012; Matus et al., 2014; Shoji et al., 1993). Also frequent on the slopes of the volcanoes are fertile Nitisols that are excellent for coffee and tea production: the Kikuyu Red Clays characterise one of the most intensive agricultural regions in East Africa (Kapkiyai et al., 1999; Lal, 2010). Nitisols develop mainly from basic iron-rich rocks such as basalt and have a dark red colour and a well-developed structure. However, the large reactive aluminium and iron contents in both Andosols and Nitisols frequently result in a large phosphate-fixation capacity and therefore a low phosphorous availability for plant growth (Kisinyo et al., 2013; Nandwa & Bekunda, 1998; Nziguheba et al., 2016). Considerable extensions of Ferralsols in Uganda result from high weathering intensity due to high rainfall

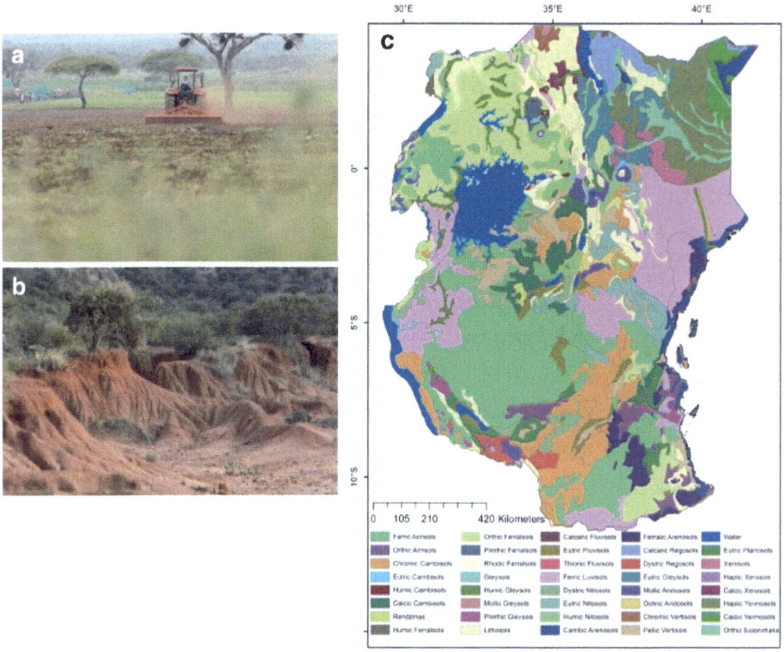

Fig. 1.6 Soils vary from the highly productive supporting industrial farming in Loitoktok (**a**) to highly degraded soils at the foothills of the Pare Mountains (**b**) The diversity and distribution of soil types in East Africa derived from the Soil Atlas of Africa, World Reference Base classification (**c**). All photographs: Rob Marchan

(Gray, 2011; Shi et al., 2015; Wasige et al., 2014). Ferralsols can support rainforest ecosystems and have a fast turnover of organic matter and nutrients; however, they are fragile soils, and can be easily degraded following deforestation (Bamutaze, 2015; Nyombi et al., 2010; Oyana et al., 2015).

Lake Victoria Basin, Serengeti and the Mara Plains, and the West of Tanzania are characterised by more acidic plutonic and sedimentary lithology bordered to the east by the Eastern branch of the Rift Valley and the Ngorongoro Conservation Area (Dawson, 2008; Kasanzu, 2016; Schlüter, 2008; Selway, 2015) where Planosols are frequent and mainly found across the Lake Victoria Basin (Lufafa et al., 2003; Rwetabula et al., 2012). These are poorly drained soils with a high clay content that can impede drainage, resulting in poorly structured topsoil (IUSS, 2015). Also in the Lake Victoria Basin are vast areas of clay-rich Vertisols (Sakané et al., 2011) and Luvisols (Lufafa et al., 2003) that develop on the river valley floors. These productive clay-rich soils form deep, wide cracks upon drying due to the presence of shrinking and swelling clays. In the extensive grasslands of the Serengeti and Mara Plains, the predominant soils are organic and lime rich Phaeozems (Blake et al., 2008). In Western Tanzania the higher acidity is reflected by extensive areas of less developed Cambisols, with Vertisols forming in the river valleys related to sedimentary deposits formed by weathered volcanic materials (Sakané et al., 2011). Towards the Western Indian Ocean coastline, the topography drops in elevation and a large, predominantly sandy plain dissected by deep valleys and sandy surfaces running downwards to the coast. The coarse-grained material gives rise to Arenosols and red coloured Ferralsols (Fig. 1.6) (Gichangi et al., 2015; Mganga et al., 2015), while the evaporation of saline groundwater originates from the widespread occurrence of Solonetz (Omuto, 2013). In Eastern Tanzania where the topography is steeper, Luvisols are developed on the metamorphic and sedimentary rocks; for example, Pare and the Usambara Mountains (ISRIC, 2006; Mathew et al., 2016; Msanya et al., 2002; Pachpute et al., 2009) (Fig. 1.6). Farther south, the undulating topography and the lithological variation of the parent materials resulted in extensive areas of Fluvisols and dark, clayey Vertisols developed in alluvial sediments of the river systems of central Tanzania, such as the highly productive Rufiji River floodplains and the Ruaha River wetlands (Armanios & Fisher, 2014; Jones et al., 2013; Mbungu & Kashaigili, 2017). Fertile Nitisols and more acidic Acrisols occur on gentle slopes, their distribution depending on underlying lithology (Lyon et al., 2015). Continuing southwards, increasing aridity and the predominance of grasslands and savannah (Dondeyne et al., 2003, 2004) gives rise to extensive Lixisols and Cambisols (Msanya et al., 2002). Hence, the soils of East Africa are increasingly varied and change over short distance. This foundation largely determines what the land can be used for, the levels of productivity, and ensuing land use management (Fig. 1.6).

1.5 East African Ecosystem Composition and Distribution

Although being fundamentally influenced by human interaction (Chapter 3), the broad patterns of vegetation distribution and ecosystem transition of East Africa largely reflects the climatic regime; the amount and seasonality of precipitation being particularly important in determining ecosystem composition, structure, and distribution (Figs. 1.7 and 1.8). Additionally, topography, alongside edaphic and disturbance

Fig. 1.7 Lowland ecosystems of East Africa (**a**). Coastal mangrove forest on intertidal mudflats, Unguja, Zanzibar (**b**). Mangrove grades into Coral rag vegetation—Chumbe Island (**c**). Extensive coconut palm groves cover much of the coastal belt (**d**). Coastal Forest, previously much more extensive, can still be found in protected areas such as the Shimba Hills, Kenya (**e**). Desert extend across much of northern Kenya and Uganda such as the Chalbi Desert in northern Kenya (**f**). Continuous savannah with intermittent trees or different composition, dominated by different species of *Acacia* extend through much of East Africa (**f**, **g**, **i**, **k**). Edaphic influences can locally be very important in controlling the presence of local woody species as demonstrated by these views of gallery forest Engaruka (**h**) and Tsavo (**l**). Savannah pastoral landscape, close to Ngorongoro (**j**) (all photographs: Rob Marchant)

Fig. 1.8 Mountainous biomes of East Africa (**a**) view from Engaruka towards surrounding highlands (**b**). High alpine vegetation on the rim of Ngorongoro (**c**). Montane vegetation within the crater of Mount Lognonot, Kenya (**d**). Extensive Montane forest in the Aberdares National Park (**e**). *Olea* spp. on the rim of Mt Marsabit, northern Kenya (**f**). Most of the montane ecosystems across East Africa have been heavily impacted in the past and are in a regenerating state, often with extensive stands of montane bamboo *Arundinaria alpina* such as on Mount Elgon (**g**). Montane forest species competing for light on Mt Elgon (**h**). Many of the montane forests are covered in *Usnea* that add to their efficiency to strip moisture out of non-precipitating clouds—Uluguru Mountains. At higher altitudes (**k**) or where the topography leads to locally dry conditions (**i**) continuous alpine grasslands form. At higher altitudes, such as on Mount Kilimanjaro *Scenecio* spp. can be locally dominant (**k**). Montane forests can be dominated by large species such as this stand of *Podocarpus* on Mt Elgon (**j**). Many of the mountains across East Africa are extinct volcanoes and often have wetlands in their caldera such as Lake Paradise on Mt Marsabit (**l**). All photographs: Rob Marchant

regimes, such as fire and grazing, impart important local controls, particularly in shaping the nature of grassland, savannah, and forest boundaries. Like any clarification scheme this can be resolved into multiple divisions or collapsed into major ecosystem types. Six major vegetation communities (phytochoria) are delineated in East Africa (Lind et al., 1974; White, 1983): Coastal rainforest, Mangrove, Ericaceous scrub, Montane Forest, Savannah grasslands, and Wooded savannah (Figs. 1.7 and 1.8). The distribution of these major ecosystem types is largely driven by the interaction between climatic regime, topography, geology, soil, and disturbance.

1.5.1 The Coastal and Lowland Vegetation

A relatively thin strip of mangrove forest is present along the tidal coastal area (Fig. 1.7a) (Punwong et al., 2013a, 2013b, 2013c) that quickly grades into coral rag, coconut groves, and coastal rainforests (Fig. 1.7b, c, and d). The latter, now highly fragmented by long history of impacts, are a globally recognised hotspot for their amazing biological diversity (Burgess et al., 1998). The East African coastal forest runs from Somalia in the north to Mozambique in the south and constitutes one of the 35 global biodiversity hotspots (Mittermeier et al., 2009) due to the high diversity of endemic plant and animal species (Burgess & Clarke, 2000; Burgess et al., 1998; Habel et al., 2017). For example, 44% of plants are endemic and 40% of plant genera are confined to a single forest patch with many new species frequently being discovered. Bird diversity, like plant diversity, is particularly high in the coastal forest and also comprises numerous threatened species such as the Fischers turaco and the Uluguru violet-backed sunbird. Being wet and warm, the coastal forests are also highly diverse in amphibians and reptiles such as the Usambara forest gecko and the Usambara short horned chameleon, with new species still being discovered. Similarly, insects are incredibly diverse in the coastal forest with some 35% of all Kenya's butterflies occurring in the Shimba Hills (Fig. 1.7d).

Despite its biological relevance, this region has undergone a long history of anthropogenic destruction and disturbance, particularly since the colonial era (Chapter 5). This use over the past millennia has transformed the endogenous forest cover to a set of small remnant patches (Burgess et al., 1998). Today, only approximately 3170 km^2 of East Africa's coastal forest still exists; around 5% of their original extent with forests confined to forest reserves or under community-protected '*kayas*' in Kenya (Ming'ate & Bollig, 2016; Tabor et al., 2010). Although natural habitats over much of this Biodiversity Hotspot have been greatly altered by urban development, exploitation of natural products, and agriculture (Mittermeier et al., 2011), several protected areas exist on Kenya's coast (Amin et al., 2019; Tabor et al., 2010). Some of the areas are better known than others. At 416 km^2 Arabuko Sokoke forest is the largest remaining forest patch of dry coastal forests of Kenya (Habel et al., 2017). The largest remnants of more mesic coastal forests are in the Shimba Hills National Reserve that was initially gazetted in 1903 with further extensions in 1956 and 1967. The biodiversity status of Boni-Dodori forest in northern Kenya is poorly known due to political insecurity and limited access to the area close to Somali (Amin et al., 2019). Possibly due to this isolation and political instability, Boni-Dodori is of major conservation importance to the coastal biodiversity hotspot, with indications that it remains relatively undisturbed and supports complete communities of predators and herbivores (Amin et al., 2019). The Boni-Dodori forest complex represents the only remaining section of the Kenyan coastline retaining a significant frontage of undisturbed natural habitats, ranging from coral reef to lagoons, mangroves, coastal forest, and grasslands; all supporting threatened biodiversity. Its protection and conservation are all the more urgent given the land-grabs, land conversion, and felling of indigenous hardwoods associated with, and driven by, the planned development of a major seaport at Lamu and cross-country pipeline development (Morris & Amin, 2012)—issues to be explored in Chapter 7.

Decreasing precipitation with distance from the coast leads to the remnant patches of coastal forest quickly grading into savannah (Fig. 1.7d). The composition and structure of East African savannahs have been particularly shaped by a long and interwoven history of interaction with animal and anthropogenic factors, including fire and herbivore type and density as primary drivers. Given their extensive distribution across East Africa, savannahs deserve a special place in this book. Savannah occurs where climatic conditions, edaphic controls and disturbance regimes interact to determine woody plant growth, density and species composition, grassland, and length of disturbance intervals (Lehmann et al., 2014). Savannahs are characterised by the continuous cover of grasses and an open canopy of drought-, fire- or browser-resistant trees. Savannahs are dominated by species of *Acacia* and *Commiphora* that comprise a heterogeneous tree and shrub layer, and an herbaceous matrix dominated by C_4 grasses (Fig. 1.7i). Savannahs are often characterised by areas of dense thicket interspersed with more open, park-like vegetation and maintained by edaphic, fire and/or herbivore influence (Fig. 1.7f, g, i, and k). For example, edaphic savannahs (e.g. the grass savannah of south-eastern Serengeti or Tsavo) (Fig. 1.7l) form on free-draining substrates where the growth of hydrophilous trees is restricted. Fires, both natural and anthropogenic, occur frequently and have led to the dominance of plants adapted to survive both regular and periodic burning (Carcaillet et al., 2002). Humans have interacted with savannahs for millennia, contributing substantially to their contemporary composition and distribution (Reid, 2012). The third main control on savannahs is herbivory.

While the East African savannah is home to the world's greatest diversity of ungulates where numerous species coexist (Arsenault & Owen-Smith, 2002; Hopcraft et al., 2010; Sinclair et al., 2010). The impact of herbivory on primary production is captured by grazing optimization theory (Georgiadis et al., 1989), whereby herbivores promote or suppress tree cover depending on their size, density, and mobility. For example, small browsers suppress shrubs and large browsers alter the structure and density of woodlands (Bond, 2008; Bond & Keeley, 2005; Hempson et al., 2015; Midgley & Bond, 2015). Moreover, elephants and humans are keystone species in regulating savannah patch dynamics (Western & Maitumo, 2004). Large and mobile herbivore populations, particularly African savannah elephants (*Loxodonta africana*), can destroy tree cover, causing the loss of habitat for browsing species (Guldemonde & Van Aarde, 2008), and can modify tree species composition (Rugemalila et al., 2016). While mature savannah trees are more resilient to these impacts, in that they can withstand prolonged droughts and grazing, juvenile specimens will be acutely affected. By debarking and knocking over trees elephants 'open' the woodland and allow grasses to flourish. This, in turn, attracts more grazing animals of greater variety into the area. Species of shrubs and trees with defensive thorns and hooks establish themselves in clearings, leading to thorny thickets that exclude grazers. Common trees and shrubs to savannah ecosystems include *Grewia* spp., *Lannea* spp., *Salvadora persica*, and *Terminalia* spp. Among these, larger trees are often scattered such as *Adansonia digitata, Balanites aegyptica, Kigelia africana*, and *Melia volkensii* (Gillson, 2015). At higher altitudes, from around 1600 m asl, savannah woodland grades into lower montane forests dominated by semi-drought-deciduous trees and shrubs (Fries & Fries, 1948; Herlocker & Dirchl, 1972). These forests are primarily composed of taxa such as: *Celtis* spp., Urticaceae, Myrtaceae, *Croton* spp., *Holoptelea* spp., *Prunus africana, Podocarpus milianjanus*, and *Ilex mitis* (Bussmann, 1994).

1.5.2 Highland Vegetation

Montane vegetation belongs to three broad ecosystem types: Afroalpine, Afromontane, and Guineo-Congolian (White, 1983) based upon floristic composition and distribution. There is a noticeable altitudinal overlap between the groups, as well as significant variation within the groups, leading to numerous subdivisions. Hedberg (1951) classified vegetation into a series of belts (regionally recognisable vegetation associations) and zones (locally based altitudinal bands); in ascending order these are the montane forest (Fig. 1.8c-g, h), the ericaceous belt (Fig. 1.8k) and the afroalpine belt (Fig. 1.8i, k). These broad vegetation associations are characterised by complex plant distributions that are not simply restricted by elevation but reflect dispersal and disturbance patterns (Hemp, 2006). Montane forest (3300–1600 m) comprises broad-leaved, hardwood trees, and less frequently conifers (Hamilton, 1982). Several different zones are recognised, different scholars use different terminology for similar zones, different "transitional altitudes" are delimited between the zones, and many taxa are distributed across these zones (Chapman & White, 1970; Hedberg, 1951, 1954; Lind et al., 1974; Livingstone, 1967; White, 1983). Indeed, Boughey (1955) identified thirty-nine different synonyms for montane forest. To escape this confusion, the classification for East African montane vegetation first proposed by Hamilton (1982) is adopted where montane forest is divided into lower and upper altitudinal zones, the adjectives moist and dry being used where appropriate.

At lower altitudes, the forest has structural and floristic similarities to lowland forest, although it is more species rich than the dry lower montane forest (Chapman & White, 1970). The vegetation is dominated by *Celtis* spp., *Juniperus procera* and *Olea capensis* ssp. *hochstetterii*. For the moist lower montane forests, frequent species on the lower slopes of valleys include *Entandrophragma excelsum*, *Neoboutonia macrocalyx*, *Parinari excelsa*, and *Syzygium cordatum*. In the mid-altitudes *Cassipourea ruwensorensis*, *Chrysophyllum albidum*, *Drypetes albidi*, *Ilex mitis*, *Strombosia scheffleri*, and *Zanthoxylum* spp. can be found. And finally, within the higher altitude upper montane forest, in areas that are edaphically drier, common taxa include *Faurea saligna*, *Hagenia abyssinica*, *Nuxia congesta*, *Olea capensis* ssp. and *Podocarpus milanjianus*. It is also worth noting that these are not fixed altitudinal lists, as taxa that are usually associated with dry environments in high-altitude areas can also be present at lower altitudes but in wetter locations (Marchant & Taylor, 2000). For example, *Podocarpus milanjianus* (Fig. 1.8j) is present on the north-west side of Lake Victoria at 1200 m, but it is a species usually found in a dry upper montane environment (Lind et al., 1974). Within a number of mountainous areas (e.g. Mount Elgon, the Aberdares and the Rukiga Highlands) bamboo dominated (*Sinarundinaria* spp.) forests can also be found. The Arundinoid grass *Arundinaria alpina* dominates the bamboo 'zone' where it can form mono-specific stands (Fig. 1.8f). Within East Africa bamboo is not thought to be a distinct vegetation type in its own right but is rather considered to represent a successional stage within montane forest (Hemp, 2005, 2006; Marchant & Taylor, 2000). This interpretation is further supported by the co-occurrence of taxa indicative of the regeneration of lower montane forest within bamboo stands (Marchant & Taylor, 2000). The upper altitudinal limit of the upper montane forest varies from place to place, it can reach up to 4000 m asl alongside stream courses on Mount Kenya (Coe, 1967) while also being as low as 2300 m asl on the Eastern Arc Mountains of Tanzania (Lovett & Pócs, 1993).

In the lower parts of the Ericaceous belt (3300 to 2700 m), which can experience nightly conditions of frost, overheating, and physiological drought during the day, montane forest taxa can be present. More regular rainfall, or occult precipitation, can encourage tree growth at higher altitudes, as can be seen in topographically protected, humid valley slopes (Coe, 1967; Coetzee, 1967). Therefore, the floristic composition of the lower part of the belt is very sensitive to the local climatic regime. Ericaceous belt vegetation is characterised by a microphyllous, thorny habit; other xeromorphic features are also common that allow it to withstand the cold dry environment. Three main communities are recognised (Hedberg, 1954): subalpine arborescent Ericaceous and *Senecio* forest, Afroalpine, and Afromontane Desert. Ericaceous and *Senecio* forest (Fig. 1.8k) is dominated by *Artemisia afra*, *Cliffortia nitidula*, *Erica arborea*, *Philippia johnstonii*, and *Stoebe kilimandscharica*. Subalpine ericaceous and mixed shrub communities are dominated by species of *Alchemilla* and *Helichrysum*. There is relatively close floristic similarity between the composition of ericaceous belt vegetation on different mountains of East Africa; the following genera being common; *Alchemilla*, *Artemisia*, *Cliffortia*, *Deschampsia*, *Helichrysum*, *Philippia*, and *Stoebe* (Hedberg, 1951; Hemp, 2006). Although having relatively poor species diversity, the Afroalpine flora (>3800 m) is sufficiently distinct from the surrounding lower floras to warrant its own zonation (Hedberg, 1964). No less than 80% of the taxa are endemic to the high mountains of East Africa, indicating that as a vegetation type it has long been isolated from other African mountains and more temperate areas (Hedberg, 1964). *Alchemilla* spp., *Helichrysum* scrub, and *Senecio* spp. are the dominant taxa in this zone (Hamilton, 1969; Hedberg, 1951), and other genera present include *Carduus*, *Festuca*, and *Lobelia*. As one moves into the lower altitudes, Afroalpine vegetation intergrades with microphyllous thicket (Hedberg, 1964) where *Philippia* and *Erica* dominate. These can form dense forest or open scrub depending on local edaphic and climatic factors (Harmsen et al., 1991). Above the Afroalpine vegetation is a nival zone defined by the sparse vegetation cover and rocky barren areas that characterise the Afromontane deserts of the highest peaks (Fig. 1.8k) of Kilimanjaro, Mount Elgon, Mount Kenya, and the Ruwenzori Mountains.

1.6 Disturbance Factors Influencing Ecosystem Composition

The preceding description is based on climate being the main factor determining species distributions. Superimposed on this dominant control are local disturbance factors such as fire, agricultural practice, and grazing regimes, all of which express themselves in the composition and distribution of vegetation. Given the long-term human inhabitation in East Africa (Chapter 3), it is not unexpected that there has been extensive clearance and/or modification of the natural vegetation. Disturbance regimes span from total conversion of natural vegetation to intensive agriculture and pasture, through to mixed agroforestry and seasonally exploited migratory rangelands. Much of the deforestation has focused on the mid-altitudes and very little primary montane forests remain between 1500 and 2500 m. Montane Forest is now largely restricted to protected areas such as Bwindi Impenetrable Forest, Mount Elgon, Mount Kenya, Udzungwa, Kilimanjaro, and the Ruwenzori National Parks; most of the remaining montane vegetation has been modified to varying degrees by people. The resultant vegetation often has components indicative of disturbance, such as ruderal species of *Chenopodium*, *Dodonaea*, *Plantago*, *Rumex*, and *Vernonia*. Tree genera indicative of similar disturbance can be locally more common, such as species of *Alchornea*, *Croton*, *Dombeya*, *Hagenia*, *Harungana*, *Macaranga*, and *Polyscias*. However, care must be taken when translating the presence of these taxa as disturbance indicators as they can all occur naturally within montane vegetation, albeit at relatively low values.

Fire can play a crucial part in local landscape management in the semi-dry regions of East Africa. Fire is an important ecological driver at the forest-savannah ecotone, promoting open grassland ecosystems at the expense of forest (e.g. Bond & Keeley, 2005; Cochrane et al., 1999), even in regions where the environmental (precipitation) regime would support forest (i.e. 'unstable savannah', Sankaran et al., 2005). Satellite and long-term field data have shown that fires in areas of grass-dominated savannah are controlled principally by the availability of fuel (Bond & Keeley, 2005). Although somewhat counter-intuitively, wetter conditions allow the production and accumulation of herbaceous litter, which means there is more fuel available for the fires; in contrast, drier conditions have the reverse effect limiting the amount of fuel (Archibald et al., 2010). Over large areas of East Africa, the observed spatial and interannual variability in biomass burning is primarily controlled by precipitation (relative drought) and land use (deliberate ignition, landscape fragmentation) (Archibald & Roy, 2009; Carcaillet et al., 2002; Colombaroli et al., 2014). In climates that support rain-fed agriculture (e.g. large areas of western Uganda), intensive human impact associated with high demographic pressure has often resulted in passive fire suppression due to landscape fragmentation and cropland under near-continuous rotation (Andela & van der Werf, 2014). In the drier climates characterising much of Kenya and Eastern Tanzania, a combination of pastoralism and dispersed subsistence agriculture instead results in more frequent and extensive burning; used as a management tool to clear land or promote the nutritional quality of grazing (Laris, 2002; Laris & Wardell, 2006) by removing unpalatable grasses and ticks (Ekblom et al., 2019).

1.7 Cultural and Land Use Diversity

Humans constitute the dominant control on modifying ecosystem composition and distribution and will continue to do so into the future. East Africa is characterised by a diverse range of people and has long been a locus of cultural and socio-economic changes and a major conduit and contact zone between diverse human populations: Bantu arriving from the east, Niolotes from the north and extensive trading populations from the coast. As social interventions evolved, social organisation formed definitive groupings or tribes; 'tribe' being a social/political group that people would first and foremost associate with as their cultural and ethnic group. East Africa has a rich mosaic of over 300 different tribes, each with their own language, culture, and traditions. Even though the different tribes have undergone intermarriage, transition, and splintering, there remains strong cultural affinity and identity. This diverse cultural backdrop can be separated into four broad categories that have shared livelihoods or land use practices: Hunter-gather, Coastal Swahili, the Pastoral Nilotic, and Agricultural Bantu. These groups are highly dynamic as pastoralist cultures transition to agriculture ones, mobile hunter-gatherers to settled agriculture.

Hunter-gatherers comprise a relatively minor and increasingly marginalised group that form some of the last remaining hunter-gatherers populations in the world. For example, the Hadza in north-central Tanzania that live around Lake Eyasi are descendants of ancient foraging communities. The Twa are an ancient population who live in the western forests of south-west Uganda with more extensive, but still very much minority, populations farther East. The East African coast is home to the Swahili ('People of the Coast'), descendants of Bantu-Arab traders who share a common language and mutual traditions. Although generally not regarded as a single tribal group, the Swahili have for centuries had their own distinct societal structures and can be considered a single group that follow the Islamic calendar and traditions.

The largest ethnic group across East Africa are the variety of Bantu agriculturalists that are spread out across much of the region and comprise a highly diverse group. For example, Uganda's largest tribal group, the Baganda, comprises almost 20% of Uganda's population and are the source of the country's name ('Land of the Baganda'). Similarly, the Kikuyu, who comprise about 22% of Kenya's population, are the country's largest tribal group, have their heartland surrounding Mt Kenya. Other major Bantu-speaking groups include the Meru who occupy the North-Eastern slopes of Mt Kenya and represent 6% of Kenya's population. Pare inhabit the Pare mountains in North-Eastern Tanzania, where they migrated several centuries ago from the Taita Hills area of Southern Kenya. The Sukuma, Bantu speakers from Southern Lake Victoria, comprise almost 15% of Tanzania's total population.

The Niolitic group have a proto homeland in the Nile Valley to the north of East Africa and include a diverse group of Buran and Maa-speaking populations and, although traditionally focused around cattle keeping, some large groups have transition to agricultural livelihoods. For example, the Luo who live on the north-eastern shores of Lake Victoria began their migration to the area from Sudan around the fifteenth century. Although their numbers are relatively small in Tanzania, in Kenya they comprise about 13% of the population and are the country's third-largest tribal group. The Kalenjin are one of Kenya's largest groups, comprising some 12% of the country's population, they have transitioned from pastoral to agricultural livelihood base. The Karamojong, living in Karamoja in north-eastern Uganda, are one of East Africa's most insulated tribes. As with the Samburu, Maasai and other Nilotic pastoralist peoples, life for the Karamojong centres on cattle, which are kept at night in the centre of the family living compound and graze by day on the surrounding land. Cattle are the main measure of wealth, ownership is a mark of adulthood, and cattle raiding, and warfare are central parts of the culture. Similarly, the Maasai are pastoral nomads who have actively resisted change, and many still follow the same traditional lifestyle that centres on their cattle, which, along with their land, are considered sacred. Cows provide many of their needs: milk, blood, and meat for their diet, and hides and skins for clothing, although sheep and goats also play an important dietary role, especially during the dry season. Closely related to the Maasai, and speaking the same language, the Samburu occupy the semi-arid area north of Mt Kenya and make up around 0.5% of Kenya's population. Similarly, the Turkana are a Nilotic people who live in the North-Western Kenya desert where they migrated from Southern Sudan and North-Eastern Uganda. Although the Turkana only emerged as a distinct tribal group during the early to mid-nineteenth century, they are notable today for their strong sense of tribal identification although they are closely related, linguistically and culturally, to Uganda's Karamojong.

Herding (i.e. keeping domestic animals without sole reliance on livestock for subsistence) preceded plant cultivation as the main livelihood system (Marshall & Hildebrand, 2002). East African herding systems rely upon cattle (*Bos taurus*), sheep (*Ovis aries*), goat (*Capra hircus*), camels (*Camelus dromedaries*), and/or donkeys (*Equus asinus*), with relative proportions of these animals at archaeological sites varying widely (Chritz et al., 2019). There is still considerable uncertainty regarding the origins, evolution, and timing of cultural transitions in East Africa; this uncertainty is particularly driven by the archaeological and historical records being both patchy and partial, as we will explore in Chapter 2.

References

Adler et al. (2021). IPCC Mountain Special Report.

Amin, R., Wacher, T., Bowkett, A. E., Ogwoka, B., Morris, M., & Agwanda, B. R. (2019). Africa's Forgotten Forests: The Conservation Value of Kenya's Northern Coastal Forests for Large Mammals. *Journal of East African Natural History, 107*(2), 41–61.

Andela, N., & Van Der Werf, G. R. (2014). Recent Trends in African Fires Driven by Cropland Expansion and El Nino to La Nina Transition. *Nature Climate Change, 4*(9), 791–795.

Archibald, S., & Roy, D. P. (2009). Identifying Individual Fires from Satellite-Derived Burned Area Data. *Geoscience and Remote Sensing Symposium, 2009 IEEE International, IGARSS 2009, 3,* III–160.

Archibald, S., Scholes, R. J., Roy, D. P., Roberts, G., & Boschetti, L. (2010). Southern African Fire Regimes as Revealed by Remote Sensing. *International Journal of Wildland Fire, 19*(7), 861–878.

Armanios, D. E., & Fisher, J. B. (2014). Measuring Water Availability with Limited Ground Data: Assessing the Feasibility of an Entirely Remote-Sensing-Based Hydrologic Budget of the Rufiji Basin, Tanzania, Using TRMM, GRACE, MODIS, SRB, and AIRS. *Hydrological Processes, 28*(3), 853–867.

Arsenault, R., & Owen-Smith, N. (2002). Facilitation Versus Competition in Grazing Herbivore Assemblages. *Oikos, 97*(3), 313–318.

Bamutaze, Y. (2015). Geopedological and Landscape Dynamic Controls on Productivity Potentials and Constraints in Selected Spatial Entities in Sub-Saharan Africa. In R. Lal, B. R. Singh, D. L. Mwaseba, D. Kraybill, D. O. Hansen, & L. O. Eik (Eds.), *Sustainable Intensification to Advance Food Security and Enhance Climate Resilience in Africa* (pp. 21–44). Springer International Publishing.

Barry, R. C., & Chorley, R. J. (2009). *Atmosphere, Weather and Climate.* Routledge.

Behera, S. K., Luo, J. J., Masson, S., Delecluce, P., Gualdi, S., Navarra, A., & Yamagata, T. (2005). Paramount Impact of the Indian Ocean Dipole on the East African Short Rains: A CGCM Study. *Journal of Climate, 18,* 4514–4527.

Behera, S. K., Luo, J. J., Masson, S., Yamagata, T., Delecluse, P., Gualdi, S., & Navarra, A. (2003). Impact of the Indian Ocean Dipole on the East African Short Rains: A CGCM Study. *CLIVAR Exchanges, 27,* 43–45.

Bergonzini, L., Richard, Y., Petit, L., & Camberlin, P. (2004). Zonal Circulations Over the Indian and Pacific Oceans and the Level of Lakes Victoria and Tanganyika. *International Journal of Climatology, 24*(13), 1613–1624.

Blake, G. R., Steinhardt, G. C., Pombal, X. P., Muñoz, J. C. N., Cortizas, A. M., Arnold, R. W., Schaetzl, R. J., Stagnitti, F., Parlange, J. Y., Steenhuis, T. S., Chesworth, W., Mualem, Y., Morel-Seytoux, H. J., & Spaargaren, O. (2008). Phaeozems. In W. Chesworth (Ed.), *Encyclopedia of Soil Science* (pp. 538–539). Springer Netherlands.

Bond, W. J. (2008). What Limits Trees in C_4 Grasslands and Savannas? *Annual Review Ecology and Evolution Systems, 39*, 641–659.

Bond, W. J., & Keeley, J. E. (2005). Fire as a Global 'Herbivore': The Ecology and Evolution of Flammable Ecosystems. *Trends in Ecology and Evolution, 20*(7), 387–394.

Boughey, A. S. (1955). The Vegetation of the Mountains of Biafra. *Proceedings of the Linnean Society of London, 165*(2), 144–150.

Braile, L. W., Keller, G. R., Wendlandt, R. F., Morgan, P., & Khan, M. A. (2006). The East African Rift System. *Developments in Geotectonics, 25*, 213–231.

Burgess, N. D., & Clarke, G. P. (2000). *Coastal forests of Eastern Africa*. IUCN-The World Conservation Union, Publications Services Unit.

Burgess, N. D., Fjeldså, J., & Botterweg, R. (1998). Faunal Importance of the Eastern Arc Mountains of Kenya and Tanzania. *Journal of East African Natural History, 87*(1), 37–58.

Bussmann, R. W. (1994). *The Forests of Mount Kenya (Kenya): Vegetation, Ecology, Destruction and Management of a Tropical Mountain Forest Ecosystem.* Unpublished Ph.D. dissertation, Bayreuth University.

Cai, W., Borlace, S., Lengaigne, M., Van Rensch, P., Collins, M., Vecchi, G., Timmermann, A., Santoso, A., McPhaden, M. J., Wu, L., & England, M. H. (2014). Increasing Frequency of Extreme El Niño Events Due to Greenhouse Warming. *Nature Climate Change, 4*(2), 111–116.

Calais, E., Ebinger, C., Hartnady, C., & Nocquet, J. M. (2006). Kinematics of the East African Rift from GPS and Earthquake Slip Vector Data. *Geological Society, London, Special Publications, 259*, 9–22.

Carcaillet, C., Almquist, H., Asnong, H., Bradshaw, R. H. W., Carrion, J. S., Gaillard, M. J., Gajewski, K., Haas, J. N., Haberele, S. G., Hadorn, P., Muller, S. D., Richard, P. J. H., Richoz, I., Rosch, M., Sanchez, M. F., von Stedingk, H., Stevenson, A. C., Talon, B., Tardy, C., Tinner, W., Tryterud, E., Wick, L., & Willis, K. J. (2002). Holocene Biomass Burning and Global Dynamics of the Carbon Cycle. *Chemosphere, 49*, 845–863.

Chapman, J. D., & White, F. (1970). *The Evergreen Forest of Malawi*. Holywell.

Chritz, K. L., Cerling, T. E., Freeman, K. H., Hildebrand, E. A., Janzen, A., & Prendergast, M. E. (2019). Climate, ecology, and the spread of herding in eastern Africa. *Quaternary Science Reviews, 204*, 119–132.

Cochrane, M. A., Alencar, A., Schulze, M. D., Souza, C. M., Nepstad, D. C., Lefebvre, P., & Davidson, E. A. (1999). Positive Feedbacks in the Fire Dynamic of Closed Canopy Tropical Forests. *Science, 284*(5421), 1832–1835.

Coe, R. S. (1967). Paleo-Intensities of the Earth's Magnetic Field Determined from Tertiary and Quaternary Rocks. *Journal of Geophysical Research, 72*(12), 3247–3262.

Coetzee, J. A. (1967). Pollen Analytical Studies in East and Southern Africa. *Palaeoecology of Africa, 3*, 1–146.

Colombaroli, D., Ssemmanda, I., Gelorini, V., & Verschuren, D. (2014). Contrasting Long-Term Records of Biomass Burning in Wet and Dry Savannas of Equatorial East Africa. *Global Change Biology, 20*, 2903–2914.

Costa, K., Russell, J., Konecky, B., & Lamb, H. (2014). Isotopic Reconstruction of the African Humid Period and Congo Air Boundary Migration at Lake Tana, Ethiopia. *Quaternary Science Reviews, 83*, 58–67.

Cuní-Sanchez, A., Pfeifer, M., Marchant, R., & Burgess, N. D. (2016). Ethnic and Locational Differences in Ecosystem Service Values: Insights from the Communities in Forest Islands in the Desert. *Ecosystem Services, 19*, 42–50.

Dawson, J. B. (2008). Geophysical Evidence for the Structure of the Crust and Upper Mantle of the Tanzania Craton and the Gregory Rift Valley. *Geological Society, London, Memoirs, 33*, 113–120.

Dinku, T., Ceccato, P., & Connor, S. J. (2011). Challenges of Satellite Rainfall Estimation Over Mountainous and Arid Parts of East Africa. *International Journal of Remote Sensing, 32*(21), 5965–5979.

Dondeyne, S., Ngatunga, E. L., Cools, N., Mugogo, S., & Deckers, J. (2003). Landscapes and Soils of Southeastern Tanzania: Their Suitability for Cashew. In C. P. Topper & L. J. Kasuga (Eds.), *Knowledge Transfer for Sustainable Tree Crop Development: A Case History of Tanzania Integrated Cashew Management Programme* (pp. 229–239). Bio Hybrid Agriculture Systems.

Dondeyne, S., Wijffels, A., Emmanuel, L. B., Deckers, J., & Hermy, M. (2004). Soils and Vegetation of Angai Forest: Ecological Insights from a Participatory Survey in South Eastern Tanzania. *African Journal of Ecology, 42*(3), 198–207.

Ekblom, A., Shoemaker, A., Gillson, L., Lane, P., & Lindholm, K. J. (2019). Conservation Through Biocultural Heritage—Examples from Sub-Saharan Africa. *Land, 8*(1), 5.
Fries, R. E., & Fries, T. C. (1948). *Phytogeographical Researches on Mt. Kenya and Mt. Aberdare, British East Africa.* Almqvist and Wiksell.
Funakawa, S., Yoshida, H., Kilasara, M., Sugihara, S., Kosaki, T., & Watanabe, T. (2012). Soil Fertility Status and Its Determining Factors in Tanzania. In M. C. Hernandez-Soriano (Ed.), *Soil Health and Land Use Management.* InTech. https://doi.org/10.5772/29199
Georgiadis, N. J., Ruess, R. W., McNaughton, S. J., & Western, D. (1989). Ecological Conditions That Determine When Grazing Stimulates Grass Production. *Oecologia, 81*(3), 316–322.
Gichangi, E. M., Gatheru, M., Njiru, E. N., Mungube, E. O., Wambua, J. M., & Wamuongo, J. W. (2015). Assessment of Climate Variability and Change in Semi-Arid Eastern Kenya. *Climatic Change, 130*(2), 287–297.
Gillson, L. (2015). Evidence of a Tipping Point in a Southern African Savanna? *Ecological Complexity, 21,* 78–86.
Gray, C. L. (2011). Soil Quality and Human Migration in Kenya and Uganda. *Global Environmental Change, 21*(2), 421–430.
Guldemond, R., & Van Aarde, R. (2008). A Meta-Analysis of the Impact of African Elephants on Savanna Vegetation. *Journal of Wildlife Management, 72*(4), 892–899.
Habel, J. C., Casanova, I. C. C., Zamora, C., Teucher, M., Hornetz, B., Shauri, H., Mulwa, R. K., & Lens, L. (2017). East African Coastal Forest Under Pressure. *Biodiversity and Conservation, 26*(11), 2751–2758.
Hamilton, A. C. (1969). The Vegetation of Southwest Kigezi. *Uganda Journal, 33*(2), 175–199.
Hamilton, A. C. (1982). *Environmental History of East Africa: A Study of the Quaternary.* Academic Press.
Harmsen, R., Spence, J. R., & Mahaney, W. C. (1991). Glacial Interglacial Cycles and Development of the Afroalpine Ecosystem on East African Mountains II. Origins and Development of the Biotic Component. *Journal of African Earth Sciences (and the Middle East), 12*(3), 513–523.
Hedberg, O. (1951). Vegetation Belts of East African Mountains. *Svensk Botanisk Tidskrift, 45,* 140–195.
Hedberg, O. (1954). A Pollen-Analytical Reconnaissance in Tropical East Africa. *Oikos, 5*(2), 137–166.
Hedberg, O. (1964). *Afroalpine Plant Ecology.* Almqvist and Wiksell.

Hemp, A. (2005). Climate Change-Driven Forest Fires Marginalize the Impact of Ice Cap Wasting on Kilimanjaro. *Global Change Biology, 11*(7), 1013–1023.

Hemp, A. (2006). Continuum or Zonation? Altitudinal Gradients in the Forest Vegetation of Mt. Kilimanjaro. *Plant Ecology, 184*(1), 27–42.

Hempson, G. P., Archibald, S., Bond, W. J., Ellis, R. P., Grant, C. C., Kruger, F. J., Kruger, L. M., Moxley, C., Owen-Smith, N., Peel, M. J., & Smith, I. P. (2015). Ecology of Grazing Lawns in Africa. *Biological Reviews, 90*(3), 979–994.

Herlocker, D. J., & Dirchl, H. J. (1972). *Vegetation of Ngorongoro Conservation Area, Tanzania.* Canadian Wildlife Service, Report Series No. 19.

Hijmans, S. E., Cameron, J. L., Parra, P. G., & Jones, A. J. (2005). Very high resolution interpolated climate surfaces for global land areas. *International Journal of Climatology, 25*, 1965–1978.

Hopcraft, J. G. C., Olff, H., & Sinclair, A. R. E. (2010). Herbivores, Resources and Risks: Alternating Regulation Along Primary Environmental Gradients in Savannas. *Trends in Ecology and Evolution, 25*(2), 119–128.

Indeje, M., Semazzi, H. M. F., & Ogallo, J. L. (2000). ENSO Signal in East African Rainfall Seasons. *International Journal of Climatology, 20,* 19–46.

ISRIC. (2006). *Soils of Tanzania and Their Potential for Agriculture Development*. Draft Report. Milingano Agricultural Research Institute, Department of Research and Training, Ministry of Agriculture, Food Security and Co-Operatives, Tanga, Tanzania, p. 34.

IUSS Working Group WRB. (2015). World Reference Base for Soil Resources 2014, Update 2015.

Jones, A., Breuning-Madsen, H., Brossard, M., Dampha, A., Deckers, J., Dewitte, O., Gallali, T., Hallett, S., Jones, R., Kilasara, M., Le Roux, P., Micheli, E., Montanarella, L., Spaargaren, O., Thiombiano, L., Van Ranst, E., Yemefack, M., & Zougmoré, R. (2013). *Soil Atlas of Africa.* European Commission.

Kapkiyai, J. J., Karanja, N. K., Qureshi, J. N., Smithson, P. C., & Woomer, P. L. (1999). Soil Organic Matter and Nutrient Dynamics in a Kenyan Nitisol Under Long-Term Fertilizer and Organic Input Management. *Soil Biology and Biochemistry, 31*(13), 1773–1782.

Karlén, W., Fastook, J. L., Holmgren, K., Malmstrom, M., Matthews, J. A., Odada, E., Risberg, J., Rosquist, G., Sandgren, P., Shemesh, A., & Westerberg, L. O. (1999). Glacier Fluctuations on Mount Kenya Since 6000 cal. Years BP: Implications for Holocene Climatic Change in Africa. *Ambio, 28,* 409–418.

Kasanzu, C. H. (2016). Apatite Fission Track and (U-Th)/He Thermochronology from the Archean Tanzania Craton: Contributions to Cooling Histories of Tanzanian Basement Rocks. *Geoscience Frontiers* [online]. https://doi.org/10.1016/j.gsf.2016.09.007

Kenworthy, J. M. (1966). Temperature Conditions in the Tropical Highland Climates of East Africa. *East African Geographical Review, 1966*(4), 1–11.

Kisinyo, P. O., Othieno, C. O., Gudu, S. O., Okalebo, J. R., Opala, P. A., Maghanga, J. K., & Ogola, B. O. (2013). Phosphorus Sorption and Lime Requirements of Maize Growing Acid Soils of Kenya. *Sustainable Agriculture Research, 2*(2), 116.

Lal, R. (2010). Enhancing Eco-Efficiency in Agro-Ecosystems Through Soil Carbon Sequestration. *Crop science, 50*(Supplement 1), 120–131.

Laris, P. (2002). Burning the Seasonal Mosaic: Preventative Burning Strategies in the Wooded Savanna of Southern Mali. *Human Ecology, 30*(2), 155–186.

Laris, P., & Wardell, D. A. (2006). Good, Bad or 'Necessary Evil'? Reinterpreting the Colonial Burning Experiments in the Savanna Landscapes of West Africa. *The Geographical Journal, 172*(4), 271–290.

Lehmann, C. E., Anderson, T. M., Sankaran, M., Higgins, S. I., Archibald, S., Hoffmann, W. A., Hanan, N. P., Williams, R. J., Fensham, R. J., Felfili, J., & Hutley, L. B. (2014). Savanna Vegetation-Fire-Climate Relationships Differ Among Continents. *Science, 343*(6170), 548–552.

Lind, E. M., Morrison, M. E., & Hamilton, A. C. (1974). *East African Vegetation*. Longman.

Livingstone, D. (1967). Postglacial Vegetation of the Ruwenzori Mountains in Equatorial Africa. *Ecological Monographs, 37*, 25–52.

Loomis, S. E., Russell, J. M., Verschuren, D., Morrill, C., De Cort, G., Damsté, J. S. S., Olago, D., Eggermont, H., Street-Perrott, F. A., & Kelly, M. A. (2017). The Tropical Lapse Rate Steepened During the Last Glacial Maximum. *Science Advances, 3*(1), e1600815.

Los, S. O., Street-Perrott, F. A., Loader, N. J., Froyd, C. A., Cuní-Sanchez, A., & Marchant, R. A. (2019). Sensitivity of a Tropical Montane Cloud Forest to Climate Change, Present, Past and Future: Mt. Marsabit, N. Kenya. *Quaternary Science Reviews, 218*, 34–48.

Lovett, J. C., & Pócs, T. (1993). *Assessment of the Condition of the Catchment Forest Reserves: A Botanical Appraisal, Iringa Region*. Catchment Forestry Report 93.3. Dar es Salaam: Forestry and Beekeeping Division, Ministry of Tourism, Natural Resources and Environment, p. 300.

Lufafa, A., Tenywa, M. M., Isabirye, M., Majaliwa, M. J. G., & Woomer, P. L. (2003). Prediction of Soil Erosion in a Lake Victoria Basin Catchment

Using a GIS-Based Universal Soil Loss Model. *Agricultural Systems, 76*(3), 883–894.

Lyon, B. (2014). Seasonal Drought in the Greater Horn of Africa and Its Recent Increase During the March-May Long Rains. *Journal of Climate, 27,* 7953–7975.

Lyon, S. W., Koutsouris, A., Scheibler, F., Jarsjö, J., Mbanguka, R., Tumbo, M., Robert, K. K., Sharma, A., & Velde, Y. (2015). Interpreting Characteristic Drainage Timescale Variability Across Kilombero Valley, Tanzania. *Hydrological Processes, 29*(8), 1912–1924.

Mace, G. M., Norris, K., & Fitter, A. H. (2012). Biodiversity and Ecosystem Services: A Multilayered Relationship. *Trends in Ecology and Evolution, 27*(1), 19–26.

Macgregor, D. (2015). History of the Development of the East African Rift System: A Series of Interpreted Maps Through Time. *Journal of African Earth Sciences, 101,* 232–252.

Marchant, R. A., & Hooghiemstra, H. (2004). Rapid Environmental Change in Tropical Africa and Latin America About 4000 Years Before Present: A Review. *Earth Science Reviews, 66,* 217–260.

Marchant, R. A., & Taylor, D. M. (2000). Numerical Analysis of Modern Pollen Spectra and in situ Montane Forest-Implications for the Interpretation of Fossil Pollen Sequences from Tropical Africa. *New Phytologist, 146,* 505–515.

Marchant, R., Mumbi, C., Behera, S., & Yamagata, T. (2007). The Indian Ocean Dipole–The Unsung Driver of Climatic Variability in East Africa. *African Journal of Ecology, 45*(1), 4–16.

Marshall, F., & Hildebrand, E. (2002). Cattle Before the Crops: The Beginnings of Food Production in Africa. *Journal of World Prehistory, 16,* 99–143.

Maslin, M. A., Brierley, C. M., Milner, A. M., Shultz, S., Trauth, M. H., & Wilson, K. E. (2014). East African Climate Pulses and Early Human Evolution. *Quaternary Science Reviews, 101,* 1–17.

Mathew, M. M., Majule, A. E., Marchant, R., & Sinclair, F. (2016). Variability of Soil Micronutrients Concentration Along the Slopes of Mount Kilimanjaro, Tanzania. *Applied and Environmental Soil Science.* https://doi.org/10.1155/2016/9814316

Matus, F., Rumpel, C., Neculman, R., Panichini, M., & Mora, M. L. (2014). Soil Carbon Storage and Stabilisation in Andic Soils: A Review. *Catena, 120,* 102–110.

Mbaluka, J. K., & Brown, F. H. (2016). Vegetation of the Koobi Fora Region Northeast of Lake Turkana, Marsabit County, Northern Kenya. *Journal of East African Natural History, 105*(1), 21–50.

Mbungu, W. B., & Kashaigili, J. J. (2017). Assessing the Hydrology of a Data-Scarce Tropical Watershed Using the Soil and Water Assessment Tool: Case of the Little Ruaha River Watershed in Iringa, Tanzania. *Open Journal of Modern Hydrology, 7*(2), 65.

Mganga, K. Z., Musimba, N. K. R., Nyariki, D. M., Nyangito, M. M., & Mwang'ombe, A. W. (2015). The Choice of Grass Species to Combat Desertification in Semi-Arid Kenyan Rangelands Is Greatly Influenced by Their Forage Value for Livestock. *Grass and Forage Science, 70*(1), 161–167.

Midgley, G. F., & Bond, W. J. (2015). Future of African Terrestrial Biodiversity and Ecosystems Under Anthropogenic Climate Change. *Nature Climate Change, 5*(9), 823–829.

Ming'ate, F. L. M., & Bollig, M. (2016). Local Rules and Their Enforcement in the Arabuko-Sokoke Forest Reserve Co-Management Arrangement in Kenya. *Journal of East African Natural History, 105*(1), 1–19.

Mittermeier, R. A., Turner, W. R., Larsen, F. W., Brooks, T. M., & Gascon, C. (2009). Global Biodiversity Conservation: The Critical Role of Hotspots. In F. E. Zachos & J. C. Habel (Eds.), *Biodiversity Hotspots—Distribution and Protection of Conservation Priority Areas*. Springer.

Mittermeier, R. A., Turner, W. R., Larsen, F. W., Brooks, T. M., & Gascon, C. (2011). Global Biodiversity Conservation: The Critical Role of Hotspots. In *Biodiversity Hotspots* (pp. 3–22). Springer.

Morris, M., & Amin, R. (2012). An Update on the Threats to Afrotheria in Northern Coastal Kenya. *Afrotherian Conservation* [online], *9*, 6–7. IUCN. http://afrotheria.net/newsletter.html

Msanya, B. M., Magoggo, J. P., & Otsuka, H. (2002). Development of Soil Surveys in Tanzania (Review). *Pedologist, 46*(2), 79–88.

Mutai, C. C., & Ward, M. N. (2000). East African Rainfall and the Tropical Circulation/Convection on Inter-Seasonal to Interannual Timescales. *Journal of Climate, 13*, 3915–3939.

Nandwa, S. M., & Bekunda, M. A. (1998). Research on Nutrient Flows and Balances in East and Southern Africa: State-Of-The-Art. *Agriculture, Ecosystems and Environment, 71*(1), 5–18.

Nelson, D. M., Verschuren, D., Urban, M. A., & Hu, F. S. (2012). Long-Term Variability and Rainfall Control of Savanna Fire Regimes in Equatorial East Africa. *Global Change Biology* [online], *18*, 3160–3170. https://doi.org/10.1111/j.1365-2486.2012.02766.x

Nicholson, S. E. (1994). Recent Rainfall Fluctuations in Africa and Their Relationship to Past Conditions Over the Continent. *The Holocene, 4,* 121–131.

Nicholson, S. E. (1996). Environmental Change Within the Historical Period. In W. A. Adams, A. S. Goudie, & A. R. Orme (Eds.), *The Physical Geography of Africa* (pp. 60–87). Oxford University Press.

Nicholson, S. E. (2000). The Nature of Rainfall Variability Over Africa on Time Scales of Decades to Millennia. *Global and Planetary Change, 26,* 137–158.

Nicholson, S. E. (2001). Climatic and Environmental Change in Africa During the Last Two Centuries. *Climate Research, 17*(2), 123–144.

Nicholson, S. E. (2017). Climate and climatic variability of rainfall over eastern Africa. *Reviews of Geophysics, 55*(3), 590–635.

Nicholson, S. E., Nash, D. J., Chase, B. M., Grab, S. W., Shanahan, T. M., Verschuren, D., Asrat, A., Lezine, A. M., & Umer, M. (2013). Temperature Variability Over Africa During the Last 2000 Years. *The Holocene, 23,* 1085–1094.

Nyombi, K., Van Asten, P. J., Corbeels, M., Taulya, G., Leffelaar, P. A., & Giller, K. E. (2010). Mineral Fertilizer Response and Nutrient use Efficiencies of East African Highland Banana (Musa spp., AAA-EAHB, cv. Kisansa). *Field Crops Research, 117*(1), 38–50.

Nziguheba, G., Zingore, S., Kihara, J., Merckx, R., Njoroge, S., Otinga, A., Vandamme, E., & Vanlauwe, B. (2016). Phosphorus in Smallholder Farming Systems of Sub-Saharan Africa: Implications for Agricultural Intensification. *Nutrient Cycling in Agroecosystems, 104*(3), 321–340.

Oettli, P., & Camberlin, P. (2005). Influence of Topography on Monthly Rainfall Distribution Over East Africa. *Journal of Climate Research, 28,* 199–212.

Ogallo, L. J., Jonawiak, J. E., & Halpert, M. S. (1988). Teleconnection Between Seasonal Rainfall over East Africa and Global Sea-Surface Temperature Anomalies. *Journal of Meteorological Society of Japan, Series II*(66), 807–822.

Omeny, P., Ogallo, L., Okoola, R., Hendon, H., & Wheeler, M. (2008). East African Rainfall Variability Associated with Madden-Julian Oscillation. *Journal of Meteorological and Related Sciences, 2,* 109–118.

Omuto, C. T. (2013). Chapter 11: Major Soil and Data Types in Kenya. In P. Paron, D. O. Olago, & C. T. Omuto (Eds.), *Kenya: A Natural Outlook: Geo-Environmental Resources and Hazards, 16,* 123.

Owiti, Z., Ogallo, L., & Mutemi, J. (2008). Linkages Between Indian Ocean Dipole and East African Seasonal Rainfall Anomalies. *Journal of Meteorological and Related Sciences, 2,* 3–17.

Oyana, T. J., Kayendeke, E., Bamutaze, Y., & Kisanga, D. (2015). A Field Assessment of Land Use Systems and Soil Properties at Varied Landscape Positions in a Fragile Ecosystem of Mount Elgon, Uganda. *African Geographical Review, 34*(1), 83–103.

Pachpute, J. S., Tumbo, S. D., Sally, H., & Mul, M. L. (2009). Sustainability of Rainwater Harvesting Systems in Rural Catchment of Sub-Saharan Africa. *Water Resources Management, 23*(13), 2815–2839.

Platts, P. J., Omeny, P. A., & Marchant, R. (2015). AFRICLIM: High-Resolution Climate Projections for Ecological Applications in Africa. *African Journal of Ecology, 53*(1), 103–108.

Punwong, P., Marchant, R., & Selby, K. (2013a). Holocene Mangrove Dynamics and Environmental Change in the Rufiji Delta, Tanzania. *Vegetation History and Archaeobotany, 22*(5), 381–396.

Punwong, P., Marchant, R., & Selby, K. (2013b). Holocene Mangrove Dynamics in Makoba Bay, Zanzibar. *Palaeogeography, Palaeoclimatology, Palaeoecology, 379,* 54–67.

Punwong, P., Marchant, R., & Selby, K. (2013c). Holocene Mangrove Dynamics from Unguja Ukuu, Zanzibar. *Quaternary International, 298,* 4–19.

Reid, R. (2012). *Savannas of Our Birth: People, Wildlife, and Change in East Africa.* University of California Press.

Rugemalila, D. R., Holdo, R., & Anderson, T. M. (2016). Precipitation and Elephants, Not Fire, Shape Tree Community Composition in Serengeti. *Biotropica, 48,* 476–482.

Rwetabula, J., De Smedt, F., & Rebhun, M. (2012). Simulation of Hydrological Processes in the Simiyu River, Tributary of Lake Victoria, Tanzania. *Water SA, 38*(4), 623–632.

Saji, H. N., Goswami, B. N., Vinayachandran, P. N., & Yamagata, T. (1999). A Dipole Mode in the Tropical Indian Ocean. *Nature, 401,* 360–363.

Sakané, N., Alvarez, M., Becker, M., Böhme, B., Handa, C., Kamiri, H. W., & Möseler, B. M. (2011). Classification, Characterisation, and Use of Small Wetlands in East Africa. *Wetlands, 31*(6), 1103–1116.

Sankaran, M., Hanan, N. P., Scholes, R. J., Ratnam, J., Augustine, D. J., Cade, B. S., & Ardo, J. (2005). Determinants of Woody Cover in African Savannas. *Nature, 438*(7069), 846–849.

Selway, K. (2015). Negligible Effect of Hydrogen Content on Plate Strength in East Africa. *Nature Geoscience, 8,* 543–546.

Schlüter, T. (2008). *Geological Atlas of Africa* (pp. 110–113). Springer.

Shanko, D., & Camberlin, P. (1998). The Effects of the Southwest Indian Ocean Tropical Cyclones on Ethiopian Drought. *International Journal of Climatology, 18*(12), 1373–1388.

Shi, D., Chen, G., Xiong, L., Huang, Y., Liu, K., Xing, Y., Liu, Z., & Huang, Y. (2015). Properties of Basic Soils in Different Types and Improvements and Uses in Burundi. *Agricultural Science and Technology, 16*(4), 733.

Shoji, S., Dahlgren, R., & Nanzyo, M. (1993). Terminology, Concepts and Geographic Distribution of Volcanic Ash Soils. *Developments in Soil Science, 21,* 1–5.

Sinclair, A. R. E., Metzger, K., Brashares, J. S., Nkwabi, A., Sharam, G., & Fryxell, J. M. (2010). Trophic Cascades in African Savanna: Serengeti as a Case Study. In *Trophic Cascades: Predators, Prey and the Changing Dynamics of Nature*, 255–274.

Street, F. A., & Grove, A. T. (1975). Environmental and Climatic Implications of Late Quaternary Lake-Level Fluctuations in Africa. *Nature, 261,* 385–390.

Summerfield, M. A. (2005). A Tale of Two Scales, or the Two Geomorphologies. *Transactions of the Institute of British Geographers, 30*(4), 402–415.

Tabor, K., Burgess, N. D., Mbilinyi, B. P., Kashaigili, J. J., & Steininger, M. K. (2010). Forest and Woodland Cover and Change in Coastal Tanzania and Kenya, 1990 to 2000. *Journal of East African Natural History, 99*(1), 19–45.

Thorn, J., Marchant, R., Cuni-Sanchez et al. (2021). Climate Change Perspectives from Mountains in Tanzania. In Review Land. In press.

Trauth, M. H., Maslin, M. A., Deino, A. L., Strecker, M. R., Bergner, A. G., & Dühnforth, M. (2007). High-and Low-Latitude Forcing of Plio-Pleistocene East African Climate and Human Evolution. *Journal of Human Evolution, 53*(5), 475–486.

Trauth, N., Schmidt, C., Vieweg, M., Maier, U., & Fleckenstein, J. H. (2014). Hyporheic Transport and Biogeochemical Reactions in Pool-Riffle Systems Under Varying Ambient Groundwater Flow Conditions. *Journal of Geophysical Research: Biogeosciences, 119*(5), 910–928.

Walling, D. E. (1999). Linking Land Use, Erosion and Sediment Yields in River Basins. In J. Garnier & J. M. Mouchel (Eds.), *Man and River Systems* (pp. 223–240). Springer Netherlands.

Wasige, J. E., Groen, T. A., Rwamukwaya, B. M., Tumwesigye, W., Smaling, E. M. A., & Jetten, V. (2014). Contemporary Land Use/Land Cover Types

Determine Soil Organic Carbon Stocks in South-West Rwanda. *Nutrient Cycling in Agroecosystems, 100*(1), 19–33.

Western, D., & Maitumo, D. (2004). Woodland Loss and Restoration in a Savanna Park: A 20-Year Experiment. *African Journal of Ecology, 42*(2), 111–121.

White, F. (1983). *The Vegetation of Africa, a Descriptive Memoir to Accompany the UNESCO/AETFAT/UNSO Vegetation Map of Africa* (3 Plates, Northwestern Africa, Northeastern Africa, and Southern Africa, 1: 5,000,000).

Yang, W. C., Seager, R., Cane, M. A., & Lyon, B. (2015). The Annual Cycle of East African Precipitation. *Journal of Climate, 28,* 2385–2404.

2

Origins and Migration: Environmental and Cultural Change Over the Last 300,000 Years in East Africa

2.1 Introduction

To understand transitions back in time we must draw on paleoenvironmental research that can access geoarchives and other natural sedimentary sources of information on the Earth's history to examine signals of environmental change, often using environmental indicators and proxy measurements (Marchant et al., 2018; Smol et al., 2001). The interpretative power of paleoenvironmental records increases dramatically when they are combined with archaeological data to begin exploring past interactions (Marchant & Lane, 2014). Archaeologists and palaeoecologists are currently addressing the correlations and interrelated connections between environmental change and changes in land use across East Africa, indeed across the world, through the land cover 6K project. As the intensity of crop growth, irrigation, wood harvesting, and use of fire the fingerprints of anthropogenic land cover change have become more transformative, the spatial and temporal linkages between East African societies and environmental change interrelations become more and more intertwined. There is the increasing utility of this information of past land cover change on contemporary issues such as ecosystem and

R. Marchant, *East Africa's Human Environment interactions*,
https://doi.org/10.1007/978-3-030-88987-6_2

climate modelling, biogeochemical cycling, managing ecosystem services, and conservation (Chapter 7). As more paleoenvironmental and archaeological records become available, it is apparent that the only constant in the East African environment, ecosystems, and cultures is change.

Research in East Africa has focussed on revealing cultural and environmental prehistory (Butzer, 1968; Marchant & Lane, 2014), as such it is an ideal test bed for investigating environmental-cultural dynamics, for discussing the interrelationships of these and for assessing how these have evolved to shape the present (Chapter 1) and can consequently be used to inform the future. Across East Africa, when investigating how climate change has impacted ecosystems and how those ecosystems have responded, sedimentary records have proven to be invaluable over a range of time scales (Graham et al., 2003). Fluctuations in past lake levels (e.g. Cohen et al., 1997; Gasse, 2000, 2002; Kiage & Liu, 2006; Street-Perrott et al., 2000; Verschuren et al., 2002) and ecosystem changes (Elenga et al., 2000; Finch et al., 2017; Hamilton, 1982; Mumbi et al., 2008, 2014; Rucina et al., 2009) have been well documented and demonstrate the sensitivity of East African ecosystems, in particular to moisture variability (Bollig, 2016; Vincens et al., 1999) and the ensuing interaction with drivers such as fire (Colombaroli et al., 2014; Finch et al., 2017) and plant-animal interactions (Kariuki et al., 2021). Drought and flood episodes throughout geological history have been more dramatic, and more persistent, than those captured by the relatively short instrumental records available; the latter failing to capture the full spectrum of climatic variability across the region (Gasse, 2000). Despite the clear value of the palaeo-archive, understanding abrupt climate change in the past remains challenging. Understanding climate change remains of paramount importance and, to date, the focus has been principally on the impacts of increasing global temperatures, ice sheet melting, and rising levels of atmospheric carbon dioxide (IPCC, 2021). However, there is great complexity in the processes and associated impacts of climate change, which are the results of interactions between changes in solar activity, atmospheric composition, atmospheric circulation, land-surface conditions, ocean currents, and human activities (Marchant,

2021; Marchant & Lane, 2014). Within the Global Change community there is an increasing recognition of the importance of the tropics for shedding light on complex climatic and environmental histories, as well as a growing understanding of how these histories interact with people through time. This volume draws on a realisation of the importance of providing new insights into ecosystem histories by collating available archaeological and palaeoecological evidence for cultural and environmental change across East Africa.

2.2 Detection of Environmental Change Across East Africa

In East Africa, the most abundant records of past environmental variability have been established from the analysis of lacustrine, palustrine, and peat geoarchives (Fig. 2.1).

As sediments accumulate in these terrestrial depositional systems, the properties of the sediment stratigraphy maintain environmental signals that provide a history of ecosystem variables, such as precipitation, biological productivity, and vegetation type (Romans et al., 2016). Variability in these environmental signals can be caused by natural and anthropogenically modified processes; thus, they provide us with insights into long-term environmental processes, abrupt ecosystem perturbations and associated consequences, and human-environment interactions. Other sources of paleoenvironmental information include coastal deposits, Indian Ocean marine sediments, corals, cave deposits, and tree rings (Stahl, 2009). The environmental signals embedded within these geo- and bio-archives provide us with information on how ecosystems and their processes changed at multiple spatial and temporal scales. There are multiple sources for examining the geological and environmental history of East Africa.

The East African landscape is characterised by numerous lakes, swamps, mires, and peatlands (Figs. 2.1 and 2.2). The large basins are influenced by broad-scale regional processes such as climate and climate-mediated erosion rates and the sediment basins are often much older, extending beyond the Quaternary (Butzer et al., 1972; Johnson et al.,

Fig. 2.1 Sources of palaeoenvironmental information. (**a**) Glacial ice archive on Kilimanjaro (photo: Rob Marchant). (**b**) Maua mire, at nearly 4000 m asl south-eastern Kilimanjaro (photo: Rob Marchant). (**c**) At over 2100 m asl, Rumuiku Swamp sits within a volcanic crater on eastern Mount Kenya (photo: Rob Marchant). (**d**) Palustrine swamp, a source of organic-rich sediments in the relatively arid areas of Amboseli (photo: Rob Marchant). (**e**) Cyperaceae-dominated wetland northern Kenya (photo: Rob Marchant). (**f**) Mangrove sediments, Zanzibar (photo: Rob Marchant). (**g**) Fluvially incised exposed soil facies northern Tanzania (photo: Rob Marchant). (**h**) Cut timbers at Mpingo showing tropical tree growth rings (photo: Neil Burgess). (**i**) A stalagmite collected from Cold Air Cave, South Africa can provide a palaeoclimatic record (photo: Karin Holmgren)

2002; Muiruri et al., 2021a). The smaller, shallow lakes of the Western Rift respond more rapidly to hydroclimatic changes and can therefore dry out completely, often leading to discontinuous late-Holocene sediment records (Verschuren et al., 2002). Small lake basins are often much younger than their larger counterparts, forming at the end of volcanic processes these can provide sediment records from the Pleistocene or earlier (Courtney-Mustaphi et al., 2016; Ficken et al., 2002; Maslin, 2016; Rucina et al., 2009). Each lake will respond as high or low stands as the hydrological budget changes due to a combination of broad scale, as well as local scale, climatic factors that influence hydrology (Sect. 1.2).

Fig. 2.2 Sources of palaeoenvironmental geoarchives can be quite varied. (**a**) Mubwindi Swamp, Bwindi Forest, Uganda. (**b**) Lake Momella, Arusha National Park. (**c**) Lake Victoria, views from Speke Bay Tanzania. (**d**) Mangrove sediments, Zanzibar. (**e**) Amboseli Swamp. (**f**) Lake Amboseli during a period of flooding. (**g**) Lake Naivasha during a relatively high stand (as demonstrated by the drowned forest). (**h**) A salt lake Serengeti National Park). (**i**) Paradise Lake, an extinct volcanic crater in Mount Marsabit National Park (**j**) Sediments can accumulate within small depressions in Kopi—Serengeti National Park. (**k**) Deva Deva Swamp, Uluguru Mountains, Tanzania. (**l**) Lake Magadi, Ngorongoro crater, Tanzania (All photographs: Rob Marchant)

An important mechanism of the formation of sedimentary archives at the highest elevations is the retreat of the montane ice caps since the Last Glacial Maximum, which led to the development of small lakes and mires in deglaciated valleys and tarns (Courtney-Mustaphi et al., 2017; Karlén et al., 1999; Perrott, 1982; Rietti-Shati et al., 1998). Other sources of paleoenvironmental history for East Africa come from coastal mangroves, corals, glacial deposits, geomorphology, dendrochronology, speleothems, soils, marine sediments, and peat deposits (Fig. 2.1). As the number of sediment-based records from East Africa is steadily increasing the community can move from a largely descriptive base to an interpretative one. This exciting transition is only possible due to the long tradition of palaeoecological research, extending back to some of the pioneering work of the late 1950s and early 1960s, that laid the foundations for an increasingly dense assemblage of palaeoecological sites that clearly demonstrate the highly responsive nature of East African ecosystems to past climatic variability (Finch et al., 2014; Marchant et al., 2018).

2.3 Detection of Cultural Changes in East African Geoarchives

From the earliest recorded presence of humans in East Africa, through until the present day, human-induced ecosystem change has gone from being relatively minor and restricted (Shipton et al., 2018), to becoming widespread and a dominate factor behind ecosystem composition and distribution (Marchant et al., 2018). Increasing human pressure on natural resources, radiating out from urban centres that have increasingly wide spheres of influence (Ahrends et al., 2010), ensures that, as with every corner of the world, all East African ecosystems have been impacted on by humans.

Information on past human activities often derives from sedimentary sequences based upon the assumption that major changes in human activities will impact upon the sediments by altering catchment conditions. Such accounts are particularly sensitive to recording the environmental effects of human populations, especially in sedimentary basins, since these both contain resources that attract human populations (water, deep soils) and allow for the accumulation of thick sediment sequences. Detecting anthropogenic signals and diagnostic characteristics in paleoenvironmental data is possible through sedimentologic analysis (e.g., sediment magnetism, particle sizes, isotopes) and proxy indicators (e.g., pollen types, plankton and mollusc remains) associated with known types of human land use or land cover change, archaeological or documentary evidence. However, human actions are not always commuted to the sedimentary record—an absence of human indicators does not mean humans were absent. Similarly, humans are not solely responsible for disturbance in a catchment. For example, we should be wary of assuming that the onset of soil erosion can be readily equated with vegetation clearance (e.g. Heckmann, 2011; for discussion see, e.g., Lane,

2009). There are several possible reasons for unexpectedly weak signals of human activity in pollen records. First, incompatible, or loose chronologies may mean that researchers are looking for evidence in the wrong part of the sedimentary column. Second, some past societies may indeed have trodden lightly on the landscape, particularly where population densities were low and/or populations were highly mobile.

Studies of sedimentary sequences from East Africa in general have tended to rely on relatively few proxy indicators of human impact such as certain pollen and diatom types, and the abundance of charcoal (Finch et al., 2017; Hawthorne et al., 2017; Lejju et al., 2005; Rucina et al., 2010). Fire, for example, is now recognised as an important driver of vegetation dynamics with humans clearly playing a major role in the distribution of vegetation communities (Haberle & Chepstow-Lusty, 2000; Haberle et al., 2001). Combining these proxies enables us to chart anthropogenic modification and be used to partition the degree of change exerted by human activities over natural drivers. The principal anthropogenic influences on the landscapes of East Africa are a reduction in arboreal cover and changes in fire activity. These interpretations of anthropogenic change are inferred from the pollen record (a reduction in arboreal pollen) and other proxies, such as peaks in microscopic charcoal (indicative of burning). This is exemplified from numerous studies such as those from Ahakagyezi (Hamilton et al., 1986; Taylor, 1990) and Muchoya Swamps (Marchant & Taylor, 1998; Taylor, 1990) in Uganda, Sacred Lake (Street-Perrott et al., 1997) in Kenya, and Lake Tanganyika (Cohen et al., 1997; Msaky et al., 2005) and Lake Masoko (Vincens et al., 2007b) in Tanzania (for locations see Fig. 1.5). These tools are not without their challenges. Pollen, one of the mainstays of detecting environmental change and human interaction, can be a relatively blunt tool to trace human impacts. Similar changes in the composition of vegetation and modified burning regimes are also evident in mid-elevation records from Mount Shengena in the Eastern Arc Mountains (Finch et al., 2017), reflecting the intensification of human activities rather than the mere presence of people in the landscape. Thus, it is often assumed in East African palynology that the introduction of agriculture, in the absence of direct indicators such as pollen from domesticated plants, is marked by evidence of forest clearance and burning (e.g., Finch et al.,

2017; Rucina et al., 2010). This assumption may hold for some activities, but shifting cultivators often make use of natural forest gaps or move with sufficient regularity to allow forest recovery (e.g. Fairhead & Leach, 1996), and indeed some forms of agriculture could be said to promote tree cover such as low intensity cultivation of oil palm (*Elaeis guineensis*), or the protection of trees to provide shade to crops grown within so-called 'forest gardens' (e.g. Stump & Tagseth, 2009). Of equal relevance to discussions of the detectability of change is the issue that human-induced alterations in some vegetation types, such as humid rainforest, may simply remain hidden in sedimentary records because the change was not sufficient to induce a change in the proxy. A good example of this is 'closed' (or amplifier) lakes in semi-arid parts of the region. Although these are the preferred locations for reconstructing past variations in effective precipitation, the nature of the ecosystems surrounding these lakes mean that the paleoenvironmental proxies recorded in lake sediments may be very insensitive to change. Such systems might therefore be sensitive to relatively small fluctuations in rainfall, but they may not record human impacts of low intensity. Furthermore, the relevant part of the sediment column might be missing, either because of non-deposition or post-depositional erosion. Thus, sedimentary evidence, as with other sources of data, should be used with caution—an approach that will be adopted within this book.

Simplistic interpretations of human-induced ecosystem transition are often challenged when there are multiple drivers interacting (Heckmann et al., 2014; Marlon et al., 2013, 2016; Salzmann et al., 2002). These challenges are compounded as the location of palaeoecological and archaeological sites is biassed to certain locations and periods (Fig. 2.3). Analysis of some palaeoecological archives from African contexts has led to the interpretation that a reduction in pollen abundance from forest tree taxa is accompanied by an increase in grass pollen and is associated with clearance (e.g., Vincens et al., 2007a). This could be indicative of a preference for agroforestry trees (e.g., *Arundinaria alpina*), the presence of cultivated crops (e.g. castor bean *Ricinus communis*), and/or disturbance, shown through the presence of pollen from ruderal plants indicating clearance for agriculture or pastoral activities. Poaceae pollen grains > 80 μm are assumed to represent the non-native crop

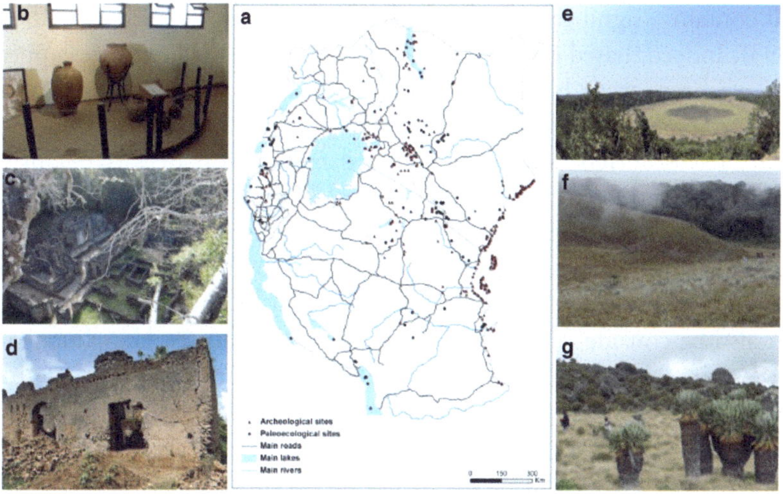

Fig. 2.3 Location of key Palaeoecological and Archaeological sites across East Africa (**a**). There is a clear concentration of Swahili sites along the coast that have yielded many trade goods (**b**) such as Gedis (**c**) and sites on Zanzibar (**d**). These are increasing complemented by a growing number of palaeoecological sites such as Lake Paradise, Marsabit (**e**), Deva Deva Swamp, Uluguru, (**f**) and Maua Swamp, Kilimanjaro (**g**) (All photographs: Rob Marchant)

maize (Eubanks, 1997), while grains 60–80 μm fall in the 'cereal' category but cannot be attributed with certainty to maize (Fearn & Liu, 1995; Tsukada & Rowley, 1964) and are hence attributed to indigenous cereals such as finger millet (*Eleusine coracana*) and/or sorghum (*Sorghum bicolor*). Although it is not possible to positively identify pollen from most cultivated plants in the palaeoecological record with light microscopy alone (cf. Msaky et al., 2005), which makes detailed studies of land use from only pollen unachievable. This is exemplified at the archaeological site of Munsa, where it proved to be impossible to distinguish between Sorghum and Millet from wild grasses (Lejju, 2009; Lejju et al., 2006).

A reliance on charcoal as an indicator of human presence within landscapes can also be problematic as it ignores natural vegetation fire dynamics. For example, records from the Ahakagyezi and Muchoya swamp sediments in Southwest Uganda extend into the last glacial period and contain a significant amount of charcoal throughout (Taylor, 1990, 1993). Indeed, based on the charcoal record from Muchoya, fires appear to have been every bit as common at the end of the last Ice Age as they are today, yet anthropogenic forest clearance based on pollen variability has occurred only since the third millennium BP (Marchant & Taylor, 1998). Down-core fluctuations in charcoal abundance at Rusaka Swamp also show a similar pattern (Mohammed et al., 1996). High levels, or peaks, in microscopic charcoal often occur with high percentages of grass pollen; this can be an indication of land use; however, it can also be considered as a drought event (Marchant & Hooghiemstra, 2001). Fire regimes represent one area where human-environment interactions are deeply entangled. Even before the development of farming, humans used fire as a tool for manipulating the landscapes they inhabited. Patterns of landscape-scale fire use vary according to land use regime (Archibald et al., 2012; Brockett et al., 2001; Gil-Romera et al., 2011; Parr & Anderson, 2006); the very different ecological outcomes that may be difficult to differentiate in paleoenvironmental records and require testing in modelling and field-experiment frameworks (Hawthorne et al., 2017; Marlon et al., 2016).

In certain situations, particularly in swamps and peatlands where the pollen signal is likely to be relatively local (compared to lakes), dung fungal spores can be used to aid in differentiating between a human or climatic cause for the environmental signature present. Certain fungal spores, and assemblages of these spores, can be indicative of human activity e.g., settlement sites, as well as the presence of grazing herbivores. Importantly, these fungal assemblages differ from those of undisturbed natural ecosystems (van Geel et al., 2011). Thus far, there has been relatively limited use of fossil fungal spores in paleoenvironmental studies in East Africa, the main exceptions to this being work carried out in the vicinity of the archaeological site of Munsa in Uganda (Lejju, 2009; Lejju et al., 2006), on the Laikipia plateau of Kenya (Muiruri et al., 2021b), and in coastal Kenya (Szymanski, 2017). Similarly, there has been a lack of archaeobotanical research in the region until recently (Crowther et al., 2016a, 2017; Giblin & Fuller, 2011; Walshaw, 2015) is nevertheless also a significant factor. While it has long been argued that low population densities and/or high levels of mobility would also make pastoralist groups hard to see archaeologically (Smith, 2008), a growing body of evidence shows that pastoralist activities can be detected, often through micromorphological and/or geochemical approaches as well as through analysis of microfaunal communities (Boles & Lane, 2016; Lane, 2016; Muchiru et al., 2009; Petek, 2015; Shahack-Gross et al., 2008). Past generations of East African pastoralists left highly visible traces across the landscape, in the form of abundant stone tools and animal bone middens, as well as less visible but detectable traces of dung and ceramic scatters.

The challenge of making accurate inferences about human history should not be underestimated—it requires the integration of archaeological, historical, genetic, and linguistic data alongside paleoenvironmental evidence. As with the synthesis of information derived from the accumulated sediments, a range of techniques are combined to reconstruct how cultures and societies have changed through time. This range of evidence varies from direct analysis of past occupation layers revealed by archaeological investigations and associated artefacts such as pottery, to documentary evidence. For example, traditionally in archaeology, ceramic artefacts have been used to establish patterns of regional and temporal succession of pottery styles, which are then used to trace the movements and interactions of different cultural groups (Ashley & Grillo, 2015). Archaeologists have long realised the limits of equating 'pots to peoples' (Lane, 2015a) and thus of using material culture as a proxy for identifying impact and, by extension, migration (Ashley et al., 2016).

Molecular approaches have been increasingly useful for examining past population movements and interactions in East Africa, including geochemical sourcing of artefacts (e.g., Goldstein & Munyiri, 2017; Merrick & Brown, 1984), stable isotope analyses (e.g., Chritz et al., 2015; Janzen, 2015; Kiura, 2008), modern DNA from people (e.g., Henn et al., 2008; Ranciaro et al., 2014; Tishkoff et al., 2007, 2009) and livestock (e.g., Gifford-Gonzalez & Hanotte, 2011; Hanotte et al., 2002). A relatively new direction of research has been the application of DNA from human artefacts (Llorente et al., 2015; Skoglund et al., 2017) and sediments (McGlynn et al., 2013). An additional source of information comes from studies of the genetics of living specimens to reveal histories of domestication and the rate, direction, and timing of spread (Rossel, 1995). The cultivated crops commonly found in East Africa have distinct origin centres (reviewed by Crowther et al., 2017; Fuller & Hildebrand, 2013), with pearl millet (*Pennisetum glaucum*) and cowpea (*Vigna unguiculata*) originating in West Africa, finger millet most likely coming from the Ethiopian highlands, and sorghum possibly from North-Eastern Sudan (Harlan, 1993; Harlan et al., 1976). Exotic crops such as Asian rice (*Oryza sativa*), banana, Asian yam (*Dioscorea alata*), and taro arrived via Indian Ocean exchanges (Boivin et al., 2013, 2014; Crowther et al., 2016a). Microfossils of maize have been recorded from the Pare Mountains (Tanzania) and Lake Naivasha (Kenya) from approximately 250 yr BP (Finch et al., 2017), on the Laikipia Plateau (Kenya) by c. 200 yr BP (Taylor et al., 2005), and Munsa Swamp after c. 170 yr BP (Lejju et al., 2005), suggesting that this crop arrived on the Indian Ocean coast from the New World before rapidly spreading inland, presumably along established caravan trade routes (Chapter 4). Tracing the histories of such plants is crucial as they can be closely associated with human populations (e.g., Crowther et al., 2016b). Evidence cited for food production and type of subsistence falls primarily within four categories; (i) botanical and faunal remains from archaeological sites, (ii) vegetation records obtained from perennially wet areas, (iii) settlement characteristics including deposits of, for example, animal dung, and (iv) genetic evidence from living crops and their wild ancestors (Neumann & Hildebrand, 2009; cf. Piperno et al., 2002).

Combined, these analyses show that the past cultural record of East Africa is not characterised by neatly bound developmental stages. Rather the opposite where the human landscape of East Africa has been characterised by highly fluid economic and social mosaics: landscapes of interaction between people with distinct subsistence strategies and cultural traditions (Crowther et al., 2017; Kusimba & Kusimba, 2005). Similarly, to the information for paleoenvironmental insights, as more data location and range of investigations develop it is now clear that there is tremendous heterogeneity across space and time that often defies previous categorical definitions and is making us rethink the nature of past human-environmental interactions and how such past interactions shape contemporary issues (Chapter 7).

2.4 Chronological and Methodological Considerations

It is possible to link archaeological and paleoenvironmental records (Fig. 1.3) through independent chronology, such as provided by radiocarbon dating. Throughout the text, ages are expressed in calibrated years before present (yr BP) except for specific recent or historical-period dates, which are given as Common Era (CE). Bulk sediment samples are often radiocarbon dated, meaning that the processes of sediment accumulation and erosion need to be fully understood. Also, the sediments may be subject to the 'old-carbon effect', returning determinations that are considerably older than the event in question. Similarly, charcoal that is frequently a target for dating raises the question of the 'old wood effect', especially within the forested landscapes where trees can potentially live for many hundreds of years. Secondly, there is the issue of the precision of the dates as the fluctuating nature of the radiocarbon calibration curve means that dating plateaux exist where dates span hundreds of years (Blaauw et al., 2011; Reimer et al., 2004, 2013; Stanley et al., 2003).

These issues of accuracy and precision can be addressed through careful sample selection, combining multiple dating methods, and using other stratigraphic information to constrain the calibrated date using Bayesian statistical techniques (Bayliss, 2009). One of the best paleoenvironmental examples from East Africa where this has occurred is the high-resolution record from Lake Challa on the Kenyan/Tanzanian border (Blaauw et al., 2011) where radiocarbon determinations were obtained on 168 bulk sediment samples, with ^{210}Pb-derived age determinations to test for any old carbon offsets. The calibrated dates were then constrained using Bayesian statistics to produce age estimates for the Holocene that ranged from approximately 50–230 yr (Blaauw et al., 2011). The utility of Bayesian statistics to constrain calibrated radiocarbon dates is clearly seen in the work of Crowther et al. (2016b) that examined the introduction of Southeast Asian crops to East Africa at 18 sites from the East African mainland coast, nearshore islands (e.g. Pemba, Zanzibar), the Comoros and Madagascar. Forty-eight macrobotanical samples (charred seeds) were selected for dating with the resulting dates modelled using Bayesian statistics to account for the stratigraphic relationships between samples, and hence to produce a narrower estimated date range for when the crop introductions occurred. Notwithstanding these developments, geochronological and age-depth model uncertainties are common, making correlations problematic at best (Blaauw et al., 2007; Gearey et al., 2009; Parnell et al., 2008), or in some cases simply wrong (Flantua et al., 2016).

2.5 Environmental and Human and Cultural History

Although the amount of data from archaeological studies has greatly increased, many gaps remain. In some cases, these gaps reflect foci in academic research, and in others result from pragmatic issues like ease of site access from major roads or settlements; factors that equally affect the distribution of paleoenvironmental data (Fig. 2.3). The coverage of palaeoecological and archaeological surveys is certainly not comprehensive, as reflected in large blank areas on maps of site distribution (Fig. 2.3), and is biassed towards better-investigated regions, time periods, and forms of evidence (like pottery or stone tools), with a feedback effect whereby discoveries in one study area tend to prompt additional research. The Pastoral Neolithic (PN) period, for example, is particularly well known from the Central Rift Valley of Kenya, whereas the Early Iron Age is better documented in the Lake Victoria Basin and along the Indian Ocean coast; whether this reflects true distributions (and is thus informative about human choices), or results from sampling patchiness, in space or time, is uncertain. For the context of this book there are some introductory insights provided from the early human origins but there is a particular focus on transitions through the current glacial to interglacial period and the last 6000 years or so that comprises the arrival of food production across the region.

2.5.1 Foundations of Modern Humans

Presently, humans across the globe constitute the dominant control on modifying ecosystem composition and distribution and will continue to impact on the planetary system into the future. However, the way in which ecosystem composition has changed over time and space, and how both human-induced and environmentally induced land use changes have interacted, remains largely unresolved. To shed light on the nature of these interactions we must look to our human origins. An important contributing factor to all stages of hominid evolution that took place in Africa during the Plio-Pleistocene was that of the environmental backdrop and how this fluctuated in response to climatic change; this affected both hominid evolution and how humans have shaped and interacted with the environment (e.g., Demenocal, 1995; Deocampo et al., 2002; Maslin, 2016; Muiruri et al., 2021b). The longer-term evolution of early hominin species is outside of the focus here so we will concentrate on the era of modern humans—*Homo sapiens sapiens*. Recent discoveries and modern genetics all point to *Homo sapiens* evolving as a new species of *Homo* in Africa over 300,000 years ago (James et al., 2019). In keeping with any time-bounded event, the timing and antiquity of this shifts as new finds become available from diverse locations that date the first modern human population divergence to sometime between 350,000 and 260,000 years ago (Schlebusch et al., 2017). Much of human evolution appears to have taken place in savannah ecosystems, with the Rift Valley system of Eastern Africa, and cave sites in Morocco and South Africa providing many of the key sites and fossils (Clark et al., 2003; McDougall et al., 2005). The character and chronology of early *H. sapiens* fossils, together with their geographic distribution across Africa, suggests that evolution may have progressed independently in different regions, in populations that were often semi-isolated for millennia by distance and/or ecological barriers, such as hyper-arid regions or tropical forests. Following the evolution of modern humans, they subsequently colonised the rest of the globe as well as adapting to the environments across the African continent. Since the migration of modern humans 'out of Africa', numerous population movements have played a role in shaping patterns of linguistic and genetic variation within the continent itself (Campbell & Tishkoff, 2008). This antiquity of our species across Africa is one of the reasons why our species is so linguistically and culturally diverse (Sect. 1.7) across East Africa: more than 2000 languages are recorded for the whole continent, comprising 30% of the world's languages (Gordon & Grimes, 2005).

Given global interest in the 'origin of our species', Plio-Pleistocene hominin fossils from Africa have been the subject of considerable interest in developing a detailed understanding of our origins, and also possible links between human cultural and/or physical evolution and environmental change (Maslin et al., 2015; Potts et al., 2020). The numerous Hominin sites across East Africa include the foundational work on various *Homo* lineages that lead to modern human. Particularly notable is the work by the Leakey's from the 1930s at Olduvia Gorge (Leakey, 1971) and the description of 'Lucy' (*Homo habilis*) dated to c. 2 million years and Turkana Boy (*Homo ergaster*) from Northern Kenya that was c 1.6 million years old (Dunsworth, 2010). Building on this early work, several research projects have been exploring human environmental relationships surrounding the large East African lakes that were formed during the Miocene some 7 million years ago. Following their formation through the rifting process provide an environmental archive that can be interpreted alongside the hominin fossils scatter at key sites across the region. The Paleolakes Drilling Project has drilled into the sediments in five lake basins across the region which contain thick sequences of lacustrine, fluvial, and terrestrial sediments that can be correlated with nearby sites with significant hominin fossils of differing ages—some going back 7 million years (Muiruri et al., 2021a). These long-term sediment deposits document how the glacial to interglacial oscillations from the Quaternary period have impacted on East African ecosystems and the associated fauna, including humans. Again, we want to focus on the most recent of these 22 or so glacial-interglacial periods that each lasted around 100,000 years each and oscillated between a 'glacial period' (when environments were characterised by reductions in temperatures of 4 °C, precipitation by c 30%, sea levels by some 100 m lower and levels of atmospheric CO_2 almost half at c. 200 ppmV relative to today) and a 'interglacial period' where environments were similar to todays, albeit with levels of atmospheric CO_2 at c. 280 ppmV and not the 418.94 ppmV of July 2021. As we will see throughout the book, these changes have a fundamental impact on the composition, distribution, and structure of East African ecosystems. Before we focus on the most recent of these glacial-interglacial climate oscillations, it is worth emphasising that the glacial state is the norm for the past 2.2 million years when East Africa, as with the rest of the planet, has spent around 90% of the time characterised by lower temperature, precipitation, sea level, and carbon dioxide before going through the relatively short lived interglacial periods, like the one we are currently in, before returning to the 'normal' glacial state. This provides useful context for the current challenges as this is really the first time East Africa is moving from a warmer, wetter, high atmospheric CO_2, high sea level environment to a future that is predicted to see increases in all of these—clearly these will bring unprecedented changes to climates, environment, and how the ecosystems, people, and animals interact with these variations. These new environments will bring transformative change requiring a myriad of adaptation strategies—challenges that will be explored in more detail in Chapter 7. Fundamentally, this backdrop of highly dynamic and changing environments outlines where an understanding of these past changes not only in the physical and biological systems, but also in the nature of societal interaction, is so important. To provide this context we can begin at the last interglacial period before moving into increasingly resolved detail as we transition towards the present day.

Around 125,000 yr BP was the last interglacial, when the world was as warm as present and sea levels were within a few metres of those of today as polar ice caps were at a minimum (Martinson et al., 1987). Warmer oceans meant higher evaporation and generally, more precipitation on land led to environments that allowed populations to expand, leaving traces of this expansion in the numbers of Middle Stone Age (MSA) sites (Deacon, 2001). The Magadi data indicate that these varying environments would have changed the resource base (water, plant/animal foods) available to hominins and, in turn, created selective pressures in the area (Kübler et al., 2015). Understanding human's response to environmental transitions and if the innovative, flexible behaviour and ecological adaptiveness exhibited by modern humans allowed past communities to adjust to their surrounding environment is of utmost importance. Alongside this, we need to chart how these interactions changed as climates did. Several sites have produced evidence for novel activities and adaptation, reflected in the appearance of backed stone tools and beads over the last 50–60,000 years (Ambrose, 1998; Shipton et al., 2018). Panga ya Saidi (PYS), a fantastic cave occupation situated 15 km from the present-day shoreline in the Zanzibar-Inhambane coastal forest mosaic, contains a continual record of human activity back to 78,000 yr BP. The consistent presence of the forest ecotone throughout the last c. 80,000 years is matched by evidence of increasing occupation intensity at PYS, perhaps suggesting a growing human presence in the region (Shipton et al., 2018) as populations expanded and did not join the 'out of Africa' migration. The intermittent, non-continuous, and multi-directional appearance of complex technological and social traits in the PYS record, further documents how humans are a flexible species capable of adopting diverse survival strategies (Shipton et al., 2018). How such social and technological transformations laid the ground for major dispersal events remains to be investigated but is highly likely that the flexible and diverse interaction with the environment is likely to have been mirrored by populations across the East African landscape.

2.5.2 Last Glacial Period: Forest Refugia and Imprints on Today's Landscape

There have been twenty-one glacial-interglacial cycles during the Pleistocene (Hamilton & Taylor, 1991). The precise timing of the last glacial 'maximum' (LGM) differs depending on the type of evidence used and the area from where the evidence was collected, although there is a consensus that it occurred about 21,000 yr BP. Despite quite large variations in the temperatures estimated for the LGM, there was a cooling during the LGM of –3 °C ± 1.9 °C (Bonnefille et al., 1992). The large standard error associated with this figure indicates the heterogeneity of temperature changes during the LGM reconstructed from different proxies and different transfer function approaches. These also reflect the diversity of past ecosystem responses that provides useful context for future projected climate changes. These range from 10 to 14 °C colder relative to present day with significant drying through to climatic conditions along the eastern zone of the Tanzanian/Kenyan coast where it may have been permanently warm and humid throughout the LGM (Dinesen et al., 1994; Finch et al., 2017; Platts et al., 2015). There is continued debate on the relative impact on vegetation by these changes in temperature (Colinvaux et al., 2000; Farrera et al., 1999; Flenley, 1998; Loomis et al., 2017; Van Zinderen-Bakker & Coetzee, 1972), precipitation (Bonnefille et al., 1990; Vizy & Cook, 2007), atmospheric CO_2 concentration (Boom et al., 2001; Jolly & Haxeltine, 1997; Street-Perrott et al., 1997), UV-B radiation flux (Flenley, 1996), wildfires (Ivory & Russell, 2016; Wooller et al., 2000), and herbivory (Ivory & Russell, 2016), let alone the interaction of all of these and how they combine to determine the composition and distribution of East African ecosystems.

Until the 1960's the idea that tropical rainforests had been unaffected by these past climatic oscillations was accepted by most biologists. As evidence became available that documented the dynamic nature of these forests in response to climatic change, an understanding that climatic and vegetation upheavals during the Pleistocene fragmented the previously continuous ranges of many species into isolated refugia. This simple model has evolved as more evidence is amassed concerning present and former biotic distributions, and patterns within these distributions, to explain the differential forest composition and centres of species richness and biodiversity (Haffer, 1997). Clearly, climatic variations associated with past glacial periods would have imparted major changes in the composition and distribution of the vegetation. During the LGM lowland mesic forests contracted throughout East Africa but, by contrast, many montane forest taxa moved to lower elevations with an estimate that cloud-base height decreased by 500 m in response to a 4 °C cooling, causing a 20–100% increase in forest area despite a 30% decrease in rainfall, a 22% decrease in atmospheric humidity and a substantial reduction of atmospheric CO_2 levels (Los et al., 2019). Regional disagreements occur about the lower boundary of montane forest at the LGM (Izumi & Lézine, 2016). In some locations the upper limit of the Afromontane Forest changed substantially between glacial and interglacial stages, but the lower limit remained relatively unchanged (Lézine et al., 2019).

Certain areas are known to have experienced greater reductions in temperature and precipitation levels than others. Locations where climates were not as variable as other areas would have experienced relatively minor changes to the vegetation composition and distribution due to local climatic anomalies, and topographic and edaphic buffering. These differential rates of vegetation change, following spatially heterogeneous climatic change, resulted in the fragmentation of species into ecological refugia where forest survived the extensive cold and dry climatic conditions characteristic of glacial periods. Many of these areas that remained relatively climatically stable are now biological hotspots, containing comparatively high degrees of endemism and high levels of diversity. However, much criticism has been levied at this simple model, particularly as forests are non-linear, dynamic systems that are unique and evolving, i.e. they have a dynamic in time and space. Although we often think of ecosystem moving en-masse it should be emphasised that there were individualistic species responses to locally prevalent combinations of cooling, changed rainfall, CO_2 induced physiological drought, fewer fog days, enhanced wildfire incidence, and grazing pressures (Rucina et al., 2010).

Species exhibit three responses to climatic change; acclimatisation, adaptation, or migration, with a species having to response in one way, or a combination of ways, to remain viable. Relatively few species have gone extinct during the Quaternary period, and of those that did, large proportions were unable to migrate, being restricted to islands and patchy habitats (Markham, 1996). The presence of refugia is attributed to the ability of species to persist through periods of environmental change within areas of unique climatic regime. Moreover, it is by virtue of their persistence within these environmental parameters that these localities are deemed relatively stable and may offer future stability. Although it is possible, where the biogeographical and palaeocological information regarding past and present species distribution is of a sufficiently high spatial and temporal resolution to map the location of past forest refugia across Africa (Fig. 2.4), the location future forest refugia may not be placed in the same locations as those in the past. The specifics of climatic change experienced during the Late Pleistocene are likely to have been very different to those changes suggested in the future, as

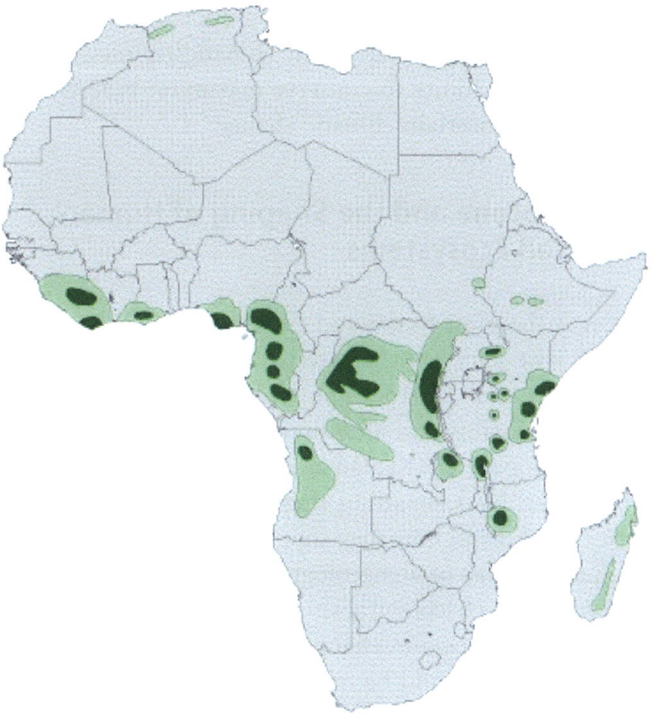

Fig. 2.4 Location of past forest refugia across Africa depicting the potential extent of the core (dark green) and peripheral area (light green). Present-day forests that have been located within refugia over geological timescales are ancient forests. Forest refugia occupy areas of relatively stable past climatic regimes that could provide the best opportunity to conserve maximum biodiversity during a period of future uncertain climatic change. Although of crucial importance to future biodiversity conservation the nature of predicted future climate change is very different to the climate variability experienced in the past

implied by the IPPC (2021). Nevertheless, why certain areas were more stable in the past may well be commuted to the future. Therefore, determining the location and extent of forest refugia is of crucial importance to biodiversity conservation, as is indeed knowing what the driver behind these areas of relatively stable climate were. The key factor in maintaining a moist climate has been the relatively constant temperature of the Indian

Ocean during the LGM (Burgess & Clarke, 2000; Marchant, 2021). By occupying areas of relatively stable past climatic regimes, forest refugia provide the best opportunity to conserve maximum biodiversity during the future period of uncertain climatic change.

2.5.3 The Holocene and the Shaping of Human Impacted Ecosystems

The Holocene is the most recent and current geological period, starting around 10,000 years ago; it is characterised by a relatively warm and wet climate (Shanahan et al., 2015). The Holocene is particularly identifiable by climatic variability and intensifying anthropogenic modifications towards the 'Anthropocene' (Kaplan et al., 2011; Ruddiman et al., 2015; Stott et al., 2000). Understanding these transitions in where societies settled, what they did and how they supported their livelihoods is largely based on different forms of archaeological evidence drawn from a range of different types of archaeological site (Fig. 2.5). The evidence for environmental and anthropogenic change will be presented in a series of time-steps; however, it should be reiterated that the past environmental and archaeological (Fig. 2.3) record does not fit easily into a series of stages, as increasing complex mosaics of foragers and food producers defined the landscape (Kusimba & Kusimba, 2005). Although the transitions tend to be presented as moving from hunter-gathering subsistence, to pastoral, and then to settled agriculture, this is not a succession and the lines between these different subsistence bases are blurred and transitory. For example, fishing remains a common strategy during the last two millennia BP as seen at Turkana Basin sites with Turkwel ceramics like Lopoy and Apeget (Robbins, 1980, 1984) and at Victoria Basin sites with Urewe ceramics like Wadh Lang'o and Usenge 3 (Lane et al., 2007; Prendergast, 2008). In all these sites, faunal remains attest to mixed herding-hunting-fishing economies, with herding varied widely in importance. While iron-using agropastoral extended across much of East Africa c. 1300–900 yr BP, the arid Turkana Basin appears to have been largely unaffected by the spread of both iron working and farming,

Fig. 2.5 Examples of archaeological sites. (**a**) Kuumbi cave on Zanzibar (photo: Ceri Shipton). (**b**) Rock art from Kondoa rock art World Heritage Site, Central Tanzania. Rock art associated with Holocene hunter-gatherers rituals of the Sandawe and the Hadzabe that date to 5000–20,000 years ago (photo: Emmanuel Bwasiri). (**c**) Early Holocene shell beds at Lothagam in northern Kenya (photo: Larry Robbins). (**d**) The Great Mosque of Kilwa Kisiwani, along the Swahili coast of Tanzania (photo: Stephanie Wynne-Jones). (**e**) The 'Bwogero' earthworks at Ntuusi, Uganda (photo: Andrew Reid). (**f**) Sirikwa holes, Kenya, dating from 850 to 550 yr BP (photo: John Sutton). (**g**) The abandoned irrigated agricultural site of Engaruka, Tanzania, occupied from c. 600 to 150 BP with former habitation platforms in foreground and former field system in the plain beyond (photo: Daryl Stump). (**h**) Traditional irrigation reservoir (Ndiva), North Pare Mountains, Tanzania (photo: Daryl Stump). (**i**) Contemporary boma (pastoral enclosure) site in northern Tanzania (photo: Suzi Richer)

and today the economic mainstays remain drylands mobile pastoralism (including camel pastoralism) and fishing.

2.5.3.1 The Early Holocene

Palaeoecological records from East Africa suggest that aridity increased from the early Holocene as temperatures rose; Barker et al. (2011), proposing that annual temperatures were around 2–3 °C higher in the Holocene compared to the LGM. This increase in arid conditions can be seen in a pollen record from Lake Tanganyika, which shows a change in the catchment area of the lake from being forested to a more open grassland from 5000 yr BP (Msaky et al., 2005; Tierney et al., 2010). The level of Lake Turkana decreased between ca. 8000 and 4500 yr BP, ultimately resulting in a regression of c. 90 m between 5200 and 4500 yr BP (Bloszies et al., 2015). While it continued to maintain this lowered level for much of the mid- to late Holocene, there is considerable variation in the magnitude of this reconstructed change due to the application of different proxies with different dating controls (Bloszies et al., 2015; Garcin et al., 2012; Morrissey & Scholz, 2014; Owen et al., 1982). In another large lake, the diatom record from Lake Victoria indicates a phase of low precipitation between c. 5000 and 4000 yr BP (Nakintu & Lejju, 2016).

Detailed characterisation of this period of geological history is challenging as a number of sites throughout East Africa are characterised by sedimentary gaps or hiatuses around this time. Gaps in sedimentation are often regarded as a lack of information, although the presence of sedimentary gaps is one of the strongest paleoenvironmental indicators within a sedimentary sequence! The main problem in defining sedimentary gaps is determining whether they result from a lack of sediment accumulation or from post-depositional erosion of the sediment. In the case of Mubwindi Swamp in Southwest Uganda, the early to mid-Holocene gaps are thought to result from erosion, particularly as sediment erosion can be observed at the swamp (Marchant & Taylor, 2000; Marchant et al., 1997). The temporal occurrences of the hiatuses in the Mubwindi Swamp sequence reiterate this suggestion as they occur during periods of increased effective moisture throughout the region (Bonnefille, 1987), thus providing the mechanism through an increased hydrological budget necessary to erode in situ sediments. Consequently, reinitiating of peat accumulation during the late Holocene (around 4000 yr BP) may have commenced because of gradual aridification or increased seasonality (Sect. 2iv). Throughout Western East Africa there appears to be a slow but sustained increase in the depth of sediment accumulated down the altitudinal sequence: no peat formed at the higher

Kuwasenkoko Swamp, Rwanda (Hamilton, 1982) at 2340 m asl, or at Muchoya Swamp, Uganda, at 2260 m asl (Jolly et al., 1997), but an accumulation of 7–10 m of peat was recorded at Ahakagyezi Swamp, Uganda (Taylor, 1990) at the slightly lower elevation of 1830 m.

On the coast of East Africa, sea levels rose dramatically following low stands characteristic of the last glacial period when sea levels were approximately 100 m below present (Prendergast et al., 2016). Sea levels reached a mid-Holocene high stand of up to 3.5 m above present mean sea level by 4500 yr BP, followed by subsequent falls to the present level in the late Holocene (Punwong et al., 2013, 2017). Sea level change is critically important to coastal societies, and it is likely that there are archaeological sites now underwater that predate the early to mid-Holocene sea level rise. Sea level variability impacts low-lying coastal areas such as deltas and mangrove forests and would have led to all the coastal islands off East Africa becoming separated from the mainland—such as those across the Zanzibar archipelago, Mafia, and Lamu.

Around 6000 yr BP the East African landscape was occupied by fishers and foragers associated with Late Stone Age (LSA) lithic technological traditions. While often viewed as a common society, there is tremendous diversity among Holocene LSA sites, both in terms of technological variability and the settlement and subsistence strategies of their occupants. A key challenge to understanding this variability is the poor temporal resolution that characterises much of Holocene archaeology in the region. There are few radiocarbon dates falling within the 6000–4500 yr BP range; whether this reflects sampling patchiness, or a decrease and/or dispersal in populations, remains to be understood (Beyin et al., 2017). In the Turkana Basin, fishing communities are documented from as early as the Pleistocene–Holocene transition and are associated with almost exclusively aquatic fauna, bone points or harpoons, and at some sites, small numbers of ceramics (for reviews, see Barthelme, 1977, 1981, 1985; Phillipson, 1977; Prendergast & Beyin, 2017; Robbins, 1972; Stewart, 1989; Yellen, 1998). Recently, human remains recovered at Nataruk were interpreted as evidence of violent intergroup conflict among fisher-foragers of this era (Lahr et al., 2016). Given that lake levels dropped dramatically at c. 6000 yr BP (Garcin et al., 2012; Wright, 2014) one might expect such settlements to disappear, although aquatic resource exploitation continued in the Turkana Basin in the past millennia (Barthelme, 1985; Lahr et al., 2016; Robbins, 1980; Yamasaki et al., 1972), through to the present day (e.g., Sobania, 1988).

Around Lake Victoria there are numerous sites dating to c. 8000–2000 yr BP all sharing a type of ceramic known as Kansyore (Fig. 2.6), as well as an aquatic subsistence focus and a quartz/quartzite-based lithic technology (Seitsonen, 2010). Kansyore sites are found across a wider region of Uganda, Kenya, and Tanzania, nearly always near the shorelines or palaeo-shorelines of Lake Victoria, or along its tributaries (Dale & Ashley, 2010; Prendergast, 2010). Kansyore pottery has also been found in contexts associated with fishing in North-Western Uganda (Schmidt, 2016), and in rock shelters in grasslands and near a soda lake (Lake Eyasi) in Northern Tanzania; these finds are associated with foraging but have little or no association with fishing (Mehlman, 1989; Prendergast, 2011; Prendergast et al., 2017). In these instances, occupations with Kansyore ceramics appear indistinct from roughly co-occurring aceramic LSA occupations at these and other sites in the region. Across East Africa, the Kansyore chronology is problematic, and notably most sites fall on either side of the c. 6000–4500 yr BP window (Prendergast et al., 2014). The paucity of radiocarbon dates for both fishing and terrestrial hunting sites in the 6000–4500 yr BP window is notable (Ambrose, 1998) (Fig. 2.7). At other sites, however, so few dates are available that it is difficult to know whether the gap c. 6000–4500 yr BP represents a sampling error or a true hiatus.

Away from lakes and rivers, foragers occupied rock shelters, and presumably undetected open-air sites, in the plains and highlands of East Africa throughout the Holocene (Fig. 2.7). The best documented areas for Holocene LSA occupations are the Central Rift Valley of Kenya (Ambrose, 1982, 1984); Lukenya Hill in Southern Kenya (Bower & Nelson, 1978; Gramly, 1975; Kusimba, 2002); the South-Eastern plains and coastal hinterland of Kenya (Abungu & Muturo, 1993; Helm et al., 2012; Kusimba et al., 2005; Shipton et al., 2013); and the highlands and some lowland lake basins of North-Central Tanzania (Inskeep, 1962; Masao, 1979; Mehlman, 1989; Odner, 1971a). Superficially, many of

Fig. 2.6 Examples of archaeological evidence. (a) Kanysore Ware, reconstruction by Ceri Ashley (photo: Paul Lane). (b) Barbed bone point (harpoon) in situ in early–mid Holocene shell beds at Lothagam, Kenya (photo: Larry Robbins). (c) A worked bone implement from the site of Luxmanda, Tanzania, interpreted by Langley et al. (2019) as a matting needle, since it exhibits signs of wear associated with working hides or plant material (photo: Mary Prendergast). (d) Urewe Ware from Lolui Island, Uganda (photo: Oliver Boles). (e) Excavated iron smelting furnace, Mwanga, Tanzania, this example dated to 650 to 510 cal BP, scales in 10 cm increments (photo: Daryl Stump, 2010). (f) Wound glass beads from 14–fifteenth centuries CE, from excavations at Songo Mnara (photo: Steph Wynne-Jones). (g) Hair Cell Phytolith from Songo Mnara, Tanzania in association with a wattle and daub structure, scale is 20 microns (photo: Hayley McParland). (h) Micrograph of soils from within an historically irrigated field at Engaruka, Tanzania (photo: Carol Lang). (i) Non-charred (left) and charred (right) *Portulaca oleraceae* seeds. This species is a beneficial agricultural weed in East Africa with shallow roots that help retain soil moisture in dry environments (photo: Senna Thornton-Barnett)

these occupations seem similar: they are documented in rock shelters, sometimes with paintings that may or may not be associated; they share LSA lithic technologies, which have been little studied (but see Ambrose, 1984; Barut, 1994; Mehlman, 1989; Wilshaw, 2016); and their subsistence strategies include hunting of mainly large and medium sized game

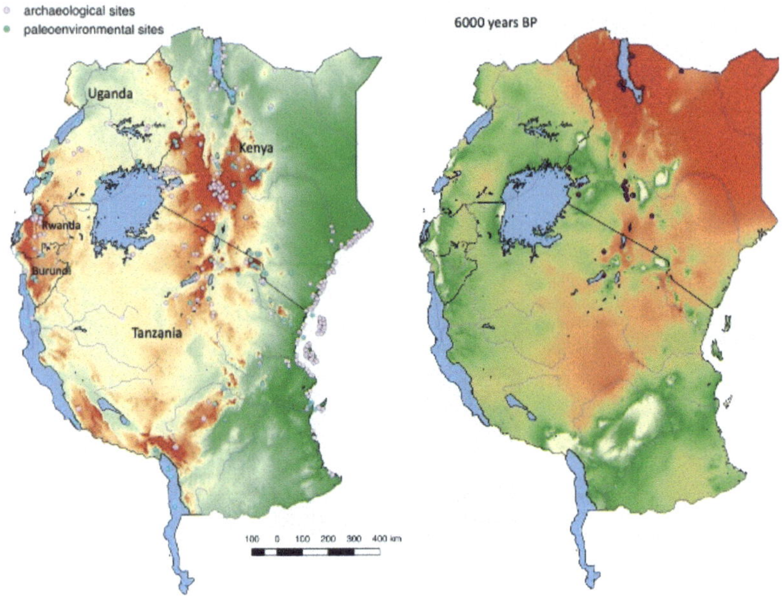

Fig. 2.7 Location of key Palaeoecological and Archaeological sites across East Africa (left) used for the synthesis. At 6000 years BP there were relatively few palaeoecological and archaeological sites due to the antiquity, the relatively sparsely populated nature of the region and many sedimentary sites having a mid-Holocene hiatus in sediment accumulation. Those sites that do date to this period (right) record a largely subsistence economy based on the exploitation of seasonal resources and fish (Maps produced by Oli Boles and modified by the author)

(Mehlman, 1989; Prendergast, 2011), though in more forested locations, trapping of smaller game was important (e.g., Ambrose, 1984; Marean et al., 1992). While other components of the modern hunter-gatherer diet do not preserve (e.g. honey) or have not been systematically sought (e.g. plants), it is likely that animals were part of a much broader diet. Although there is considerable uncertainty in characterising how these early human populations interacted across different user communities or interacted with the flora and fauna, principally due to the lack of data (Fig. 2.7), what we can say is that on current evidence it seems that populations at this time were living lightly on the landscape and employing subsistence strategies that did not involve major vegetation clearance.

2.5.3.2 Mid Holocene Environmental Shifts and the Arrival of Food Production

The start of this period is marked by a relatively abrupt event that shifted climate towards arid conditions characterised by reduced rainfall and/or a more seasonal environment (deMenocal et al., 2000). There is ample evidence for sharp lake level drops at approximately 4000 yr BP (e.g., Gasse, 2000; Kiage & Liu, 2006; Marchant & Hooghiemstra, 2004) from several sites around the region. Quite typical is the Lake Bogoria (Kenya) pollen sequence, which from 4500 yr BP shows a sharp decline in high-altitude forest pollen from taxa such as *Hagenia, Hypericum, Stoebe,* and Ericaceae, and increases in taxa such as *Podocarpus, Juniperus, Acacia,* and *Dodonaea,* which are more drought tolerant (Vincens, 1986). Similarly, on Mount Kenya and Mount Elgon sharp increases in *Podocarpus* occurred after 4500 yr BP and 3500 yr BP, respectively (Hamilton, 1982; Street-Perrott & Perrott, 1988) marking the onset and establishment of dry conditions across the region. The low abundance of Panicoid grass cuticles at Sacred Lake on Mount Kenya during the mid to late Holocene is also thought to indicate a shift in the seasonality of rainfall and/or fire (Wooller et al., 2000, 2003). Another record from Mount Kenya investigating the isotopic character of the sediments from Lake Rutundu indicates a very rapid rise in C_3 grasses about 4500 yr BP that is relatively short lived (Ficken et al., 2002) and similarly may derive from a sudden change in the precipitation regime. In Rwanda and Burundi, Jolly et al. (1997) suggest that montane forest taxa that are more drought tolerant, for example *Podocarpus,* increased in abundance after 4100 yr BP. These widely recorded ecosystem responses correspond with a visible dust layer 30 mm thick in the ice core record from Kilimanjaro, Tanzania that, based on its chemical signature, has been interpreted as reflecting a drier phase with less vegetation cover from 4500 to 3500 yr BP (Thompson et al., 2002). Lake level records derived from Lake Rukwa (Tanzania) also indicate that a dry phase occurred at around 3000 yr BP (Talbot & Livingstone, 1989). The outflow from Lake Kivu (Rwanda) to Lake Tanganyika (Burundi) was interrupted from about 3500 yr BP with lake levels lowered by about 75 m; this altered the salt balance in Lake Tanganyika and changed the resultant diatom flora (Haberyan & Hecky, 1987). Again, this phase of aridity can be seen in the Lake Kivu catchment where, around 3200 yr BP, a reduction in forest taxa occurred that coincided with an expansion of the savannah, without any indication that people were responsible for this change (Ssemmanda & Vincens, 1993).

2.5.3.3 Archaeological Insights into the Timing and Spread of Pastoralism

As with the origins, the timing, and mechanisms of adopting metal technologies by East Africa's various pastoralist communities remains poorly understood as the archaeological record concerning the trajectory of pastoralism is both patchy and partial (Lane, 2011a). Unfortunately, this has led to quite a bit of conjecture as to the timing and nature of pastoral development. Unlike other areas around the world, mobile livestock herding preceded sedentary farming in much of East Africa (Marshall & Hildebrand, 2002). Concomitant with the mid-Holocene drop in lake levels, beginning as early as c. 5000 yr BP, this area also witnessed the arrival of the first livestock herders of East Africa. The slow spread of pastoralism has been attributed to zoonotic diseases, aridity, and the presence of indigenous foragers or other social factors (Gifford-Gonzalez, 2000; Marshall et al., 2011). The appearance of mobile herding around Lake Turkana coincides with dropping lake levels (Grillo & Hildebrand, 2013; Wright et al., 2015) and, as noted above, is possibly accompanied by a reduction in fishing sites and diversification of fishing strategies (Stewart, 1989), though this may be an artefact of uneven archaeological sampling (Fig. 2.8).

Although fish remains continue to appear at sites with livestock, exclusively barbed bone points and aquatic diets appear to all but disappear by the mid-Holocene arid period as mixed subsistence strategies seem to be the norm. Beginning c. 5000 yr BP, East Africa witnesses the first appearance, and an initial slow spread, of food production, as well as a continuation of fishing and foraging activities described in Sect. 2.5.3.1. This marks the beginning of the Pastoral Neolithic (PN)

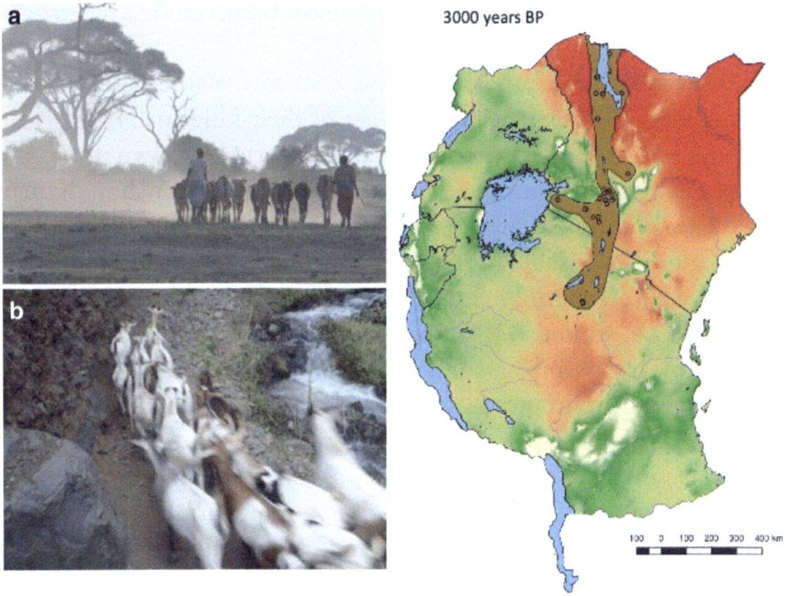

Fig. 2.8 The first major transition in food production, away from subsistence livelihoods, was the expansion of pastoralism and the keeping of domestic livestock, particularly focused on cattle (**a**), sheep and goats (**b**). Although the timing and direction of this pastoral spread is continually being refined the spread appears to originate from the north into East Africa funnelled along the rift valley (All photographs: Rob Marchant)

that started around 5000 yr BP (for reviews see: Bower, 1991; Gifford-Gonzalez, 2005; Lane, 2013a, 2013b; Marshall et al., 2011; Wright, 2014). Increasing aridity in the Sahara during the mid-Holocene may have been the driving force leading to the Southward movement of pastoralists into East Africa (Maslin, 2016). The spread of herding is linked to the emergence of new stone tool and ceramic remains that first appear in the Turkana Basin c. 4500–4000 yr BP at the site of Dongodien (GaJi4), and possibly at neighbouring GaJi2 (Goldstein, 2019; Hildebrand et al., 2018; Marshall et al., 1984), these artefacts are associated with 'Nderit' ceramics and LSA lithic technology. Elsewhere around Lake Turkana, from c. 5000 to 4000 yr BP, direct evidence of herding is present but scarce, and Nderit ceramics have been taken as a proxy for pastoralist communities.

The spread of food production also coincided with the creation of elaborate megalithic cemeteries, the first of their kind in Eastern Africa and some of the earliest monumental architecture known from sub-Saharan Africa that also offer important perspectives on how humans reacted to significant changes in the world around them. Early herders in the region built elaborate megalithic cemeteries ~5000 yr BP overlooking Lake Turkana (Sawchuk et al., 2019). Archaeological-rich horizons dating to around 4100 yr BP in North-Eastern Lake Turkana yielded a diverse assemblage of artefacts suggesting a mixture of subsistence strategies (hunting and gathering, fishing, and herding animals) rather than a single type of subsistence practice (Ashley et al., 2017). Increased research at megalithic mortuary sites around Lake Turkana, many of which also have Nderit pottery, more firmly link them to mobile pastoralism (Grillo & Hildebrand, 2013; Hildebrand & Grillo, 2012); at the site of Jarigole, clay livestock figurines may offer additional insight to these early food producers (Nelson, 1995). Importantly, these megalithic mortuary sites, which were constructed relatively rapidly around the Turkana Basin, do not appear to have extended beyond this area despite finds of Nderit ceramics in Central and Southern Kenya and possibly into Northern Tanzania (Gifford-Gonzalez, 1998). The diversity of this range of finds suggests that the Lake Turkana basin was a major centre of ritual activity and a locus of long-distance exchange from the fifth millennium BP (Gifford-Gonzalez, 1998; Hildebrand et al., 2018). This is not just restricted to the shoreline but has a much wider fingerprint: Stiles and Munro-Hay (198) note hundreds of simple cairns across the region to the east of Lake Turkana and date many of these to c. 3500 yr BP. Grave cairns, one of which was reported as containing a quern resembling stone bowls found adjacent to Lake Turkana (Barthelme, 1977), were documented as rimming the Chalbi Desert (Stiles & Munro-Hay, 1981). These cairns are likely to be associated with the early Pastoral Neolithic (Davies, 2013), indeed they are still constructed by the Gabbra, Boran, and Rendille today.

South of Lake Turkana, mobile herders appear in isolated occurrences from 4000 yr BP, but there is little direct evidence of their activity

(Wright, 2014). Farther South, pastoralism in the Ewaso Basin initially appears some 4500 years ago until the early twentieth century (Lane, 2015a). At the Eburran forager site of Enkapune ya Muto in the Central Rift Valley, caprines appear in an otherwise wild faunal assemblage, beginning possibly as early c. 4500–4300 yr BP and becoming commonplace at the site by c. 3700–3200 yr BP (Ambrose, 1998). At Wadh Lang'o, a Kansyore site with later occupation phases near Lake Victoria, caprines and cattle appear after c. 4000 yr BP, again being minor components of otherwise fisher-forager assemblages (Prendergast, 2010); the same is true in the Kansyore horizons at the multi-period sites of Usenge 3 at c. 3500 yr BP (Lane et al., 2007) and Gogo Falls, where dating of domesticates is problematic (Karega-Munene, 1993; Robertshaw, 1991). Thus, the initial southward spread of livestock seems to be patchy and involving the agency of existing fisher-forager groups (Masao, 1979; Mehlman, 1989; Mjema, 2008; Prendergast et al., 2014, 2016). Like the populations around Lake Turkana, the hunter-gatherers occupying Laikipia possessed a complex and sophisticated lithic technology and were relatively residentially mobile with extended territories. The wider area of Northern Kenya probably also included pockets of Oromo-speaking Warra Daaya and possibly Turkana and Rendille; indeed, there is also good evidence to suggest that, at least in the recent past, the boundaries between different 'ethnic', 'subsistence', and perhaps even 'linguistic' groups were fluid and that cultural intermixing as well as interaction through exchange and other social mechanisms was common. Such zones of interaction are particularly well documented in the Central Rift Valley of Kenya (Ambrose, 1984) and the Eyasi Basin in Northern Tanzania (Mehlman, 1989; Prendergast, 2011; Prendergast & Mutundu, 2009). In short, the wider area of Northern Kenya was an exceptionally dynamic environment, in which both climatic and veterinary factors made intergroup social alliances and exchange crucial (Gifford-Gonzalez, 1998) with the presence of robust regional pastoral exchange networks (Wright, 2019).

While Nderit pottery, like that of the Turkana Basin 5000 yr BP, does appear at multiple Kenyan sites (Gifford-Gonzalez, 1998, 2005), and possibly as far south as Lake Eyasi in Tanzania (Mehlman, 1989), these findings are scattered and not clearly associated with livestock. True

PN sites, marked by abundant livestock remains as well as novel ceramic styles and stone bowls, do not appear more widely until after c. 3000 yr BP (Fig. 2.8). This late date may be due to a heavy reliance at PN sites on bone apatite, a notoriously unreliable material for dating, as well as numerous problems of contextual association (Collett & Robertshaw, 1983). More recently obtained dates from PN sites in South-Eastern Kenya (Wright, 2005) and Northern Tanzania (Grillo & Prendergast et al., 2018; Prendergast et al., 2014) slightly predate c. 3000 yr BP. Redating more northerly sites and increasing the number of sites from the period is thus essential to reassess the temporal spread of herding; although there does seem to have been a 'moving frontier' of pastoralism ca. 4700–3000 yr BP that became increasingly specialised.

The third millennium BP is characterised by the expansion of specialised pastoralism across much of East Africa (Fig. 2.8), while foraging and fishing communities also continued to thrive, in some cases interacting with in-migrating herders. This widespread dispersal into the Rift Valley and grasslands of Central and Southern Kenya, and ultimately into Tanzania, is frequently viewed as a response to ameliorating climatic conditions (Marshall, 1990, 2000; Marshall et al., 2011). Both the number of pastoralist sites, and the extent of research on them (particularly in Central and Southern Kenya), have enabled a relatively good understanding of the material culture, subsistence strategies, settlement patterns, and mortuary traditions of specialised herders of this era (Lane, 2013a, 2013b). These studies have enabled two broad groupings that date to c. 3000–1200 yr BP (Ambrose, 1984; Robertshaw, 1988). Elmenteitan sites, found in geographically restricted areas of Southwestern Kenya, are defined by distinctive and relatively uniform material culture and by consistent access to obsidian sources. By contrast, Savannah Pastoral Neolithic (SPN) sites are both more geographically widespread, from Northern Kenya to Northern Tanzania, and more variable in terms of ceramics, obsidian sources, and settlement patterns. It has been suggested that SPN heterogeneity could reflect localised adaptations to increasingly unpredictable rainfall and grazing conditions (Marshall et al., 2011). Archaeological surveys and excavations in the Mara plains in Southwest Kenya also attest to the dominance of pastoralism in this area (Robertshaw, 1990) despite the continuing human and livestock disease threats posed by tsetse fly. While the spread of specialised pastoralism is the most distinctive phenomenon of the 3000–1500 yr BP window, it is also spatially restricted: to date, no unambiguous PN settlements have been found near the Indian Ocean coast, for example, nor farther west than the Nyanza area of Eastern Lake Victoria. While recent discoveries extend the distribution of the PN southward (Prendergast et al., 2016), there is still no evidence that stone-using herders penetrated into Central-Southern Tanzania, despite strong genetic and, more debatably, archaeological arguments for connections between Eastern and Southern African prehistoric herding communities (e.g., Ranciaro et al., 2014; Russell & Lander, 2015; Skoglund et al., 2017).

In summary, the period c. 4500–3000 yr BP is best characterised as one in which foragers and (in the Victoria Basin) fisher-foragers dominated the landscape before pastoralism emerged in Northern Kenya, around 5000 yr BP (Homewood, 2008) as a culturally and linguistically diverse range of societies with differences in terms of the livestock species kept (small ruminants, cattle, camels). Through grazing and the use of fire, pastoralists and their livestock played a role in shaping the ecology of rangeland landscapes, impacting the composition and nutrient content of grass species and the degree of tree cover (Lankester & Davis, 2016). The presence of wildlife on the rangelands has also presented many challenges to pastoralists. Species such as elephant and buffalo pose threats to human life and, together with smaller herbivores, cause crop damage, while carnivore species prey upon livestock. There has also been an ongoing debate around the importance of grazing competition between wild and domestic species (Butt & Turner, 2012). The spread of specialised pastoralism may have resulted in substantial but localised effects on vegetation and microfaunal communities, as indicated by several ecological and ethnoarchaeological studies (e.g., Muchiru et al., 2008, 2009; Shahack-Gross et al., 2003, 2004, 2008; Weissbrod, 2013; see also Boles, 2017; Boles & Lane, 2016). These studies highlight how pastoralist settlements can shape local ecodynamics such as soil nutrient redistribution as a product of grazing-corralling cycles with the implication that, for archaeologists, such effects offer investigative proxies for the ephemeral material residues associated with mobile communities. While methodologies for recognising and interpreting these ecological effects in the deep past are still in development, heterogeneity in soils and vegetation communities has been successfully linked to pastoralism at Sugenya in South-Western Kenya, c. 2000 yr BP (Shahack-Gross et al., 2008). Thus, pastoralists and wildlife have co-existed and competed for millennia in the rangelands of East Africa using similar and complementary adaptations. However, at the end of the 19th Century, shifting socio-political factors began to dramatically reshape the East African landscape (Enghoff, 1990). The opening up of grasslands in Kenya and Tanzania after c. 3500 yr BP, as Rift Valley lakes shrank or disappeared altogether and forests receded, may have helped facilitate the spread of pastoralism, and this seems a more likely starting point for the spread of the PN (Lane, 2013a). One assumes a marginal and highly localised impact upon vegetation that might be associated with grazing where population levels were relatively low. However, further archaeological, and palaeoecological work has the potential to change this narrative (Muiruri et al., 2021a).

2.5.3.4 The Pastoral Iron Age

In the Rift Valley of Kenya and Tanzania, evidence of early farming is nearly absent, other than at the Pastoral Iron Age (PIA) site of Deloraine at c. 1200 yr BP. This site is remarkable for an apparently unique ceramic tradition as well as having, until recently, one of the few direct signs of agriculture in the form of finger millet grain (Ambrose, 1984; Harlan, 1992; Sutton, 1993). This, and neighbouring PIA sites such as Hyrax Hill (Leakey et al., 1943; Kyule, 1997), are interpreted as possibly emerging from later Elmenteitan communities, subsequently giving rise to a distinct form of permanent settlements known as 'Sirikwa holes' or 'hollows' after the depressions formed at their centre. The distribution of Sirikwa sites covered much of the Western Highlands of Kenya, extending into the central part of the Eastern Rift Valley as far as Lake Nakuru, with over 200 individual Sirikwa sites documented; that together span the period from c. 750 to 150 yr BP (Davies, 2012; Sutton, 1971, 1984). Excavations have shown that Sirikwa sites comprised one or more individual homesteads consisting of a central livestock enclosure around which were grouped a series of houses and other structures (interpreted by some as guard huts), with a single entrance into the enclosure that could be closed off (presumably at night and during times of conflict to protect the livestock and occupants from animal and human predators). The food-producing economy of these sites varied; in the drier areas of the Rift Valley, such as at Hyrax Hill and Lanet, more emphasis was placed on livestock herding, whereas at sites in the wetter Western Highlands, such as Chemagel, more emphasis may have been placed on crop production (Kyule, 1997; Posnansky, 1957; Sutton, 1993). Although, it is important to note that much more systematic archaeobotanical, bioarchaeological, and geoarchaeological research is needed at these sites to determine the range of crops exploited, the relative balance of herding versus cultivation, and the overall nature of land use around these sites.

2.5.3.5 The Onset of Iron Age and the Arrival of Bantu Agriculturalists and Cereal Crops

The environmental backdrop of a highly variable climate continues through the Holocene across East Africa. Variation in the carbonate content of sediment cores from the northern part of the Lake Turkana basin documents several arid events occurring between c. 2200 yr BP and 1700 yr BP (Halfman et al., 1994). A lake level reconstruction derived from Lake Tanganyika based on fossil ostracod assemblages suggests that lake levels were low from 2200 to 2000 yr BP (Alin & Cohen, 2003). Conversely, a high-resolution record of diatom-inferred conductivity (dissolved-ion content) from Northern Lake Tanganyika (Stager et al., 2009) shows decreased mean water-residence time over the past 3300 years, implying higher lake levels and a general trend of climatic wetting throughout this period. In Central Kenya the deep Crescent Island Crater basin within Lake Naivasha last stood dried out at c. 2200 yr BP (Verschuren, 2001). At Sacred Lake, on the lower slopes of Mount Kenya, a study of plant leaf waxes indicates increased aridity shortly before 1700 yr BP (Konecky et al., 2014). In Western

Uganda, peak Mg levels within calcite deposited in Lake Edward point to a distinct century-scale long phase of intense evaporation around c. 1900 yr BP (Russell & Johnson, 2005); this evidence for pronounced aridity has also been recorded at a nearby Ugandan crater lake (Russell et al., 2007). We can see evidence for climatic change from Lake Challa near Kilimanjaro, where a shift to relatively low values of the Branched to Isoprenoid Tetraether (BIT) index from bacterial cell-wall lipids have suggested a period of reduced rainfall between 2000 and 1800 yr BP (Buckles et al., 2016). On the coast of East Africa, a higher sea level was recorded after 3200 cal yr BP until around 2000 cal yr BP before falling below the present level at approximately 1400 yr BP (Punwong et al., 2017); this fall in sea level also coincides with a period of reduced rainfall seen at Lake Challa. Following this episode of apparently intense drought or extended dry seasons, much of East Africa appears to have experienced a transition to slightly wetter conditions between c. 1800 and 1500 yr BP. At Lake Naivasha this is substantiated by the refilling of the previously dry basin, followed by a relative high stand lasting until about 1400 yr BP (Verschuren, 2001). Similarly, reduced Mg values of calcite in Lake Edward for the period 1800 to 1600 yr BP (Russell & Johnson, 2005) suggest wet conditions on the western shoulder of the East African plateau, while at Lake Tanganyika, a contemporaneous humid phase is dated to between c. 1750 and 1450 yr BP (Stager et al., 2009). Despite local variation in the strength of the signal, a period of extreme aridity does seem to have centred around 2000 yr BP. Clearly this period of pronounced climate variability had considerable regional distinctions at a time when people may have initiated or accentuated the signals of environmental change.

This period is marked by the arrival of food producers and the onset of the Iron Age in Eastern Africa, extending over the last c. 2200 years, and, as a cultural term, designates groups of people who were metal-using agriculturalists (Huffman, 1982). African crops were first brought into view by Vavilov (1926) who, based on the diversity of cultivated plants, defined eight centres where agriculture could have emerged. Vavilov considered the East African highlands to be a potential cradle of agriculture, where he thought barley (*Hordeum vulgare*), coffee (*Coffea arabica*), sorghum (*Sorghum bicolor*), and pearl millet (*Pennisetum glaucum*) originated in Ethiopia (Barnett, 1999: 60; Edwards, 1991). A wide range of artefacts shows that populations were metal-using food producers particularly growing bullrush millet (*Pennisetum americanum*) and finger millet (*Eleusine coracana*) (Huffman, 1982). It is widely accepted that the early agricultural communities were part of a much larger expansion of Bantu-speaking agriculturalists that spread rapidly through the region. These farming migrants whose origins lie in equatorial West-Central Africa, the hypothesised homeland of Bantu languages, form this so-called 'Iron Age package' comprising Bantu languages, cereal grains domesticated in Africa, and iron technology. However, this was not a single, tightly packed population that spread as one into East Africa; it is likely that there was some physical migration of people, disseminations of words, ideas, technologies, and crops, but this would have varied across time and location (de Maret, 2013; Crowther et al., 2017; Grollemund et al., 2015). The spread of Bantu language, pottery, food production, and metallurgy throughout Eastern Africa was likely the result of several, often separate, processes of immigration diffusion, and invention (Kusimba & Kusimba, 2005) rather than a single colonisation of the region by migrant farmers with metal technology. Over the last 2000 years, East Africa's most populated areas were complex multi-ethnic, multi-economic regions which we call mosaics. Mosaics typically included several communities practising different economies, religions, inventions, and vocations, bound together by friendships and clientship's, alliances, knowledge, and concepts of personal and social identity (Kusimba & Kusimba, 2005; Mapunda, 2000). Although there are problems associated with dating, the beginning of metallurgy was well established by 2000 yr BP (Chirikure, 2018). It should be emphasised iron working was not solely for those communities growing crops; Pastoral Iron Age (PIA) (Iles & Lane, 2015) is an important phase where Iron technologies became widely practised by pastoral communities. For example, Larick (1986) identified seventeen smelting sites in Samburu and Southern Marsabit Districts, which he classified into three broad groups: mountain base sites, plains oasis sites and mountain ridge sites.

Many East African Iron Age sites are characterised as using Urewe ceramics (e.g., Ashley, 2010; Childs & Herbert, 2005; Schmidt, 1980;

Schmidt & Childs, 1995); a pottery style which appears at sites around the Victoria Basin in Uganda, Tanzania, and Kenya, as well as in more forested and highland areas in Rwanda and Burundi (Desmedt, 1991) (Fig. 2.6). The earliest of the Urewe sites appear West of Lake Victoria beginning c. 2500 yr BP (e.g., Posnansky, 1969; Reid, 1994; Van Grunderbeek, 1992), reaching the North-Eastern side (Nyanza) around c. 1800–1500 yr BP (e.g., Ashley, 2010; Lane et al., 2006; Leakey et al., 1948). In Nyanza, Urewe contexts frequently overlie Kansyore and/or Elmenteitan horizons and provide evidence for the transition and contact between foragers, herders, and these newly arrived food producers (Ashley, 2010; Lane et al., 2007; Robertshaw, 1991) with the possible periodic formation of more stable cultural and economic frontiers between them (Lane, 2004). We cannot reconstruct the nature of these transitions but it is likely there was a mix of new immigrants as well as the adoption of new techniques and technologies by the extant populations. On the West side of Lake Victoria, there is also evidence for some degree of continuity between earlier and Urewe occupations (Reid, 1994). Archaeobotanical and archaeofaunal data from Early Iron Age sites, in both the Great Lakes region (Giblin & Fuller, 2011; Lane et al., 2007; Prendergast, 2008) and along the coast in Kenya and Tanzania (Crowther et al., 2016b, 2017; Pawlowicz, 2011), are now helping to disentangle subsistence, material culture, and identity. Unfortunately, coastal Early Iron Age sites have poor bone preservation, preventing a full interpretation of subsistence strategies; sites where faunal and botanical data are present point to a strong marine diet, with occupants reliant on shellfish and fish, with only occasional hunting and no unambiguous evidence of animal husbandry nor farming (Crowther et al., 2016a). In the Victoria Basin, fauna is well preserved at several archaeological sites with Urewe ware, and in each case there is evidence of mixed hunting, fishing, and herding strategies that were previously common in this area in association with Kansyore and Elmenteitan ceramics (Lane et al., 2007; Prendergast, 2008). Historical linguistic data further suggests that by the end of this period, early Bantu-language speakers settled in the area had developed an integrated farming regime that included yams (*Dioscorea* spp.), oil palm, cowpea, and Bambara groundnut (*Vigna subterranea*) among the root crops, and had also adopted sorghum, pearl millet, and perhaps finger millet from neighbouring groups farther north (Schoenbrun, 1993, 1998). The appearance of new terms around 1500 yr BP for 'field', 'to weed' and 'to open land by clearing trees', as well as terms for 'milking and bleeding cattle', also point to the development of additional agricultural and herding practices associated with an increasing domestication and modification of the landscape (Schoenbrun, 1993).

A dualistic concept of hunter-gatherers and food producers as opposite and exclusive is not appropriate for East Africa. Traditional land use systems with little mechanisation are still practiced on a large scale, and wild or semi-domesticated plants play a central role in contemporary African subsistence (Neumann, 2005). Furthermore, the biased search for the 'oldest' domesticated plants has often led to a distorted picture of prehistoric economies. The presence of domesticated plants in the archaeological record does not signal reliance on agriculture; a single grain of domesticated sorghum does not justify calling the corresponding human population 'farmers'. Rather, the status of domesticates must be defined in a broader economic and ecological context based on complete assemblages of plant remains (Neumann, 2005). Perhaps unsurprisingly, the lack of direct archaeobotanical proof for cultivated banana (*Musa* sp.) agriculture in the rainforest raises more questions than it solves. How did the banana cross the Indian Ocean and East Africa before it entered the rain forest? Did the introduction of Asian crops act as a stimulus for the development of indigenous rain forest agriculture? What is the role of the banana in an agricultural system, which enabled the occupation of the rain forest c. 2000 yr BP? (Mbida et al., 2000; Mindzie et al., 2001; Neumann, 2005).

During the Iron Ages, the development of increasingly structured and technologically advanced societies may have significantly impacted the environment through land use for agricultural and herding activities. Escalations in human activities combined with the development of metallurgy (Schmidt & Childs, 1995) led to a rise in population (Vansina, 1995) and resulting forest clearance (Marchant & Taylor, 2000; Vincens et al., 2003). The connections between society and a changing climate during the East African Iron Age are not well known.

Increased fire activity and grassland expansion are documented beginning approximately 2500 yr BP (Colombaroli et al., 2014; Finch & Marchant, 2011). In Lake Victoria this fire history, as determined by Lvg flux, indicates a peak in biomass burning between 1850 and 2050 yr BP (Battistel et al., 2017). High-resolution charcoal fluxes in neighbouring lakes such as Lake Katinda (Colombaroli et al., 2014) demonstrate a similar increase in biomass burning magnitude and frequency in 1850–2150 yr BP coincident with a vegetation transition to open savannah. This data supports the idea of a connection between the increase in fire activity and human presence (Battistel et al., 2017). Historical evidence suggests a development of early metallurgy and ceramic production that required high-temperature fires and burning fuels that provide sufficiently high heat of combustion for processing metals and clays (Battistel et al., 2017). Bayon et al. (2012) argue that Bantu colonists caused major vegetation changes centred around ~2500 yr BP while Maley et al. (2012) ascribe the regional forest change to natural climatic factors.

2.6 Environmental-Human Interconnections: Linking Environmental and Cultural Change

Humans have long distinguished themselves by using tools and technologies to shape East Africa's ecosystems so that today, as with most parts in the world, the landscapes of East Africa are imprinted with a legacy of past anthropogenic activity. There are numerous challenges to investigating the interaction between environmental and cultural change. One of the key obstacles is apparent when we plot the palaeoecological and the archaeological sites together across Eastern Africa; we can see there is a large disconnect in the location of these (Fig. 2.3). Indeed, there are only a few sites where the palaeoecological and the archaeological insights are from the same location (Lejju et al., 2005, 2006). Unfortunately, the spatial disconnection between palaeoecological and archaeological sites mitigates against direct comparison between the evidence for environmental and cultural history.

Locations where palaeoecological and archaeological sites have been studied together can provide great insights. This is particularly exemplified from the Lake Turkana catchment where the changing nature of the lake as it responded to environmental change has impacted directly on the cultural use of the catchment. For example, extensive lowering of lake levels during the mid- to late Holocene exposed large and previously inundated lake deposits (Garcin et al., 2012) that supported rich pasture and provided a conduit for the movement of pastoral communities into the East African region (Gifford-Gonzalez, 1998). However, the radiocarbon dates from the earliest archaeological evidence for pastoralism at the sites around Lake Turkana, e.g. Dongodien (GaJj2), Koobi Fora Ridge (GaJj2) (Owen et al., 1982), and FwJj5 (Ashley et al., 2011) suggest that there was an initial lag following this pronounced environmental change and consequently a gradual take-up of pastoralism in relation to this newly available land. This potential of a temporal lag is further supported by Tierney et al.'s (2013) observation for the Tanganyika Basin whereby the terrestrial ecosystem has an inertia or resilience in responding to climatic stress, and vegetation change appears to have contributed to the abrupt hydrologic change.

Around 4000 yr BP clear indications for an abrupt arid event are apparent in the ice core record from Kilimanjaro (Thompson et al., 2002), which can be correlated to similar events across tropical Africa (Marchant & Hooghiemstra, 2004) between 4100 and 4300 yr BP. Archaeologically, throughout this period we see an expansion of pastoralist activity (Coughenour & Ellis, 1993). However, there is a significant caveat to assessing the impact that severe drought events have as the paleoenvironmental and archaeological records may not be well enough resolved (in terms of chronology and sampling) to detect these short-term, but intense, events. Due to local and regional differences, lags in response time of vegetation and water levels, and the general lack of precision in radiocarbon dating in this period, accurately detecting, and dating decadal-scale events may not be possible in the paleoenvironmental records, leaving the potential for a slightly 'blurred' interpretation of climate from sedimentary and archaeological records alone. While this may be evident with the expansion of pastoralism in

East Africa at this time, as pastoral populations expanded farther southwards from an increasingly dry Sahara, finding the direct chronological and environmental link between people and the climatic events is still elusive.

Similarly to elsewhere in the world, the transition to metal working transformed the ability of populations to modify large tracts of land both in terms of a resource needed for metal working, and for rapidly expanding and growing populations and increasing socio-political organisation (Chirikure, 2018). Indeed it has been inferred, together with palaeoecological data, that iron production resulted in major changes to vegetation and reductions in forest cover, particularly in montane areas that were highly suited to grain-based agriculture. Kiage and Liu (2006) have suggested that from 2500 yr BP a steady decline in tree cover, seen in pollen profiles, can be linked with a coeval increase in evidence for human activity and that increased sedimentation rates are indicative of erosion caused by land clearance for agriculture. Examples given include the presence of *Elaeis guineensis* (wild oil palm) pollen grains found within the Lake Masoko sediment record (Vincens, 1989); coeval deforestation at Batongo and Katenga mires in Uganda (Morrison & Hamilton, 1974; Taylor, 1990). While these are signs of deforestation, disturbance, and change, in the absence of further distinct human indicators or archaeological evidence, assigning a human cause to this disturbance is not only possible but also highly likely. The technological shift to iron working, as throughout the world, is of particular importance when discussing the relationship between people and their environment: it is arguably the first time that people affected the environment around them as much as climate had done in preceding periods. Human impact on the vegetation has often been assumed as a by-product of the need for fuel in the iron production process (Bayon et al., 2012; Bostoen et al., 2015) and the forest clearance for building and agricultural expansion to support a (rapidly) growing population (Taylor, 1990). Evidence for these causal effects is extremely limited, however, and the direct impact from iron smelting was probably minimal, the real link to deforestation being that iron arrives with farming and enables vegetation clearance for agriculture. A good example can be seen from the Pare Mountains in Northern Tanzania, where increased erosion is recorded in the sedimentary records from 2000 yr BP, correlating well with the arrival of agriculture and ironworking in the area (Heckmann, 2011; Heckmann et al., 2014). Pare was a centre of iron production from at least c. 1100 yr BP and several iron smelting and smithing furnaces in North Pare have been excavated. Archaeometallurgical research on samples recovered from these sites suggests that the industry was relatively fuel efficient, despite some variations in raw materials selection and smelting technologies over time. More critically, it is clear from the geoarchaeological studies that soil erosion had been initiated prior to the intensification of smelting activity, which nevertheless continued well into the modern era.

Traditionally, residents of Northern Kenya have relied extensively on vast social networks and economic diversification to mitigate adverse effects of climate change, and there are grassroots efforts underway to restructure that form of resilience in the region's pastoral ecology (Opiyo et al., 2015; Sørbø, 2003). Intertribal social relations have gradually become increasingly hostile between neighbouring ethnic communities, particularly during drought conditions (Schlee, 2013). Looking directly at the archaeological record of the Chalbi Desert, evidence for long-distance social networks include non-local obsidian artefacts and ceramics that were part of a broader East African pastoral Neolithic cultural complex. As much of the existing insights come from around lake margins, such as Lake Turkana, there is a clear need for future paleoenvironmental and archaeological research in inland areas (Chritz et al., 2019), given the proximity, interest, and developing challenges the area around Mount Marsabit must be a strong contender for this focus. Rangelands have lost much of their integrity, through fences, water development, controlled stocking rates, and 'restoration' that has resulted in increasingly fragmented landscapes and disintegration of pastoral communities. Certainly, the present-day drivers behind the 'perfect storm' of development, environmental policy, and social change would not have been so acute in the past, and the communities would not have exerted so much pressure on the land. Instead, there would have been challenges faced by pastoral communities and this is clearly an area where palaeo-perspectives on environmental change and response are key.

Despite extensive forest clearance, as recorded on the mountains of Burundi, Rwanda, Tanzania, and Uganda (Bonnefille & Riollet, 1988; Jolly et al., 1997; Taylor, 1990), some quite large tracts of forest remained intact. For example, a precursor of the present Bwindi Impenetrable Forest remained forested while the surrounding region was extensively cleared. Why some areas were cleared, and others not, is difficult to decipher but clearly important information for contemporary management challenges (Chapter 7). One possibility is the occurrence of some degree of protection imparted by forest dwelling communities likely to have been present at the time—such as a precursor to the modern BaTwa—who restricted expansion of agricultural land (Hart et al., 1996). Currently there has been insufficient archaeological research on this period, or relevant ethnographic work in these montane contexts, to allow us to even begin to understand how people might have perceived and used/avoided certain forested areas. However, the palaeoecological records do inform us that by approximately 2000 yr BP agriculture had spread within the Western highlands of East Africa, and that some areas of forest had been afforded 'protection'; this protection has continued to the present day, now under the guise of National Park legislation (Fig. 2.9).

In the examples above, the benefits of multi-proxy studies are clear, especially when these include evidence of both cultural activities and environmental change. Often, we are left with only one of those types of evidence and forced to make inferences to nearby studies that can be of a very different context and at a vastly differing resolution. This has led to heated debates around the causes of environmental change (e.g., Bayon et al., 2012; Maley et al., 2012). How we bring datasets together to determine the ecological consequences of human activities in different environmental settings—whether these datasets are near each other or over larger areas—is a complex issue (Marchant & Lane, 2014) and will be explored in more detail in the next chapter that focuses on the last millennium.

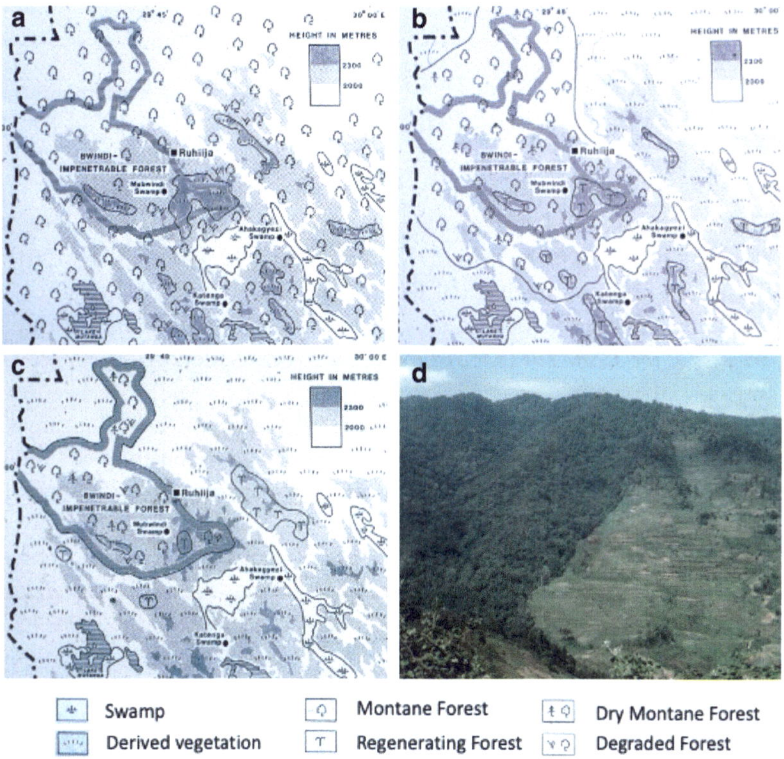

Fig. 2.9 History of forest clearance across the Rukiga Highlands. Initial forest clearance appears to be part of the spread of agriculture associated with Bantu arrival into Uganda and dates from around 2200 years BP (**a**). The focus of this early forest clearance appears to be at the highest altitudes before these were subsequently abandoned and forest clearance spread rapidly across the landscape (**c**). Regenerating forest marks the location of previously cleared forest (**b**). The present day boundary of Bwindi Impenetrable Forest National Park (**c**, **d**) is a strongly demarcated where the current day forested 'island' is surrounded by a 'sea' of agriculture (**d**) (Photograph: Rob Marchant)

References

Abungu, G. H. O., & Muturo, H. W. (1993). Coast-Interior Settlements and Social Relations in the Kenya Hinterland. In T. Shaw, P. J. J. Sinclair, B. Andah, & A. Okpoko (Eds.), *The Archaeology of Africa: Food, Metals and Towns* (pp. 694–704). Routledge.

Ahrends, A., Burgess, N. D., Milledge, S. A. H., Bulling, M. T., Fisher, B., Smart, J. C. R., Clarke, G. P., Mhoro, B. E., & Lewis, S. L. (2010). Predictable Waves of Sequential Forest Degradation and Biodiversity Loss Spreading from an African City. *Proceedings of the National Academy of Sciences of the USA, 107*(33), 14556–14561. https://doi.org/10.1073/pnas.0914471107

Alin, S. R., & Cohen, A. S. (2003). Lake-Level History of Lake Tanganyika, East Africa, for the Past 2500 Years Based on Ostracod-Inferred Water-Depth Reconstruction. *Palaeogeography, Palaeoclimatology, Palaeoecology, 199*, 31–49.

Ambrose, S. H. (1982). Archaeological and Linguistic Reconstructions of History in East Africa. In C. Ehret & M. Posnansky (Eds.), *The Archaeological and Linguistic Reconstruction of African History* (pp. 104–157). University of California Press.

Ambrose, S. H. (1984). The Introduction of Pastoral Adaptations to the Highlands of Eastern Africa. In J. D. Clark & S. A. Brandt (Eds.), *From Hunters to Farmers: The Causes and Consequences of Food Production in Africa* (pp. 212–239). University of California Press.

Ambrose, S. H. (1998). Chronology of the Later Stone Age and Food Production In East Africa. *Journal of Archaeological Science, 25*(4), 377–392.

Archibald, S., Staver, A. C., & Levin, S. A. (2012). Evolution of Human-Driven Fire Regimes in Africa. *Proceedings of the National Academy of Sciences, 109*(3), 847–852.

Ashley, C. Z. (2010). Towards a Socialised Archaeology of Ceramics in Great Lakes Africa. *African Archaeological Review, 27*(2), 135–163.

Ashley, C. Z., Antonites, A., & Fredriksen, P. D. (2016). Mobility and African Archaeology: An Introduction. *Azania: Archaeological Research in Africa, 51*(4), 417–434.

Ashley, C. Z., & Grillo, K. M. (2015). Archaeological Ceramics from Eastern Africa: Past Approaches and Future Directions. *Azania: Archaeological Research in Africa, 50*(4), 460–480.

Ashley, G. M., Ndiema, E. K., Spencer, J. Q. G., Harris, J. W. K., & Kiura, P. W. (2011). Paleoenvironmental Context of Archaeological Sites, Implications for Subsistence Strategies Under Holocene Climate Change, Northern Kenya. *Geoarchaeology, 26*(6), 809–837.

Ashley, G. M., Ndiema, E. K., Spencer, J. O. G., Harris, J. W. K., Kiura, P. W., Dibble, L., Du, A., & Lordan, P. T. (2017). Paleoenvironmental Reconstruction of Dongodien, Lake Turkana, Kenya and OSL Dating of Site Occupation During Late Holocene Climate Change. *African Archaeological Review, 34*(3): 345–362.

Barker, P. A., Hurrell, E. R., Leng, M. J., Wolff, C., Cocquyt, C., Sloane, H. J., & Verschuren, D. (2011). Seasonality in Equatorial Climate Over the Past 25 ky Revealed by Oxygen Isotope Records from Mount Kilimanjaro. *Geology, 39*(12), 1111–1114.

Barnett, T. (1999). *The Emergence of Food Production in Ethiopia.* Cambridge Monographs in African Archaeology 45. BAR International Series, 763. Archaeopress.

Barthelme, J. W. (1977). Holocene Sites Northeast of Lake Turkana: A Preliminary Report. *Azania, 12*, 33–41.

Barthelme, J. W. (1981). *Late Pleistocene–Holocene Prehistory Northeast of Lake Turkana, Kenya.* Ph.D., Department of Anthropology, University of California, Berkeley.

Barthelme, J. W. (1985). *Fisher-Hunters and Neolithic Pastoralists in East Turkana, Kenya.* BAR International Series, 254. British Archaeological Reports.

Barut, S. (1994). Middle and Later Stone Age Lithic Technology and Land Use in East African Savannas. *African Archaeological Review, 12*(1), 43–72.

Battistel, D., Argiriadis, E., Kehrwald, N., Spigariol, M., Russell, J. M., & Barbante, C. (2017). Fire and Human Record at Lake Victoria, East Africa, During the Early Iron Age: Did Humans or Climate Cause Massive Ecosystem Changes? *The Holocene, 27*(7), 997–1007.

Bayliss, A. (2009). Rolling Out Revolution: Using Radiocarbon Dating in Archaeology. *Radiocarbon, 51*(1), 123–147.

Bayon, G., Dennielou, B., Etoubleau, J., Ponzevera, E., Toucanne, S., & Bermell, S. (2012). Intensifying Weathering and Land Use in Iron Age Central Africa. *Science, 335*(6073), 1219–1222.

Beyin, A., Prendergast, M. E., Grillo, K. M., & Wang, H. (2017). New Radiocarbon Dates for Terminal Pleistocene and Early Holocene Settlements in West Turkana, Northern Kenya. *Quaternary Science Reviews* [online], *168*, 208–215. https://doi.org/10.1016/j.quascirev.2017.04.012

Blaauw, M., Christen, J. A., Mauquoy, D., van der Plicht, J., & Bennett, K. D. (2007). Testing the Timing of Radiocarbon-Dated Events Between Proxy Archives. *The Holocene, 17*(2), 283–288.

Blaauw, M., van Geel, B., Kristen, I., Plessen, B., Lyaruu, A., Engstrom, D. R., van der Plicht, J., & Verschuren, D. (2011). High-Resolution ^{14}C Dating of a 25,000-Year Lake-Sediment Record from Equatorial East Africa. *Quaternary Science Reviews, 30*(21), 3043–3059.

Bloszies, C., Forman, S. L., & Wright, D. K. (2015). Water Level History for Lake Turkana, Kenya in the Past 15,000 Years and a Variable Transition from the African Humid Period to Holocene Aridity. *Global Planetary Change, 132*, 64–76.

Boivin, N., Crowther, A., Helm, R., & Fuller, D. Q. (2013). East Africa and Madagascar in the Indian Ocean World. *Journal of World Prehistory, 26*(3), 213–281.

Boivin, N., Crowther, A., Prendergast, M., & Fuller, D. Q. (2014). Indian Ocean Food Globalisation and Africa. *African Archaeological Review, 31*(4), 547–581.

Boles, O. J. C. (2017). *Pastoralist Settlement and the Anthropogenic Savannah: The Archaeo-Ecology of Maili Sita, Kenya*. Unpublished Ph.D. dissertation, University College London.

Boles, O. J. C., & Lane, P. J. (2016). The Green, Green Grass of Home: An Archaeo-Ecological Approach to Pastoralist Settlement in Central Kenya. *Azania, 51*(4), 507–530.

Bollig, M. (2016). Adaptive Cycles in the Savannah: Pastoral Specialization and Diversification in Northern Kenya. *Journal of Eastern African Studies, 10*(1), 21–44.

Bonnefille, R. (1987). Evolution Forestière et Climatique au Burundi Durant les Quarante Derniers Milliers D'années. *C.R. Académie Science Paris, 305*(sér II), 1021–1026.

Bonnefille, R., Chalié, F., Guiot, J., & Vincens, A. (1992). Quantitative Estimates of Full Glacial Temperatures in Equatorial Africa from Palynological Data. *Climate Dynamics, 6*(3–4), 251–257.

Bonnefille, R., & Riollet, R. (1988). The Kashiru Pollen Sequence (Burundi): Palaeoclimatic Implications for the Last 40,000 Years in Tropical Africa. *Quaternary Research, 30*, 19–35.

Bonnefille, R., Roeland, J. C., & Guiot, J. (1990). Temperature and Rainfall Estimates for the Past 40,000 years in Equatorial Africa. *Nature* [online], *346*, 347–349. https://doi.org/10.1038/346347a0

Boom, A., Mora, G., Cleef, A. M., & Hooghiemstra, H. (2001). High Altitude C4 Grasslands in the Northern Andes: Relicts from Glacial Conditions? *Review of Palaeobotany and Palynology* [online], *115*, 147–160. https://doi.org/10.1016/S0034e6667(01)00056e2

Bostoen, K., Clist, B., Doumenge, C., Grollemund, R., Hombert, J. M., Muluwa, J. K., & Maley, J. (2015). Middle to Late Holocene Paleoclimatic Change and the Early Bantu Expansion in the Rainforests of Western Central Africa. *Current Anthropology, 56*, 354–384.

Bower, J., & Nelson, C. M. (1978). Early Pottery and Pastoral Cultures of the Central Rift Valley, Kenya. *Man, 18*, 554–566.

Bower, J. R. F. (1991). The Pastoral Neolithic of East Africa. *Journal of World Prehistory, 5*, 49–82. British Institute in Eastern Africa. 1976. Nyame Akuma no. 9(5).

Brockett, B. H., Biggs, H. C., & van Wilgen, B. W. (2001). A Patch Mosaic Burning System for Conservation Areas in Southern Africa Savannas. *International Journal of Wildland Fire, 10*(2), 169–183.

Buckles, L. K., Weijers, J. W. H., Verschuren, D., Cocquyt, C., & Sinninghe Damste, J. S. (2016). Short-Term Variability in the Sedimentary BIT Index of Lake Challa, East Africa Over the Past 2200 Years: Validating the Precipitation Proxy. *Climates of the Past, 12*, 1243–1262.

Burgess, N. D., & Clarke, G. P. (2000). *Coastal forests of eastern Africa*. IUCN-The World Conservation Union, Publications Services Unit.

Butt, B., & Turner, M. D. (2012). Clarifying Competition: The Case of Wildlife and Pastoral Livestock in East Africa. *Pastoralism: Research, Police, and Practice* [online], *2*(1), 9. https://doi.org/10.1186/2041-7136-2-9

Butzer, K. W. (1968). Comment in Response to 'The Origins and Development of Agriculture in East and Southern Africa'. *Current Anthropology, 9*, 489–509.

Butzer, K. W., Isaac, G. L., Richardson, J. L., & Washbourn-Kamou, C. (1972). Radiocarbon Dating of East African Lake-Levels. *Science, 175*, 1069–1076.

Campbell, M. C., & Tishkoff, S. A. (2008). African Genetic Diversity: Implications for Human Demographic History, Modern Human Origins, and Complex Disease Mapping. *Annual Review of Genomics and Human Genetics, 9*, 403–433.

Childs, T. S., & Herbert, E. W. (2005). Metallurgy and Its Consequences. In A. B. Stahl (Ed.), *African Archaeology: A Critical Introduction* (pp. 276–301). Blackwell Publishing.

Chirikure, S. (2018). *Precolonial Metallurgy and Mining Across Africa* (p. 2020). Oxford University Press.
Chritz, K. L., Marshall, F. B., Zagal, M. E., Kirera, F., & Cerling, T. E. (2015). Environments and trypanosomiasis risks for early herders in the later holocene of the Lake Victoria Basin, Kenya. *Proceedings of the National Academy of Sciences, 112*(12), 3674–3679.
Chritz, K. L., Cerling, T. E., Freeman, K. H., Hildebrand, E. A., Janzen, A., & Prendergast, M. E. (2019). Climate, ecology, and the spread of herding in eastern Africa. *Quaternary Science Reviews, 204*, 119–132.
Clark, J. D., Beyene, Y., WoldeGabriel, G., Hart, W. K., Renne, P. R., Gilbert, H., Defleur, A., Suwa, G., Katoh, S., & Ludwig, K. R. (2003). Stratigraphic, Chronological and Behavioural Contexts of Pleistocene Homo Sapiens from Middle Awash, Ethiopia. *Nature, 423*, 747–752.
Cohen, A. S., Talbot, M. R., Awramik, S. M., Dettman, D. L., & Abell, P. (1997). Lake-Level and Palaeoenvironmental History of Lake Tanganyika, Africa, as Inferred from Late Holocene and Modern Stromatolites. *Geological Society of America Bulletin, 109*, 444–460.
Colinvaux, P. A., De Oliveira, P. E., & Bush, M. B. (2000). Amazonian and Neotropical Plant Communities on Glacial Time Scales: The Failure of the Aridity and Refuge Hypotheses. *Quaternary Science Reviews, 19*, 141–169.
Collett, D. P., & Robertshaw, P. T. (1983). Pottery Traditions of Early Pastoral Communities in Kenya. *Azania, 18*, 107–126.
Colombaroli, D., Ssemmanda, I., Gelorini, V., & Verschuren, D. (2014). Contrasting Long-Term Records of Biomass Burning in Wet and Dry Savannas of Equatorial East Africa. *Global Change Biology, 20*, 2903–2914.
Coughenour, M. B., & Ellis, J. E. (1993). Landscape and Climatic Control of Woody Vegetation in a Dry Tropical Ecosystem: Turkana District, Kenya. *Journal of Biogeography, 20*, 383–398.
Courtney-Mustaphi, C. J., Gajewski, K., Marchant, R., & Rosqvist, G. (2017). A Late Holocene Pollen Record from Proglacial Oblong Tarn, Mount Kenya. *PLOS One, 12*(9), e0184925.
Courtney-Mustaphi, C. J., Githumbi, E. N., Shotter, L. R., Rucina, S. M., & Marchant, R. (2016). Subfossil Statoblasts of *Lophopodella Capensis* (Sollas, 1908) (Bryozoa: Phylactolaemata: Lophopodidae) in the Upper Pleistocene and Holocene Sediments of a Montane Wetland, Eastern Mau Forest Kenya. *African Invertebrates, 7*(1), 39–52.
Crowther, A., Faulkner, P., Prendergast, M. E., Quintana Morales, E. M., Horton, M., Wilmsen, E., Kotarba-Morley, A. M., Christie, A., Petek, N., Tibesasa, R., & Douka, K. (2016a). Coastal Subsistence, Maritime

Trade, and the Colonization of Small Offshore Islands in Eastern African Prehistory. *The Journal of Island and Coastal Archaeology, 11*(2), 211–237.
Crowther, A., Lucas, L., Helm, R., Horton, M., Shipton, C., Wright, H. T., Walshaw, S., Pawlowicz, M., Radimilahy, C., Douka, K., & Picornell-Gelabert, L. (2016b). Ancient Crops Provide First Archaeological Signature of the Westward Austronesian Expansion. *Proceedings of the National Academy of Sciences, 113*(24), 6635–6640.
Crowther, A., Prendergast, M. E., Fuller, D. Q., & Boivin, N. (2017). Subsistence Mosaics, Forager-Farmer Interactions, and the Transition to Food Production in Eastern Africa. *Quaternary International* [online]. https://doi.org/10.1016/j.quaint.2017.01.014
Dale, D., & Ashley, C. Z. (2010). Holocene Hunter-Fisher-Gatherer Communities: New Perspectives on Kansyore Using Communities of Western Kenya. *Azania, 45*(1), 24–48.
Davies, M. I. (2012). Some thoughts on a 'useable' African archaeology: Settlement, population and intensive farming among the Pokot of northwest Kenya. *African Archaeological Review, 29*(4), 319–353.
Davies, M. I. (2013). Stone cairns across eastern Africa: A critical review. *Azania: Archaeological Research in Africa, 48*(2), 218–240.
Deacon, H. J. (2001). Modern Human Emergence: An African Archaeological Perspective. In P. V. Tobias, M. A. Raath, J. Muggi-Cecchi, & G. A. Doyle (Eds.), *Humanity from African Naissance to Coming Millennia—Colloquia in Human Biology and Palaeoanthropology* (pp. 217–226). Florence University Press.
de Maret, P. (2013). Archaeologies of the Bantu Expansion. In P. Mitchell & P. J. Lane (Eds.), *The Oxford Handbook of African Archaeology* (pp. 627–643). Oxford University Press.
Demenocal, P. B. (1995). Plio-Pleistocene African Climate. *Science, 270*(5233), 53–59.
deMenocal, P., Ortiz, J., Guilderson, T., Adkins, J., Sarnthein, M., Baker, L., & Yarusinsky, M. (2000). Abrupt onset and termination of the African Humid Period: Rapid climate responses to gradual insolation forcing. *Quaternary Science Reviews, 19*, 347–361.
Deocampo, D. M., Blumenschine, R. J., & Ashley, G. M. (2002). Wetland Diagenesis and Traces of Early Hominids, Olduvai Gorge. *Tanzania. Quaternary Research, 57*(2), 271–281.
Desmedt, C. (1991). Poteries Anciennes Décorées a la Roulette dans la Région des Grands Lacs. *African Archaeological Review, 9*, 161–196.

Dinesen, L., Lehmberg, T., Svendsen, T. O., Hansen, L. A., & Fjeldsa, J. (1994). A new genus and species of perdicine bird (Phasianidae, Perdicini) from Tanzania; a relict form with Indo-Malayan affinities. *Ibis, 136*(1), 3–11.

Dunsworth, H. M. (2010). Origin of the Genus Homo. *Evolution: Education and Outreach, 3*(3), 353–366.

Edwards, S. (1991). Plants with Wild Relatives Found in Ethiopia. In J. Hawkes, J. M. J. Engels, & T. Wordede (Eds.), *Plant Genetic Resources of Ethiopia* (pp. 42–74). Cambridge University Press.

Elenga, H., Peryon, O., Bonnefille, R., Prentice, I. C., Jolly, D., Cheddadi, R., Guiot, J., Andrieu, V., Bottema, S., Buchet, G., De Beaulieu, J. L., Hamilton, A. C., Maley, J., Marchant, R., Perez-Obiol, R., Reille, M., Riollet, G., Scott, L., Straka, H., … Jonson, H. (2000). Pollen-Based Reconstruction for Southern Europe and Africa 18,000 Years Ago. *Journal of Biogeography, 27*, 154–172.

Enghoff, M. (1990). Wildlife Conservation, Ecological Strategies and Pastoral Communities. A Contribution to the Understanding of Parks and People in East Africa. *Nomadic Peoples, 25–27*, 93–107.

Eubanks, M. W. (1997). Molecular Analysis of Crosses Between Tripsacum Dactyloides and Zea Diploperennis (Poaceae). *TAG Theoretical and Applied Genetics, 94*(6), 707–712.

Fairhead, J., & Leach, M. (1996). *Rethinking the forest-savanna mosaic: Colonial science and its relics in West Africa* (pp. 105–121). James Currey Ltd.

Farrera, I., Harrison, S. P., Prentice, I. C., Ramstein, G., Guiot, J., Bartlein, P. J., Bonnefille, R., Bush, M., Cramer, W., von Grafenstein, U., Holmgren, K., Hooghiemstra, H., Hope, G., Dolly, D., Lauritzen, S. E., Ono, Y., Pinot, S., Stute, M., & Yu, G. (1999). Tropical Climates at the Last Glacial Maximum: A New Synthesis of Terrestrial Palaeoclimate Data. Vegetation, Lake-Levels and Geochemistry. *Climate Dynamics* [online], *15*, 23–856. https://doi.org/10.1007/s003820050317

Fearn, M. L., & Liu, K.-B. (1995). Maize Pollen of 3500 yr BP from Southern Alabama. *American Antiquity, 60*, 109–117.

Ficken, K. J., Wooller, M. J., Swain, D. L., Street-Perrott, A. A., & Eglinton, G. (2002). Reconstruction of a Subalpine Grass-Dominated Ecosystem, Lake Rutundu, Mount Kenya: A Novel Multi-Proxy Approach. *Palaeogeography, Palaeoclimatology, Palaeoecology, 177*(1), 137–149.

Finch, J., & Marchant, R. (2011). A Palaeoecological Investigation into the Role of Fire and Human Activity in the Development of Montane

Grasslands in East Africa. *Vegetation History and Archaeobotany, 20*(2), 109–124.

Finch, J., Marchant, R., & Courtney Mustaphi, C. J. (2017). Ecosystem Change in the South Pare Mountain Bloc, Eastern Arc Mountains of Tanzania. *The Holocene, 27*(6), 796–810.

Finch, J., Wooller, M., & Marchant, R. (2014). Tracing Long-Term Tropical Montane Ecosystem Change in the Eastern Arc Mountains of Tanzania. *Journal of Quaternary Science, 29*(3), 269–278.

Flantua, S. G. A., Blaauw, M., & Hooghiemstra, H. (2016). Geochronological Database and Classification System for Age Uncertainties in Neotropical Pollen Records. *Climate of the Past, 12*(2), 387–414.

Flenley, J. R. (1996). Problems of the Quaternary on Mountains of the Sunda-Sahul Region. *Quaternary Science Review* [online], *15*, 549–555. https://doi.org/10.1016/0277e3791(96)00017e0

Flenley, J. R. (1998). Tropical Forests Under the Climates of the Last 30,000 Years. *Climate Change* [online], *39*, 177–197. https://doi.org/10.1023/A:1005367822750

Fuller, D. Q., & Hildebrand, E. A. (2013). Domesticating Plants in Africa. In P. Mitchell & P. J. Lane (Eds.), *The Oxford Handbook of African Archaeology* (pp. 507–525). University of Oxford Press.

Garcin, Y., Melnick, D., Strecker, M. R., Olago, D., & Tiercelin, J. J. (2012). East African mid-Holocene wet–dry transition recorded in palaeo-shorelines of Lake Turkana, northern Kenya Rift. *Earth and Planetary Science Letters, 331*, 322–334.

Gasse, F. (2000). Hydrological Changes in the African Tropics Since the Last Glacial Maximum. *Quaternary Science Reviews, 19*, 189–211.

Gasse, F. (2002). Diatom-Inferred Salinity and Carbonated Oxygen Isotopes in Holocene Waterbodies of the Western Sahara and Sahel (Africa). *Quaternary Science Reviews, 21*, 737–767.

Gearey, B. R., Marshall, P., & Hamilton, D. (2009). Correlating Archaeological and Palaeoenvironmental Records Using a Bayesian Approach: A Case Study from Sutton Common, South Yorkshire, England. *Journal of Archaeological Science, 36*(7), 1477–1487.

Giblin, J., & Fuller, D. Q. (2011). First and Second Millennium AD Agriculture in Rwanda: Archaeobotanical Finds and Radiocarbon Dates from Seven Sites. *Vegetation History and Archaeobotany, 20*, 253–265.

Gifford-Gonzalez, D. (1998). Early Pastoralists in East Africa: Ecological and Social Dimensions. *Journal of Anthropological Archaeology, 17*(2), 166–200.

Gifford-Gonzalez, D. (2000). Animal Disease Challenges to the Emergence of Pastoralism in Sub-Saharan Africa. *African Archaeological Review, 17*, 95–139.

Gifford-Gonzalez, D. (2005). Pastoralism and its Consequences. In A. B. Stahl (Ed.), *African Archaeology: A Critical Introduction* (pp. 187–224). Blackwell.

Gifford-Gonzalez, D., & Hanotte, O. (2011). Domesticating Animals in Africa: Implications of Genetic and Archaeological Findings. *Journal of World Prehistory, 24*(1), 1–23.

Gil-Romera, G., Turton, D., & Sevilla-Callejo, M. (2011). Landscape Change in the Lower Omo Valley, Southwestern Ethiopia: Burning Patterns and Woody Encroachment in the Savanna. *Journal of Eastern African Studies, 5*, 108–128.

Goldstein, S. T. (2019). Lithic Technological Strategies of the Earliest Herders at Lake Turkana, Northern Kenya. *Antiquity, 93*, 1495–1514.

Goldstein, S. T., & Munyiri, J. M. (2017). The Elmenteitan Obsidian Quarry (GsJj50): New Perspectives on Obsidian Access and Exchange During the Pastoral Neolithic in Southern Kenya. *African Archaeological Review, 34*(1), 43–73.

Gordon, R. G., & Grimes, B. F. (2005). *Ethnologue: Languages of the World* (p. 1272). SIL International.

Graham, M. H., Dayton, P. K., & Erlandson, J. M. (2003). Ice Ages and Ecological Transitions on Temperate Coasts. *Trends in Ecology and Evolution, 18*, 33–40.

Gramly, R. M. (1975). Meat-Eating Sites and Cattle Brands: Patterns of Rockshelter Utilization in East Africa. *Azania, 10*(1), 107–121.

Grillo, K., & Hildebrand, E. (2013). The Context of Early Megalithic Architecture in Eastern Africa: The Turkana Basin c. 5000–4000 BP. *Azania: Archaeological Research in Africa, 48*(2), 193–217.

Grillo, K. M., Prendergast, M. E., with Contreras, D., Fitton, T., Gidna, A., Goldstein, S., Knisley, M., Langley, M., & Mabulla, A. Z. P. (2018). Pastoral Neolithic Settlement at Luxmanda, Tanzania. *Journal of Field Archaeology, 43*, 102–120.

Grollemund, R., Branford, S., Bostoen, K., Meade, A., Venditti, C., & Pagel, M. (2015). Bantu expansion shows that habitat alters the route and pace of human dispersals. *Proceedings of the National Academy of Sciences, 112*(43), 13296–13301.

Haberle, S. G., & Chepstow-Lusty, A. C. (2000). Can Climate Influence Cultural Development? A View Through Time. *Environment and History, 6*(3), 349–369.

Haberle, S. G., Hope, G. S., & van der Kaars, S. (2001). Biomass Burning in Indonesia and Papua New Guinea: Natural and Human Induced Fire Events in the Fossil Record. *Palaeogeography, Palaeoclimatology, Palaeoecology, 171*(3), 259–268.

Haberyan, K. A., & Heckey, R. E. (1987). The Late Pleistocene and Holocene Stratigraphy and Palaeolimnology of Lakes Kivu and Tanganyika. *Palaeogeography, Palaeoclimatology, Palaeoecology, 61*, 169–197.

Haffer, J. R. (1997). Alternative Models of Vertebrate Speciation in Amazonia: An Overview. *Biodiversity & Conservation, 6*, 451–476.

Halfman, J. D., Johnson, T. C., & Finney, B. P. (1994). New AMS Dates, Stratigraphic Correlations and Decadal Climatic Cycles for the Past 4 ka at Lake Turkana, Kenya. *Palaeogeography, Palaeoclimatology, Palaeoecology, 111*, 83–98.

Hamilton, A. A., & Taylor, D. (1991). History of Climate and Forests in Tropical Africa During the Last 8 Million Years. In *Tropical Forests and Climate* (pp. 65–78). Springer.

Hamilton, A. C. (1982). *Environmental History of East Africa: A Study of the Quaternary*. Academic Press.

Hamilton, A. C., Taylor, D., & Vogel, J. (1986). Early Forest Clearance and Environmental Degradation in South-West Uganda. *Nature, 320*, 164–167.

Hanotte, O., Bradley, D. G., Ochieng, J. W., Verjee, Y., Hill, E. W., & Rege, J. E. O. (2002). African Pastoralism: Genetic Imprints of Origins and Migrations. *Science, 296*, 336–339.

Harlan, J. R. (1992). Indigenous African Agriculture. In J. D. Clark (Ed.), *Cambridge History of Africa vol. 1: From Earliest Times to c. 500 BC* (pp. 624–657). Cambridge University Press.

Harlan, J. R. (1993). The Tropical African Cereals. In T. Shaw, P. Sinclair, B. Andah, & A. Opoko (Eds.), *The Archaeology of Africa: Food, Metals and Towns* (pp. 53–60). Routledge.

Harlan, J. R., De Wet, J. M. J., & Stemler, A. B. L. (1976). *Origins of African Plant Domestication* (pp. 465–478). Mouton.

Hart, T. B., Hart, J. A., Deschamp, R., Fournier, M., & Atahols, M. (1996). Changes in Forest Composition over the Last 4000 Years in the Ituri Basin, Zaire. In L. J. G. van der Maesen, X. M. van der Burgt, J. M. van Medenback, & de Rooy (Eds.), *The Biodiversity of African Plants* (pp. 545–563). Kluwer.

Hawthorne, D., Courtney Mustaphi, C. J., Aleman, J. C., Blarquez, O., Colombaroli, D., Daniau, A. L., Marlon, J. R., Power, M., Vannière, B., Han, Y., Hantson, S., Kehrwald, N., Magi, B., Yue, X., Carcaillet,

C., Marchant, R., Ayodele, O., Githumbi, E. N., & Muriuki, R. M. (2017). Global Modern Charcoal Dataset (GMCD): A Tool for Exploring Proxy-Fire Linkages and Spatial Patterns of Biomass Burning. *Quaternary International* [online]. https://doi.org/10.1016/j.quaint.2017.03.046

Heckmann, M. (2011). *Soil Erosion History and Past Human Land Use in the North Pare Mountains: A Geoarchaeological Study of Slope Deposits in NE Tanzania*. Unpublished Ph.D. dissertation, University of York.

Heckmann, M., Muiruri, V., Boom, A., & Marchant, R. (2014). Human-Environment Interactions in an Agricultural Landscape: A 1400-yr Sediment and Pollen Record from North Pare, NE Tanzania. *Palaeogeography, Palaeoclimatology, Palaeoecology, 406*, 49–61.

Helm, R., Crowther, A., Shipton, C., Tengeza, A., Fuller, D., & Boivin, N. (2012). Exploring agriculture, interaction and trade on the eastern African littoral: Preliminary results from Kenya. *Azania: Archaeological Research in Africa, 47*(1), 39–63.

Henn, B. M., Gignoux, C., Lin, A. A., Oefner, P. J., Shen, P., Scozzari, R., ... Underhill, P. A. (2008). Y-chromosomal evidence of a pastoralist migration through Tanzania to southern Africa. *Proceedings of the National Academy of Sciences, 105*(31), 10693–10698.

Hildebrand, E. A., & Grillo, K. M. (2012). Early Herders and Monumental Sites in Eastern Africa: Dating and Interpretation. *Antiquity, 86*(332), 338–352.

Hildebrand, E. A., Grillo, K. M., Sawchuk, E. A., Pfeiffer, S. K., Conyers, L. B., Goldstein, S. T., Hill, A. C., Janzen, A., Klehm, C. E., Helper, M., Kiura, P., Ndiema, E., Ngugi, C., Shea, J. J., & Wang, H. (2018). A Monumental Cemetery Built by Eastern Africa's First Herders Near Lake Turkana, Kenya. *Proceedings of the National Academy of Sciences of the United States of Africa, 115*(36), 8942–8947.

Homewood, K. M. (2008). *Ecology of African Pastoralist Societies*. James Currey.

Huffman, T. N. (1982). Archaeology and Ethnohistory of the African Iron Age. *Annual Review of Anthropology, 11*(1), 133–150.

Iles, L., & Lane, P. J. (2015). Iron Production in Second Millennium AD Pastoralist Contexts on the Laikipia Plateau, Kenya. *Azania: Archaeological Research in Africa* [online], *50*, 372–401. https://doi.org/10.1080/0067270X.2015.1079379

Inskeep, R. (1962). The Age of the Kondoa Rock Paintings in the Light of Recent Excavations at Kisese II Rock Shelter. In G. Mortelmans & J. Nenquin (Eds.), *Actes du IVe Congrès Panafricain de Préhistoire et L'étude du Quaternaire* (pp. 249–256). Musée Royal de l'Afrique Centrale.

IPCC. (2021). 7th Assessment Report.
Ivory, S. J., & Russell, J. (2016). Climate, Herbivory and Fire Controls on Tropical African Forest for the Last 60ka. *Quaternary Science Review* [online], *148*, 101–114. https://doi.org/10.1016/jquascirev.2016.07.015.
Izumi, K., & Lézine, A. M. (2016). Pollen-Based Biome Reconstructions Over the Past 18,000 Years and Atmospheric CO_2 Impacts on Vegetation in Equatorial Mountains of Africa. *Quaternary Science Reviews, 152*, 93–103.
James, W. P. T., Johnson, R. J., Speakman, J. R., Wallace, D. C., Frühbeck, G., Iversen, P. O., & Stover, P. J. (2019). Nutrition and Its Role in Human Evolution. *Journal of Internal Medicine, 285*(5), 533–549.
Janzen, A. (2015). *Mobility and Herd Management Strategies of Early Pastoralists in South-Central Kenya, 3000–1200 BP*. Unpublished Ph.D. thesis, Department of Anthropology, University of California Santa Cruz.
Johnson, T. C., Brown, E. T., McManus, J., Barry, S., Barker, P., & Gasse, F. (2002). A High-Resolution Palaeoclimate Record Spanning the Past 25,000 Years in Southern East Africa. *Science, 296*, 113–117.
Jolly, D., & Haxeltine, A. (1997). Effect of Low Glacial Atmospheric CO_2 on Tropical African Montane Vegetation. *Science, 276*, 786–788.
Jolly, D., Taylor, D., Marchant, R., Hamilton, A., Bonnefille, R., Buchet, G., & Riollet, G. (1997). Vegetation dynamics in central Africa since 18,000 yr BP: Pollen records from the interlacustrine highlands of Burundi, Rwanda and Western Uganda. *Journal of Biogeography, 24*(4), 492–512.
Kaplan, J. O., Krumhardt, K. M., Ellis, E. C., Ruddiman, W. F., Lemmen, C., & Goldewijk, K. K. (2011). Holocene Carbon Emissions as a Result of Anthropogenic Land Cover Change. *The Holocene, 21*(5), 775–791.
Karega-Munene. (1993). *Neolithic/Iron Age Subsistence Economies in the Lake Victoria Basin of East Africa*. Unpublished Ph.D. dissertation, University of Cambridge.
Kariuki, R., Western, D., Willcock, S., & Marchant, R. (2021). Exploring the Relationships Between Land Use and Ecosystem Services. *Land, 10*. https://www.mdpi.com/journal/land/special_issues/landuse_ES
Karlén, W., Fastook, J. L., Holmgren, K., Malmstrom, M., Matthews, J. A., Odada, E., Risberg, J., Rosquist, G., Sandgren, P., Shemesh, A., & Westerberg, L. O. (1999). Glacier Fluctuations on Mount Kenya Since 6000 cal. Years BP: Implications for Holocene Climatic Change in Africa. *Ambio, 28*, 409–418.
Kiage, L. M., & Liu, K. B. (2006). Late Quaternary Paleoenvironmental Changes in East Africa: A Review of Multiproxy Evidence from Palynology,

Lake Sediments, and Associated Records. *Progress in Physical Geography, 30*(5), 633–658.
Kiura, P. (2008). *Ethnoarchaeological and Stable Isotopes in the Study of People's Diets*. VDM Verlag Dr Mueller e.g.
Konecky, B., Russell, J., Huang, Y., Vuille, M., Cohen, L., & Street-Perrott, F. A. (2014). Impact of Monsoons, Temperature, and CO_2 on the Rainfall and Ecosystems of Mt. Kenya During the Common Era. *Palaeogeography, Palaeoclimatology, Palaeoecology, 396*, 17–25.
Kübler, S., Owenga, P., Reynolds, S. C., Rucina, S. M., & King, G. C. P. (2015). Animal Movements in the Kenya Rift and Evidence for the Earliest Ambush Hunting by Hominins. *Scientific Reports* [online], *5*, 14011. https://doi.org/10.1038/srep14011
Kusimba, C. M., & Kusimba, S. B. (2005). Mosaics and Interactions: East Africa, 2000 bp to the Present. In A. B. Stahl (Ed.), *African Archaeology: A Critical Introduction* (pp. 392–419). Blackwell.
Kusimba, C. M., Kusimba, S. B., & Wright, D. (2005). The Development and Collapse of Precolonial Ethnic Mosaics in Tsavo, Kenya. *Journal of African Archaeology, 3*(2), 243–265.
Kusimba, S. (2002). *African Foragers: Environment, Technology, Interactions*. Altamira Press.
Kyule, D. M. (1997). The Sirikwa Economy. *Azania: Archaeological Research in Africa, 32*(1), 21–30.
Lahr, M. M., Rivera, F., Power, R. K., Mounier, A., Copsey, B., Crivellaro, F., Edung, J. E., Fernandez, J. M., Kiarie, C., Lawrence, J., & Leakey, A. (2016). Inter-Group Violence Among Early Holocene Hunter-Gatherers of West Turkana, Kenya. *Nature, 529*(7586), 394–398.
Lane, P., Ashley, C., Seitsonen, O., Harvey, P., Mire, S., & Odede, F. (2007). The Transition to Farming in Eastern Africa: New Faunal and Dating Evidence from Wadh Lang'o and Usenge, Kenya. *Antiquity, 81*, 62–81.
Lane, P. J. (2004). The 'Moving Frontier' and the Transition to Food Production in Kenya. *Azania: Archaeological Research in Africa, 39*, 243–264.
Lane, P. J. (2009). Environmental Narratives and the History of Soil Erosion in Kondoa District, Tanzania: An Archaeological Perspective. *International Journal of African Historical Studies, 42*, 457–483.
Lane, P. J. (2011a). An Outline of the Later Holocene Archaeology and Precolonial History of the Ewaso Basin, Kenya. *Smithsonian Contributions to Zoology, 632*, 11–30.
Lane, P. J. (2011b). Slavery and Slave Trading in Eastern Africa: Exploring the Intersections of Historical Sources and Archaeological Evidence. In P. J.

Lane & K. C. MacDonald (Eds.), *Slavery in Africa: Archaeology and Memory* (pp. 281–314). Oxford University Press.

Lane, P. J. (2013a). The Archaeology of Pastoralism and Stock-Keeping in East Africa. In P. Mitchell & P. J. Lane (Eds.), *Oxford Handbook of African Archaeology* (pp. 585–601). Oxford University Press.

Lane, P. J. (2013b). Trajectories to Pastoralism in Northern and Central Kenya: An Overview of the Archaeological and Environmental Evidence. In M. Bollig, M. Schnegg, & H. P. Wotzka (Eds.), *Pastoralism in Africa: Past, Present and Future* (pp. 105–143). Berghahn.

Lane, P. J. (2015a). Ethnicity, Archaeological Ceramics, and Changing Paradigms in East African Archaeology. In F. G. Richard & K. C. MacDonald (Eds.), *Ethnic Ambiguity and the African Past* (pp. 245–271). Left Coast Press.

Lane, P. J. (2015b). Just How Long Does 'Long-Term' Have to Be? Matters of Temporal Scale as Impediments to Interdisciplinary Understanding in Historical Ecology. In C. Isendahl & D. Stump (Eds.), *The Oxford Handbook of Historical Ecology and Applied Archaeology.* Oxford University Press. https://doi.org/10.1093/oxfordhb/9780199672691.013.5

Lane, P. J. (2016). New Directions for Historical Archaeology in Eastern Africa. *The Journal of African History, 57*(2), 173–181.

Lane, P. J., Ashley, C. Z., & Oteyo, G. (2006). New Dates for Kansyore and Urewe Wares from Northern Nyanza, Kenya. *Azania, 41*, 123–138.

Langley, M. C., Prendergast, M. E., & Grillo, K. M. (2019). Organic technology in the Pastoral Neolithic: Osseous and eggshell artefacts from Luxmanda, Tanzania. *Archaeological and Anthropological Sciences, 11*(1), 1–14.

Lankester, F., & Davis, A. (2016). Pastoralism and Wildlife: Historical and Current Perspectives in the East African Rangelands of Kenya and Tanzania. *Revue Scientifique Et Technique (International Office of Epizootics), 35*(2), 473–484.

Larick, R. (1986). Iron Smelting and Interethnic Conflict Among Precolonial Maa-Speaking Pastoralists of North-Central Kenya. *African Archaeological Review, 4*, 165–176.

Leakey, M. D., Leakey, L. S. B., Game, P. M., & Goodwin, A. J. H. (1943). Report on the excavations at Hyrax Hill, Nakuru, Kenya Colony, 1937–1938. *Transactions of the Royal Society of South Africa, 30*(4), 271–409.

Leakey, M. D., Owen, W. E., & Leakey, L. S. B. (1948). *Dimple-Based Pottery from Central Kavirondo, Kenya Colony* (Coryndon Memorial Museum Occasional Papers No. 2). Deighton, Bell and Co.

Leakey, R. E. (1971). Further Evidence of Lower Pleistocene Hominids from East Rudolf, North Kenya. *Nature, 231*(5300), 241–245.
Lejju, B. J., Robertshaw, P., & Taylor, D. (2006). Africa's Earliest Bananas? *Journal of Archaeological Science, 33*(1), 102–113.
Lejju, B. J., Taylor, D., & Robertshaw, P. (2005). Late-Holocene Environmental Variability at Munsa Archaeological Site, Uganda: A Multicore, Multiproxy Approach. *The Holocene, 15*(7), 1044–1061.
Lejju, J. B. (2009). Vegetation Dynamics in Western Uganda During the Last 1000 Years: Climate Change or Human Induced Environmental Degradation? *African Journal of Ecology, 47*(s1), 21–29.
Lézine, A. M., Izumi, K., Kageyama, M., & Achoundong, G., (2019). A 90,000-Year Record of Afromontane Forest Responses to Climate Change. *Science, 363*(6423), 177–181.
Llorente, M. G., Jones, E. R., Eriksson, A., Siska, V., Arthur, K. W., Arthur, J. W., Curtis, M. C., Stock, J. T., Coltorti, M., Pieruccini, P., & Stretton, S. (2015). Ancient Ethiopian Genome Reveals Extensive Eurasian Admixture in Eastern Africa. *Science, 350*(6262), 820–822.
Loomis, S. E., Russell, J. M., Verschuren, D., Morrill, C., De Cort, G., Damsté, J. S. S., Olago, D., Eggermont, H., Street-Perrott, F. A., & Kelly, M. A. (2017). The Tropical Lapse Rate Steepened During the Last Glacial Maximum. *Science Advances, 3*(1), e1600815.
Los, S. O., Street-Perrott, F. A., Loader, N. J., Froyd, C. A., Cuní-Sanchez, A., & Marchant, R. A. (2019). Sensitivity of a Tropical Montane Cloud Forest to Climate Change, Present, Past and Future: Mt. Marsabit N. Kenya. *Quaternary Science Reviews, 218*, 34–48.
Maley, J., Giresse, P., Doumenge, C., & Favier, C. (2012). Comment on Intensifying Weathering and Land Use in Iron Age Central Africa. *Science, 337*(6098), 1040.
Mapunda, B. B. (2000). Ironworking. Journal of African History 41, 487–529. 2003 Fipa Iron Technologies and Their Implied Social History. In M. Kusimba & S. Kusimba (Eds.), *East African Archaeology: Foragers, Potters, Smiths, and Traders* (pp. 71–86). University of Pennsylvania Museum Publications.
Marchant, R., Richer, S., Boles, O., Capitani, C., Courtney-Mustaphi, C. J., Lane, P., Prendergast, M. E., Stump, D., de Cort, G., Kaplan, J. O., Phelps, L., Kay, A., Olago, D., Petek, N., Platts, P., Punwong, P., Widgren, M., Wynne-Jones, S., Ferro-Vázquez, C., … Wright, D. (2018). Drivers and

Trajectories of Land Cover Change in East Africa: Human and Environmental Interactions from 6000 Years Ago to Present. *Earth-Science Reviews, 178*, 322–378.

Marchant, R. A. (2021). *Climate of East Africa*. Oxford University Press.

Marchant, R. A., & Hooghiemstra, H. (2001). 'Letter to the Editor' Climate of East Africa 6000 ^{14}C Yr BP as Inferred from Pollen Data. By Peyron, O., Jolly, D., Bonnefille, R., Vincens, A. and Guiot, J. *Journal of Quaternary Research, 56*, 133–135.

Marchant, R. A., & Hooghiemstra, H. (2004). Rapid Environmental Change in Tropical Africa and Latin America About 4000 Years Before Present: A Review. *Earth Science Reviews, 66*, 217–260.

Marchant, R. A., & Lane, P. (2014). Past Perspectives for the Future: Foundations for Sustainable Development in E. Africa. *Journal of Archaeological Science, 51*, 12–21.

Marchant, R. A., & Taylor, D. M. (1998). A Late Holocene Record of Montane Forest Dynamics from South-Western Uganda. *The Holocene, 8*, 375–381.

Marchant, R. A., & Taylor, D. M. (2000). Numerical Analysis of Modern Pollen Spectra and in situ Montane Forest-Implications for the Interpretation of Fossil Pollen Sequences from Tropical Africa. *New Phytologist, 146*, 505–515.

Marchant, R. A., Taylor, D. M., & Hamilton, A. C. (1997). Late Pleistocene and Holocene History at Mubwindi Swamp, South-West Uganda. *Quaternary Research, 47*, 316–328.

Marean, C. W., Spencer, L. M., Blumenschine, R. J., & Capaldo, S. D. (1992). Captive Hyena Bone Choice and Destruction, the Schlepp Effect and Olduvai Archaeofaunas. *Journal of Archaeological Science, 19*, 101–121.

Markham, A. (1996). Potential Impacts of Climate Change on Ecosystems: A Review of Implications for Policymakers and Conservation Biologists. *Climate Research, 6*(2), 179–191.

Marlon, J. R., Bartlein, P. J., Daniau, A. L., Harrison, S. P., Maezumi, S. Y., & Power, M. J. (2013). Global Biomass Burning: A Synthesis and Review of Holocene Paleofire Records and Their Controls. *Quaternary Science Reviews, 65*, 5–25.

Marlon, J. R., Kelly, R., Daniau, A. L., Vannière, B., Power, M. J., Bartlein, P., Higuera, P., Blarquez, O., Brewer, S., Brücher, T., Feurdean, A., Gil-Romera, G., Iglesias, V., Maezumi, S. Y., Magi, B., Courtney Mustaphi, C. J., & Zhihai, T. (2016). Reconstructions of Biomass Burning from Sediment Charcoal Records to Improve Data-Model Comparisons. *Biogeosciences, 13*, 3225–3244.

Marshall, F. (1990). Origins of Specialized Pastoral Production in East Africa. *American Anthropologist, 92*(4), 873–894.
Marshall, F., Grillo, K. M., & Arco, K. (2011). Prehistoric Pastoralists and Social Responses to Climatic Risk in East Africa. In N. Miller, K. Moore, & K. Ryan (Eds.), *Sustainable Lifeways: Cultural Persistence in an Ever-Changing Environment* (pp. 38–73). University of Pennsylvania Museum of Archaeology and Anthropology.
Marshall, F., & Hildebrand, E. (2002). Cattle Before the Crops: The Beginnings of Food Production in Africa. *Journal of World Prehistory, 16*, 99–143.
Marshall, F., Stewart, K. M., & Barthelme, J. (1984). Early Domestic Stock at Dongodien in Northern Kenya. *Azania, 19*, 120–127.
Marshall, F. B. (2000). The Origins and Spread of Domestic Animals in East Africa. In R. M. Blench & K. C. MacDonald (Eds.), *The Origins and Development of African Livestock: Archaeology, Genetics, Linguistics, and Ethnography* (pp. 191–221). University College London Press.
Martinson, D. G., Pisias, N. G., Hays, J. D., Imbrie, J., Moore, T. C., & Shackleton, N. J. (1987). Age Dating and Orbital Theory of the Ice Ages: Development of a High Resolution 0-to-300,000-year Chronostratigraphy. *Quaternary Research, 27*, 1–29.
Masao, F. T. (1979). *The Later Stone Age and the Rock Paintings of Central Tanzania.* Franz Steiner Verlag.
Maslin, M. A. (2016). *The Cradle of Humanity: How the Changing Landscape of Africa Made Us so Smart.* University Press.
Maslin, M. A., Shultz, S., & Trauth, M. H. (2015). A Synthesis of the Theories and Concepts of Early Human Evolution. *Philosophical Transactions of the Royal Society B: Biological Sciences* [online], *370*, 20140064. https://doi.org/10.1098/rstb.2014.0064
Mbida, C. M., Van Neer, W., Doutrelepont, H., & Vrydaghs, L. (2000). Evidence for Banana Cultivation and Animal Husbandry During the First Millennium BC in the Forest of Southern Cameroon. *Journal of Archaeological Science, 27*(2), 151–162.
McDougall, I., Brown, F. H., & Fleagle, J. G. (2005). Stratigraphic Placement and Age of Modern Humans from Kibish, Ethiopia. *Nature, 433*, 733–736.
McGlynn, G., Mooney, S., & Taylor, D. (2013). Palaeoecological Evidence for Holocene Environmental Change from the Virunga Volcanoes in the Albertine Rift, Central Africa. *Quaternary Science Reviews, 61*, 32–46.

Mehlman, M. J. (1989). *Later Quaternary Archaeological Sequences in Northern Tanzania*. Unpublished Ph.D. dissertation, University of Illinois at Urbana-Champaign.

Merrick, H., & Brown, F. H. (1984). Obsidian Sources and Patterns of Source Utilization in Kenya and Northern Tanzania: Some Initial Findings. *African Archaeological Review, 2*, 129–152.

Mindzie, C. M., Doutrelepont, H., Vrydaghs, L., Swennen, R. L., Swennen, R. J., Beeckman, H., De Langhe, E., & De Maret, P. (2001). First Archaeological Evidence of Banana Cultivation in Central Africa During the Third Millennium Before Present. *Vegetation History and Archaeobotany, 10*(1), 1–6.

Mjema, E. (2008). *The Later Stone Age of Endadu Rock Shelter, Babati District, Northern Tanzania*. Unpublished M.A. dissertation, University of Dar es Salaam.

Mohammed, M. U., Bonnefille, R., & Johnson, T. C. (1996). Pollen and Isotopic Records in Late Holocene Sediments from Lake Turkana, Kenya. *Palaeogeography, Palaeoclimatology, Palaeoecology, 119*(3–4), 371–383.

Morrison, M. E., & Hamilton, A. C. (1974). Vegetation and Climate in the Uplands of South-Western Uganda During the Later Pleistocene Period: II. Forest Clearance and Other Vegetational Changes in the Rukiga Highlands During the Past 8000 Years. *The Journal of Ecology, 62*(1), 1–31.

Morrissey, A., & Scholz, C. A. (2014). Paleohydrology of Lake Turkana and Its Influence on the Nile River System. *Palaeogeography, Palaeoclimatology, Palaeoecology, 403*, 88–100.

Msaky, E. S., Livingstone, D., & Davis, O. K. (2005). Paleolimnological Investigations of Anthropogenic Environmental Change in Lake Tanganyika: V. Palynological Evidence for Deforestation and Increased Erosion. *Journal of Paleolimnology, 34*(1), 73–83.

Muchiru, A. N., Western, D. J., & Reid, R. S. (2008). The Role of Abandoned Pastoral Settlements in the Dynamics of African Large Herbivore Communities. *Journal of Arid Environments, 72*(6), 940–952.

Muchiru, A. N., Western, D. J., & Reid, R. S. (2009). The Impact of Abandoned Pastoral Settlements on Plant and Nutrient Succession in an African Savanna Ecosystem. *Journal of Arid Environments, 73*(3), 322–331.

Mumbi, C. T., Marchant, R., Hooghiemstra, H., & Wooller, M. J. (2008). Late Quaternary Vegetation Reconstruction from the Eastern Arc Mountains, Tanzania. *Quaternary Research, 69*(2), 326–341.

Mumbi, C. T., Marchant, R. A., & Lane, P. (2014, May). Vegetation Response to Climate Change and Human Impacts in the Usambara Mountains. *ISRN Forestry* [online]. https://doi.org/10.1155/2014/240510

Muiruri, V. M., Owen, R. B., Lowenstein, T. K., Renaut, R. W., Marchant, R., Rucina, S. M., ... Wang, C. (2021a). A million year vegetation history and palaeoenvironmental record from the Lake Magadi Basin, Kenya Rift Valley. *Palaeogeography, Palaeoclimatology, Palaeoecology, 567*, 110247.

Muiruri, V., Marchant, R., Rucina, S. M., Scott, L., & Lane, P. J. (2021b). Late Holocene environmental change and anthropogenic: Ecosystem interaction on the Laikipia Plateau, Kenya. *Ambio*, 1–14.

Nakintu, J., & Lejju, J. (2016). Environmental Dynamics of Lake Victoria: Evidence from a 10,000 ^{14}C yr Diatom Record from Napoleon Gulf and Sango Bay. *Journal of Environmental Science and Engineering, 5*, 626–637.

Nelson, C. M. (1995). The Work of the Koobi Fora Field School at the Jarigole Pillar Site. *Kenya Past and Present, 27*, 49–63.

Neumann, K. (2005). The romance of farming: Plant cultivation and domestication in Africa. *African Archaeology: A Critical Introduction*, 249–275.

Neumann, K., & Hildebrand, E. A. (2009). Early Bananas in Africa: The State of the Art. *Ethnobotany Research and Applications, 7*, 353–362.

Odner, K. (1971a). Usangi Hospital and Other Archaeological Sites in the North Pare Mountains, North-Eastern Tanzania. *Azania, 6*, 89–130.

Odner, K. (1971b). A Preliminary Report of an Archaeological Survey on the Slopes of Kilimanjaro. *Azania, 6*, 131–149.

Opiyo, F., Wasonga, O., Nyangito, M., Schilling, J., & Munang, R. (2015). Drought Adaptation and Coping Strategies Among the Turkana Pastoralists of Northern Kenya. *International Journal of Disaster Risk Science, 6*, 295–309.

Owen, R. B., Barthelme, J. W., Renaut, R. W., & Vincens, A. (1982). Palaeolimnology and Archaeology of Holocene Deposits North-East of Lake Turkana, Kenya. *Nature, 298*(5874), 523–529.

Parnell, A. C., Haslett, J., Allen, J. R., Buck, C. E., & Huntley, B. (2008). A Flexible Approach to Assessing Synchroneity of Past Events Using Bayesian Reconstructions of Sedimentation History. *Quaternary Science Reviews, 27*(19), 1872–1885.

Parr, C. L., & Andersen, A. N. (2006). Patch Mosaic Burning for Biodiversity Conservation: A Critique of the Pyrodiversity Paradigm. *Conservation Biology, 20*, 1610–1619.

Pawlowicz, M. C. (2011). *Finding Their Place in the Swahili World: An Archaeological Exploration of Southern Tanzania.* Unpublished Ph.D. dissertation, University of Virginia.

Perrott, R. A. (1982). A High Altitude Pollen Diagram from Mount Kenya: Its Implications for the History of Glaciation. *Palaeoecology of Africa and the Surrounding Islands, 14*, 77–83.

Petek, N. (2015). An Archaeological Survey of the Lake Baringo Lowlands 2014: Preliminary Results. *Nyame Akuma, 83*, 100–111.

Phillipson, D. W. (1977). Lowasera. *Azania, 12*, 1–32.

Piperno, D. R., Holst, I., Wessel-Beaver, L., & Andres, T. C. (2002). Evidence for the Control of Phytolith Formation in Cucurbita Fruits by the Hard Rind (Hr) Genetic Locus: Archaeological and Ecological Implications. *Proceedings of the National Academy of Sciences, 99*(16), 10923–10928.

Platts, P. J., Omeny, P. A., & Marchant, R. (2015). AFRICLIM: High-Resolution Climate Projections for Ecological Applications in Africa. *African Journal of Ecology, 53*(1), 103–108.

Posnansky, M. (1957). Excavations at Lanet, Kenya, 1957. *Azania, 2*, 89–114.

Posnansky, M. (1969). Bigo bya Mugenyi. *Uganda Journal, 33*, 125–150.

Potts, R., Dommain, R., Moerman, J. W., Behrensmeyer, A. K., Deino, A. L., Riedl, S., ... Uno, K. (2020). Increased ecological resource variability during a critical transition in hominin evolution. *Science advances, 6*(43), eabc8975.

Prendergast, M. E. (2008). *Forager Variability and Transitions to Food Production in Secondary Settings: Kansyore and Pastoral Neolithic Economies in East Africa.* Unpublished Ph.D. dissertation, Harvard University.

Prendergast, M. E. (2010). Kansyore Fisher-Foragers and Transitions to Food Production in East Africa: The View from Wadh Lang'o, Nyanza Province. *Azania: Archaeological Research in Africa, 45*(1), 83–111.

Prendergast, M. E. (2011). Hunters and Herders at the Periphery: The Spread of Herding in Eastern Africa. In H. Jousse & J. Lesur (Eds.), *People and Animals in Holocene Africa: Recent Advances in Archaeozoology* (pp. 43–58). Africa Magna Verlag.

Prendergast, M. E., & Beyin, A. (2017). Fishing in a Fluctuating Landscape: Terminal Pleistocene and Early Holocene Subsistence Strategies in the Lake Turkana Basin, Kenya. *Quaternary International* [online]. https://doi.org/10.1016/j.quaint.2017.04.022

Prendergast, M. E., Grillo, K. M., Mabulla, A. Z. P., & Wang, H. (2014). New Dates for Kansyore and Pastoral Neolithic Ceramics in the Eyasi Basin, Tanzania. *Journal of African Archaeology, 12*(1), 89–98.

Prendergast, M. E., & Mutundu, K. K. (2009). Late Holocene Zooarchaeology in East Africa: Ethnographic Analogues and Interpretive Challenges. *Documenta Archaeobiologiae, 7*, 203–232.

Prendergast, M. E., Quintana Morales, E. M., Crowther, A., Horton, M. C., & Boivin, N. L. (2017). Dietary Diversity on the Swahili Coast: The Fauna from Two Zanzibar Trading Locales. *International Journal of Osteoarchaeology* [online]. https://doi.org/10.1002/oa.2585

Prendergast, M. E., Rouby, H., Punnwong, P., Marchant, R., Crowther, A., Kourampas, N., Shipton, C., Walsh, M., Lambeck, K., & Boivin, N. L. (2016). Continental Island Formation and the Archaeology of Defaunation on Zanzibar, Eastern Africa. *Plos One, 11*(2). https://doi.org/10.1371/journal.pone.0149565

Punwong, P., Marchant, R., & Selby, K. (2013). Holocene Mangrove Dynamics and Environmental Change in the Rufiji Delta, Tanzania. *Vegetation History and Archaeobotany, 22*(5), 381–396.

Punwong, P., Sritrairat, S., Selby, K., Marchant, R., Pumijumnong, N., & Traiperm, P. (2017). An 800 Year Record of Mangrove Dynamics and Human Activities in the Upper Gulf of Thailand. *Vegetation History and Archaeobotany*. https://doi.org/10.1007/s00334-017-0651-x

Ranciaro, A., Campbell, M. C., Hirbo, J. B., Ko, W. Y., Froment, A., Anagnostou, P., Kotze, M. J., Ibrahim, M., Nyambo, T., Omar, S. A., & Tishkoff, S. A. (2014). Genetic Origins of Lactase Persistence and the Spread of Pastoralism in Africa. *The American Journal of Human Genetics, 94*(4), 496–510.

Reid, D. A. M. (1994–1995). Ntusi and the Development of Social Complexity in Southern Uganda. In G. Pwiti & R. Soper (Eds.), *Aspects of African Archaeology* (pp. 621–628). University of Zimbabwe Press.

Reimer, P. J., Baillie, M. G., Bard, E., Bayliss, A., Beck, J. W., Bertrand, C. J., & Damon, P. E. (2004). IntCal04 Terrestrial Radiocarbon Age Calibration, 0–26 cal kyr BP. *Radiocarbon, 46*(3), 1029–1058.

Reimer, P. J., Bard, E., Bayliss, A., Beck, J. W., Blackwell, P. G., Ramsey, C. B., Buck, C. E., Cheng, H., Edwards, R .L., Friedrich, M., & Grootes, P. M. (2013). IntCal13 and Marine13 Radiocarbon Age Calibration Curves 0–50,000 Years cal BP. *Radiocarbon, 55*, 1889–1887.

Rietti-Shati, M., Shemesh, A., & Karlen, W. (1998). A 3000-year climatic record from biogenic silica oxygen isotopes in an equatorial high-altitude lake. *Science, 281*(5379), 980–982.

Robbins, L. H. (1972). Archaeology in the Turkana District, Kenya. *Science, 176*(4033), 359–366.

Robbins, L. H. (1980). *Lopoy: A Late Stone-Age Fishing and Pastoralist Settlement in the Lake Turkana Basin, Kenya* (Vol. 3). Michigan State University Museum.

Robbins, L. H. (1984). Late Prehistoric Aquatic and Pastoral Adaptations West of Lake Turkana, Kenya. In J. D. Clark & S. A. Brandt (Eds.), *Hunters and Farmers* (pp. 206–212). University of California Press.

Robertshaw, P. T. (1988). Environment and Culture in the Late Quaternary of Eastern Africa: A Critique of Some Correlations. In J. Bower & D. Lubell (Eds.), *Prehistoric Cultures and Environments in the Late Quaternary of Africa* (pp. 115–126). Cambridge University Press.

Robertshaw, P. T. (1990). *Early Pastoralists of South-Western Kenya.* British Institute in Eastern Africa.

Robertshaw, P. T. (1991). Gogo Falls: Excavations at a Complex Archaeological Site East of Lake Victoria. *Azania, 26*, 63–195.

Romans, B. W., Castelltort, S., Covault, J. A., Fildani, A., & Walsh, J. P. (2016). Environmental Signal Propagation in Sedimentary Systems Across Timescales. *Earth-Science Reviews, 153*, 7–29.

Rossel, G. (1995). Musa and Ensete in Africa: Taxonomy, Nomenclature and Uses. *Azania, 30*, 130–146.

Rucina, S. M., Muiruri, V. M., Downton, L., & Marchant, R. (2010). Late Holocene Savanna Dynamics in the Amboseli Basin, Kenya. *Holocene, 20*, 667–677.

Rucina, S. M., Muiruri, V. M., Kinyanjui, R. N., McGuiness, K., & Marchant, R. (2009). Late Quaternary Vegetation and Fire Dynamics on Mount Kenya. *Palaeogeography, Palaeoclimatology, Palaeoecology, 283*(1), 1–14.

Ruddiman, W. F., Fuller, D. Q., Kutzbach, J. E., Tzedakis, P. C., Kaplan, J. O., Ellis, E. C., Vavrus, S. J., Roberts, C. N., Fyfe, R., He, F., & Lemmen, C. (2015). Late Holocene Climate: Natural or Anthropogenic? *Reviews of Geophysics, 54*, 93–118.

Russell, J. M., & Johnson, T. C. (2005). A High-Resolution Geochemical Record from Lake Edward, Uganda Congo and the Timing and Causes of Tropical African Drought During the late Holocene. *Quaternary Science Reviews, 24*(12), 1375–1389.

Russell, J. M., Verschuren, D., & Eggermont, H. (2007). Spatial Complexity of "Little Ice Age" Climate in East Africa: Sedimentary Records from Two Crater Lake Basins in Western Uganda. *Holocene, 17*, 183–193.

Russell, T., & Lander, F. (2015). 'What Is Consumed Is Wasted': From Foraging to Herding in the Southern African Later Stone Age. *Azania: Archaeological Research in Africa, 50*(3), 267–317.

Salzmann, U., Hoelzmann, P., & Morczinek, I. (2002). Late Quaternary Climate and Vegetation of the Sudanian Zone of Northeast Nigeria. *Quaternary Research, 58*, 73–83.

Sawchuk, E. A., Pfeiffer, S., Klehm, C. E., Cameron, M. E., Hill, A. C., Janzen, A., Grillo, K. M., & Hildebrand, E. A.. (2019). The Bioarchaeology of Mid-Holocene Pastoralist Cemeteries West of Lake Turkana, Kenya. *Archaeological and Anthropological Sciences, 11*, 6221–6241. https://doi.org/10.1007/s12520-019-00914-4

Schlebusch, C. M., Malmström, H., Günther, T., Sjödin, P., Coutinho, A., Edlund, H., ... Jakobsson, M. (2017). Southern African ancient genomes estimate modern human divergence to 350,000 to 260,000 years ago. *Science, 358*(6363), 652–655.

Schlee, G. (2013). Territorializing Ethnicity: The Imposition of a Model of Statehood on Pastoralists in Northern Kenya and Southern Ethiopia. *Ethnic and Racial Studies, 36*, 857–874.

Schmidt, P. R. (1980). Early Iron Age Settlements and Industrial Locales in West Lake. *Tanzania Notes and Records, 1984–85*, 77–94.

Schmidt, P. R. (2016, June 28). *Forest Clearance, Fisher People and Human Burials in Western Uganda: New Light on Environmental History.* Paper presented to the Society of African Archaeologists 23rd Biannual Conference, Toulouse, France.

Schmidt, P. R., & Childs, S. T. (1995). Ancient African Iron Production. *American Scientist, 83*(6), 524.

Schoenbrun, D. L. (1993). We are what we eat: Ancient agriculture between the great lakes. *Journal of African History, 34*(1), 1–31.

Schoenbrun, D. L. (1998). *A Green Place, a Good Place: Agrarian Change, Gender, and Social Identity in the Great Lakes Region to the 15th Century.* James Currey.

Seitsonen, O. (2010). Lithics Use at Kansyore Sites in East Africa: Technological Organisation at Four Recently Excavated Sites in Nyanza Province, Kenya. *Azania: Archaeological Research in Africa, 45*(1), 49–82.

Shahack-Gross, R., Marshall, F., Ryan, K., & Weiner, S. (2004). Reconstruction of Spatial Organization in Abandoned Maasai Settlements: Implications for Site Structure in the Pastoral Neolithic of East Africa. *Journal of Archaeological Science, 31*(10), 1395–1411.

Shahack-Gross, R., Marshall, F., & Weiner, S. (2003). Geo-Ethnoarchaeology of Pastoral Sites: The Identification of Livestock Enclosures in Abandoned Maasai Settlements. *Journal of Archaeological Science, 30*(4), 439–459.

Shahack-Gross, R., Simons, A., & Ambrose, S. H. (2008). Identification of Pastoral Sites Using Stable Nitrogen and Carbon Isotopes from Bulk Sediment Samples: A Case Study in Modern and Archaeological Pastoral Settlements in Kenya. *Journal of Archaeological Science, 35*(4), 983–990.

Shanahan, T. M., McKay, N. P., Hughen, K. A., Overpeck, J. T., Otto-Bliesner, B., Heil, C. W., King, J., Scholz, C. A., & Peck, J. (2015). The Time-Transgressive Termination of the African Humid Period. *Nature Geoscience, 8*(2), 140–144.

Shipton, C., Helm, R., Boivin, N., Crowther, A., Austin, P., & Fuller, D. Q. (2013). Intersections, networks and the genesis of social complexity on the Nyali coast of East Africa. *African Archaeological Review, 30*, 427–453.

Shipton, C., Roberts, P., Archer, W., Armitage, S. J., Bita, C., Blinkhorn, J., Courtney-Mustaphi, C., Crowther, A., Curtis, R., d'Errico, F., & Douka, K. (2018). 78,000-Year-Old Record of Middle and Later Stone Age Innovation in an East African Tropical Forest. *Nature Communications, 9*(1), 1–8.

Skoglund, P., Thompson, J. C., Prendergast, M. E., Mittnik, A., Sirak, K., Hajdinjak, M., Salie, T., Rohland, N., Mallick, S., Peltzer, A., & Heinze, A. (2017). Reconstructing Prehistoric African Population Structure. *Cell, 171*(1), 59–71.

Smith, A. B. (2008). Is Absence of Evidence, Evidence of Absence? Problems in the Archaeology of Early Herding Societies of Southern Africa. In H. Bernard & W. Wendrich (Eds.), *The Archaeology of Mobility: Old World and New World Nomadism* (pp. 264–279). UCLA.

Smol, J. P., Birks, H. J. B., & Last, W. M. (2001). *Tracking Environmental Change Using Lake Sediments Volume 3: Terrestrial, Algal, and Siliceous Indicators*. Kluwer.

Sobania, N. (1988). Fishermen Herders: Subsistence, Survival and Cultural Change in Northern Kenya. *The Journal of African History, 29*(1), 41–56.

Sørbø, G. M. (2003). Pastoral Ecosystems and the Issue of Scale. *AMBIO: A Journal of the Human Environment, 32*, 113–117.

Ssemmanda, I., & Vincens, A. (1993). Végétation et Climate dans le Bassin du lac Albert (Oganda, Zaïre) depuis 13,000 ans BP. Apport de la palynologie. *C. R. Academie Sciences Paris, 316*(2), 561–567.

Stager, J. C., Cocquyt, C., Bonnefille, R., Weyhenmeyer, C., & Bowerman, N. (2009). A Late Holocene Paleoclimatic History of Lake Tanganyika, East Africa. *Quaternary Research, 72*, 47–56.

Stahl, A. B. (2009). The archaeology of African history. *The International journal of African historical studies, 42*(2), 241–255.

Stanley, J. D., Krom, M. D., Cliff, R. A., & Woodward, J. C. (2003). Short Contribution: Nile Flow Failure at the End of the Old Kingdom, Egypt: Strontium Isotopic Evidence and Petrological Evidence. *Geoarcheology, 18*, 395–402.

Stewart, K. M. (1989). *Fishing Sites of North and East Africa in the Late Pleistocene and Holocene: Environmental Change and Human Adaptation* (Vol. 521). B.A.R. International Series.

Stiles, D., & Munro-Hay, S. C. (1981). Stone Cairn Burials at Kokurmatakore, Northern Kenya. *Azania, 16*, 151–166.

Stott, P. A., Tett, S. F. B., Jones, G. S., Allen, M. R., Mitchell, J. F. B., & Jenkins, G. J. (2000). External Control of 20th Century Temperature by Natural and Anthropogenic Forcings. *Science, 290*(5499), 2133–2137.

Street-Perrott, F. A., Holmes, J .F., Waller, M., Allen, M. J., Barber, N. G. H., Fothergill, P. A., Harkness, D. D., Ivanovich, M., Kroon, D., & Perrott, R. A. (2000). Drought and Dust Deposition in the West African Sahel: A 5500-year Record from Kajemarum Oasis, Northeastern Nigeria. *The Holocene, 10*, 293–302.

Street-Perrott, F. A., Huang, Y., Perrott, R. A., Eglinton, G., Barker, P., Khelifa, L. B., Harkness, D. D., & Olago, D. O. (1997). Impact of Lower Atmospheric Carbon Dioxide on Tropical Mountain Ecosystems. *Science, 278*(5342), 1422–1426.

Street-Perrott, F. A., & Perrott, R. A. (1988). Holocene Vegetation, Lake-Levels and Climate of Africa. In H. E. Wright, J. E. Kutzbach, T. Webb, W. F. Ruddiman, F. A. Street-Perrott, & P. J. Bartlein (Eds.), *Global Climates Since the Last Glacial Maximum* (pp. 318–356). University of Minnesota Press.

Stump, D., & Tagseth, M. (2009). The History of Precolonial and Early Colonial Agriculture on Kilimanjaro: A Review. In T. Clack (Ed.), *Culture, History and Identity: Human-Environmental Relations in the Mount Kilimanjaro Area, Tanzania* (pp. 107–124). Cambridge Monographs in African Archaeology. Archaeopress.

Sutton, J. E. G. (1971). Temporal and Spatial Variability in African Iron Furnaces. In K. Haaland & P. L. Shinnie (Eds.), *African Iron Working, Ancient and Traditional* (pp. 164–196). Norwegian University Press.

Sutton, J. E. G. (1984). Irrigation and Soil-Conservation in African Agricultural History: With a Reconsideration of the Inyanga Terracing Zimbabwe and Engaruka Irrigation Works Tanzania. *The Journal of African History, 251*, 25–41.

Sutton, J. E. G. (1993). Becoming Maasailand. In T. Spear & R. Waller (Eds.), *Being Maasai: Ethnicity and Identity in East Africa* (pp. 38–60). James Currey.

Szymanski, R. M. (2017). Detection of Human Landscape Alteration Using Nested Microbotanical and Fungal Proxies. *Environmental Archaeology, 22,* 1–17.

Talbot, M. R., & Livingstone, D. A. (1989). Hydrogen Index and Carbon Isotopes of Lacustrine Organic Matter as Lake Level Indicators. *Palaeogeography, Palaeoclimatology, Palaeoecology, 70*(1–3), 121–137.

Taylor, D.M. (1990). Late quaternary pollen records from two Ugandan mires: Evidence for environmental change in the Rukiga highlands of southwest Uganda. *Palaeogeography, Palaeoclimatology, Palaeoecology, 80*, 283–300.

Taylor, D. (1993). Environmental Change in Montane South West Uganda: A Pollen Record for the Holocene from Ahakagyezi Swamp. *The Holocene, 3*, 324–332.

Taylor, D. M., Lane, P. J., Muiruri, V., Ruttledge, A., Gaj McKeever, R., Nolan, T., Kenny, P., & Goodhue, R. (2005). Mid- to Late-Holocene Vegetation Dynamics on the Laikipia Plateau, Kenya. *The Holocene, 15*(6), 837–846.

Thompson, L. G., Mosley-Thompson, E., Davis, M. E., Henderson, K. E., Brecher, H. H., Zagorodnov, V. S., & Mashiotta, T. A. (2002). Kilimanjaro Ice Core Records: Evidence of Holocene Climate Change in Tropical Africa. *Science, 298*, 589–593.

Tierney, J. E., Russell, J. M., & Huang, Y. (2010). A Molecular Perspective on Late Quaternary Climate and Vegetation Change in the Lake Tanganyika Basin, East Africa. *Quaternary Science Reviews, 29*, 787–800.

Tierney, J. E., Smerdon, J. E., Anchukaitis, K. J., & Seager, R. (2013). Multidecadal Variability in East African Hydroclimate Controlled by the Indian Ocean. *Nature, 493*, 389–392.

Tishkoff, S. A., Reed, F. A., Friedlaender, F. R., Ehret, C., Ranciaro, A., Froment, A., ... Williams, S. M. (2009). The genetic structure and history of Africans and African Americans. *Science, 324*(5930), 1035–1044.

Tishkoff, S. A., Reed, F. A., Ranciaro, A., Voight, B. F., Babbitt, C. C., Silverman, J. S., Powell, K., Mortensen, H. M., Hirbo, J. B., Osman, M., & Ibrahim, M. (2007). Convergent Adaptation of Human Lactase Persistence in Africa and Europe. *Nature Genetics, 39*(1), 31–40.

Tsukada, M., & Rowley, J. R. (1964). Identification of Modern and Fossil Maize Pollen. *Grana, 5*(3), 406–412.

van Geel, B., Gelorini, V., Lyaruu, A., Aptroot, A., Rucina, S., Marchant, R., Damsté, J. S. S., & Verschuren, D. (2011). Diversity and Ecology of Tropical African Fungal Spores from a 25,000-Year Palaeoenvironmental Record in Southeastern Kenya. *Review of Palaeobotany and Palynology, 164*(3), 174–190.

Van Grunderbeek, M. C. (1992). Essai de Délimitation Chronologique de l'Age du Fer Ancien au Burundi, au Rwanda et dans la région des Grands Lacs. *AZANIA: Journal of the British Institute in Eastern Africa, 27*(1), 53–80.

Vansina, J. (1995). New Linguistic Evidence and "The Bantu Expansion". *Journal of African History, 36*, 173–195.

Van Zinderen-Bakker, E. M., & Coetzee, J. A. (1972). A Re-appraisal of Late-Quaternary Climatic Evidence from Tropical Africa. *Palaeoecology of Africa and the Surrounding Islands, 7*, 151–181.

Vavilov, N. I. (1926). *Studies on the Origin of Cultivated Plants (Russian and English).* State Press.

Verschuren, D. (2001). Reconstructing Fluctuations of a Shallow East African Lake During the Past 1800 yrs from Sediment Stratigraphy in a Submerged Crater Basin. *Journal of Paleolimnology, 25*(3), 297–311.

Verschuren, D., Johnson, T. C., Kling, H. J., Edgington, D. N., Leavitt, P. R., Brown, E. T., & Hecky, R. E. (2002). History and Timing of Human Impact on Lake Victoria, East Africa. *Proceedings of the Royal Society of London B: Biological Sciences, 269*(1488), 289–294.

Vincens, A. (1986). Diagramme Pollinique d'un Sondage Pleistocene Superieur-Holocene du Lac Bogoria (Kenya). *Review of Palaeobotany and Palynology, 47*(1–2), 169–179 and 183–192.

Vincens, A. (1989). Paleoenvironnements du Bassin Nord-Tanganyika (Zaire, Burundi, Tanzanie) au cours des 13 derniers mille ans: Apport de la Palynologie. *Review of Palaeobotany and Palynology, 61*(1–2), 69–88.

Vincens, A., Garcin, Y., & Buchet, G. (2007a). Influence of Rainfall Seasonality on African Lowland Vegetation During the Late Quaternary: Pollen Evidence from Lake Masoko, Tanzania. *Journal of Biogeography, 34*(7), 1274–1288.

Vincens, A., Lézine, A. M., Buchet, G., Lewden, D., & Le Thomas, A. (2007b). African Pollen Database Inventory of Tree and Shrub Pollen Types. *Review of Palaeobotany and Palynology, 145*(1), 135–141.

Vincens, A., Schwartz, D., Elenga, H., Ferrera, I., Alexandre, A., Bertaux, J., Mariotti, A., Martin, L., Meunier, J. D., Nguetsop, N., Servant, M., Servant-Vildary, S., & Wirrman, D. (1999). Forest Response to Climate

Changes in Atlantic Equatorial Africa During the Last 4000 Years BP and Inheritance on the Modern Landscapes. *Journal of Biogeography, 26*, 879–885.
Vincens, A., Williamson, D., Thevenon, F., Taieb, M., Buchet, G., Decobert, M., & Thouveny, N. (2003). Pollen-Based Vegetation Changes in Southern Tanzania During the Last 4200 Years: Climate Change and/or Human Impact. *Palaeogeography, Palaeoclimatology, Palaeoecology, 198*(3), 321–334.
Vizy, E. K., & Cook, K. H. (2007). Relationship Between Amazon and High Andes Rainfall. *Journal of Geophysical Research* [online], *112*(D7). https://doi.org/10.1029/2006JD007980
Walshaw, S. C. (2015). *Swahili Trade, Urbanization, and Food Production: Botanical Perspectives from Pemba Island, Tanzania, 600–1500.* Cambridge Monographs in African Archaeology 90 (British Archaeological Reports). Archaeopress.
Weissbrod, L. (2013). The Small Animals of Maasai Settlements: Ethnoarchaeological Investigations of the Commensalism Model. *Azania: Archaeological Research in Africa, 48*(1), 152–153.
Wilshaw, A. (2016). The Current Status of the Kenya Capsian. *African Archaeological Review, 33*, 13–27.
Wooller, M., Street-Perrott, F. A., & Agnew, A. D. Q. (2000). Late Quaternary Fires and Grassland Paleoecology of Mount Kenya, East Africa: Evidence from Charred Grass Cuticles in Lake sediments. *Palaeogeography, Palaeoclimatology, Palaeoecology, 164*, 207–230.
Wooller, M. J., Swain, D. L., Ficken, K. J., Agnew, A. D. Q., Street-Perrott, F. A., & Eglinton, G. (2003). Late Quaternary Vegetation Changes Around Lake Rutundu, Mount Kenya, East Africa: Evidence from Grass Cuticles, Pollen and Stable Carbon Isotopes. *Journal of Quaternary Science, 18*, 3–15.
Wright, D. K. (2005). New Perspectives on Early Regional Interaction Networks of East African Trade: A View from Tsavo National Park, Kenya. *African Archaeological Review, 22*(3), 111–140.
Wright, D. K. (2014). East and Southern African Neolithic: Geography and Overview. In C. Smith (Ed.), *Encyclopedia of Global Archaeology* (pp. 2281–2298). Springer.
Wright, D. K. (2019). Long-Term Dynamics of Pastoral Ecology in Northern Kenya: An Old Model for New Resilience. *Journal of Anthropological Archaeology* [online]. https://doi.org/10.1016/j.jaa.2019.101068
Wright, N. J., Fairbairn, A. S., Faith, J. T., & Matsumura, K. (2015). Woodland Modification in Bronze and Iron Age Central Anatolia: An

Anthracological Signature for the Hittite State? *Journal of Archaeological Science, 55*, 219–230.
Yamasaki, N., Kangawa, K., Kobayashi, S., Kikutani, M., & Sonenberg, M. (1972). Amino Acid Sequence of a Biologically Active Fragment of Bovine Growth Hormone. *Journal of Biological Chemistry, 247*(12), 3874–3880.
Yellen, J. (1998). Barbed Bone Points: Tradition and Continuity in Saharan and Sub-Saharan Africa. *African Archaeological Review, 15*, 173–198.

3

Trading Language, New Crops, New Relationships: Digging Anthropocene Foundations

3.1 The Last 1000 Years: Continued Environmental Variability

As in the rest of the Holocene, the past 1000 years or so have been characterised by a dynamic environment (Fig. 3.1). The sedimentary record from Lake Tanganyika (Alin & Cohen, 2003) suggests increasingly wet conditions between 1200 and 1000 yr BP, while to the north Mohammed et al. (1996) interpret the pollen record from Lake Turkana dating to this period as reflecting lower lake levels. There is substantial evidence for a widespread arid phase around 1000–800 yr BP across the East African region, citing records from lakes such as Lake Naivasha (Verschuren et al., 2000), Lake Emakat (Ryner et al., 2008), Lake Victoria (Stager et al., 2005), Lake Edward (Russell & Johnson, 2005), and Lake Kitagata (Russell et al., 2007). Ryner et al. (2008) suggest that this dry phase before 800 yr BP also coincides with drier conditions at Lake Naivasha (Lamb et al., 2003; Verschuren et al., 2000) and Lake Edward (Russell & Johnson, 2007). In addition, a prolonged dry phase from 1250 to 550 yr BP has been recorded in the sediment record from Lake Masoko (Barker et al., 2000; Gibert et al.,

R. Marchant, *East Africa's Human Environment interactions*, https://doi.org/10.1007/978-3-030-88987-6_3

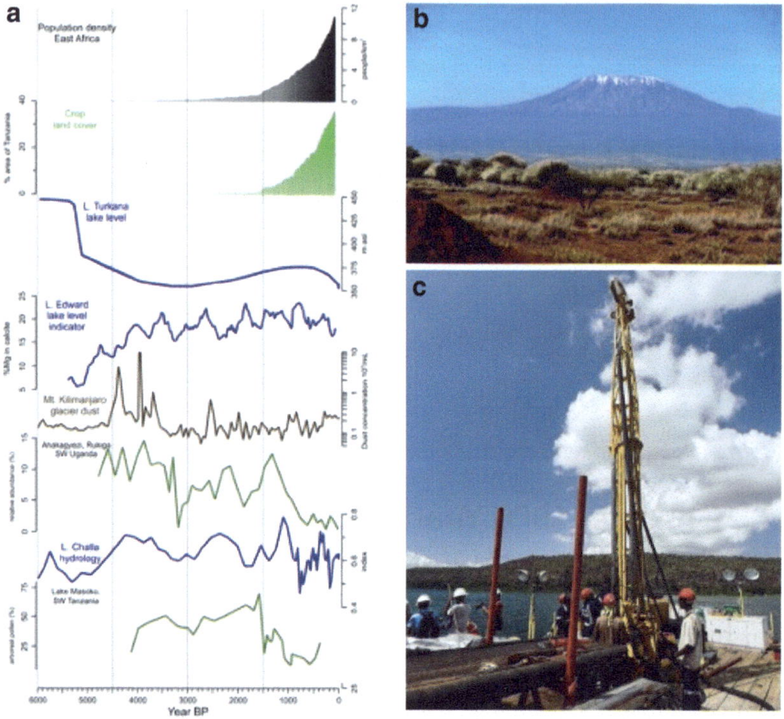

Fig. 3.1 Selected late Holocene palaeoenvironmental records from eastern Africa (**a**). The graphs are described here from top to bottom of the page: Modeled reconstructions of historical population and cropland coverage estimates (KK10; Kaplan et al., 2011; Kaplan & Krumhardt, 2011). Lake level reconstructions for interpreting hydroclimatic variability from Lake Turkana (Garcin et al., 2012) and Lake Edward (Thompson et al., 2002). Kilimanjaro northern ice field (**b**) dust concentration reconstruction as a proxy for vegetation cover in the lowland dust-entrainment source areas (Thompson et al., 2002). Afromontane pollen sums (% total pollen) as a proxy of forest cover from Ahakagyezi Swamp, Rukiga Highlands, Uganda (Taylor, 1993). Lake Challa (**c**) hydroclimate reconstruction using BIT (Verschuren et al., 2009). Re-digitised arboreal pollen relative abundance (%) as a proxy for tree cover from Lake Masoko sediment record (Thevenon et al., 2003) (All photographs: Rob Marchant)

2002; Vincens et al., 2003). The pollen data from Kasenda, Uganda (Ssemmanda et al., 2005) further supports this period of aridity, with a decline in forest taxa around 1100–1000 yr BP being attributed to this dry phase. Limited burning during the period c. 1050–950 yr BP in the Lake Simbi area, and during the longer period c. 1100–850 yr BP in the Naivasha area, may both reflect the wider region's generally dry conditions (Verschuren & Charman, 2008) with widespread fuel limitation (Colombaroli et al., 2014) across the drier savannahs of Eastern equatorial Africa. It must be emphasised that for such ecosystem transitions it is challenging to assign cause and effect as they coincide with the rise of political complexity and consequential intensification of human activities, at least in Western Uganda (Taylor & Marchant, 1996; Taylor et al., 1999). Lake sediment records from Western Uganda (Russell & Johnson, 2007; Russell et al., 2007; Mills et al., 2014) and Lake Tanganyika (Cohen et al., 1997; Alin & Cohen, 2003) record a humidity maximum from c. 800 to 500 yr BP, followed by sustained dry conditions. The high-resolution sediment record from Lake Edward suggests that regional aridity at that time was extreme (Russell & Johnson, 2007). A similar pattern is shown by the 700-year biogenic silica record from Lake Malawi (Johnsen et al., 2001), where moisture levels peaked at about 500 yr BP followed by aridification; this culminated in a severe drought centred around 200 yr BP (Crossley et al., 1984; Owen et al., 1990). Lake Bogoria in the central Kenya Rift Valley reached a high level between 800 and 600 yr BP, followed by levels that were lower than present until c. 150 yr BP (De Cort et al., 2018). To the north of Lake Bogoria, abrupt wetland formation at Loboi swamp is dated to 750 yr BP (Ashley et al., 2004; Driese et al., 2004). Wetter than present conditions between c. 750 and 150 yr BP are suggested by various paleoenvironmental records for the region, and it is possible that the consequent decrease in C_4 grasses and an increase in woody and shrubby vegetation may have encouraged an expansion of tsetse flies, reducing the overall suitability of the areas around Lakes Baringo and Bogoria for livestock herding.

Although climate variability is often attributed to current global climate impact, we can see that change has always characterised the East African landscape throughout geological history. This continued to the present day where people increasingly shaped the environment. A major dry spell centred around 350 yr BP resulted in a significant reduction in the size of Lake Baringo (Kiage & Liu, 2006). After this common, initial stage of lake transgression, notable regional spatial differences developed, particularly in the amplitude and timing of hydroclimate variability. Sites such as Lake Naivasha (Verschuren et al., 2000) and Lake Challa (Buckles et al., 2016) indicate relatively humid conditions continued throughout this period. The signal from these eastern East Africa sites are in contrast to the western part of the region that evolved towards dry conditions following an initial wet period (Tierney et al., 2013). The leaf-wax δD time series from Sacred Lake on Mount Kenya attests to wetter conditions between c. 300 and 130 yr BP (Konecky et al., 2014). Wetter conditions around 300 yr BP gave way to a widespread arid phase from 150 to 100 yr BP, which desiccated many of the lakes in the region (Bessems et al., 2008); however, from around 150 yr BP, these dry conditions were followed by a wetter phase. A general pattern of substantially increasing lake levels in the nineteenth century CE (150–50 yr BP) (Bessems et al., 2008; Verschuren et al., 2002) was followed by a period of wet, but more stable, conditions (50 yr BP to present) (Bessems et al., 2008; Russell & Johnson, 2007). Superimposed on these general patterns are shorter events, such as the period of increased rainfall across East Africa c. 90–70 yr BP (Nicholson, 1996), which may have also caused a dramatic rise in Lake Duluti, Tanzania, seen to occur around 50 yr BP, potentially turning an area of swamp into an open water lake (Öberg et al., 2012). After a period of prolonged drought up to 300 yr BP wetter conditions are also attested to at Munsa archaeological site (Robertshaw & Taylor, 2000; Robertshaw et al., 2004; Taylor et al., 1999) where forest expansion began to occur again between 300 and 100 yr BP. However, using records from Lake Baringo and two shallow crater lakes in Western Uganda, the combined evidence suggests that the nineteenth century was drier, on average, than the twentieth century (Nash et al., 2016). Following the dry period at the beginning of the nineteenth century (Bessems et al., 2008; Stager et al., 2005), when the level of Lake Simbi was low, biomass burning increased from c. 80 yr BP and peaked in the first decade of the twentieth century. The low stand noted at Lake Simbi is largely coeval with evidence of drier conditions across Eastern Africa at approximately 150–100 yr BP

(Bessems et al., 2008; Nicholson, 2000; Ryner et al., 2008; Verschuren et al., 2000). The water levels of Lake Tanganyika and Lake Rukwa, established from historical and geographical information, were low until 160 yr BP (Nicholson, 2001). In the Central Rift Valley, Kenya, a period of very low water levels and/or elevated salinity is dated to 130–110 yr BP (Verschuren et al., 2000) at which point the shallow lakes of Nakuru and Elementeita stood completely dry (De Cort et al., 2013).

The reduction in water availability between 150 and 100 yr BP likely contributed to a famine dated to 120 yr BP in Eastern Africa (Hartwig, 1979) that has been claimed to have triggered extensive population migrations and social upheaval (Anderson, 2016). Nash et al. (2016) note that historical records from Burton (1860) and Krapf (1860) describe famine and drought occurring from 130 to 110 yr BP in the Pangani Valley, Tanzania and in Mombasa, Kenya. Nash et al. (2016) suggest that this dry period began around 165 yr BP, and using this date as a point of comparison, it can be seen in Nicholson and Yin's (2001) water-balance model that rainfall within the catchment of Lake Victoria c. 165–115 yr BP was roughly 13% below the twentieth century mean. Again, this falls broadly within the 150–100 yr BP dry phase/drought period. Nonetheless, it is important to bear in mind that 'droughts' and 'famines' are not directly equivalent, the former being natural phenomena while the latter are, at least partially, socially, economically, and/or politically driven. Moreover, within the paleoenvironmental literature it is only when a dry period is associated with the potential to affect people that the discourse changes from apparently neutral terms such as 'dry periods' or 'aridity', to ones that are more socially loaded, for example, 'drought' and 'famine'.

If we turn to look at fire activity in the paleoenvironmental record, we can see a mixed picture of climate-driven fire activity and human-induced activity (Fig. 3.2). However, long-term palaeoecological records have shown that fire is generally more responsive to climate variability than has been previously thought, as most studies were based on satellite observations, which suffer from the disadvantage of only recording short-term responses rather than the longer-term climatic responses (Lehmann et al., 2014; Sankaran et al., 2005; Staver et al., 2011). Fire in the catchment of Lake Simbi, Kenya, seems to have been limited during

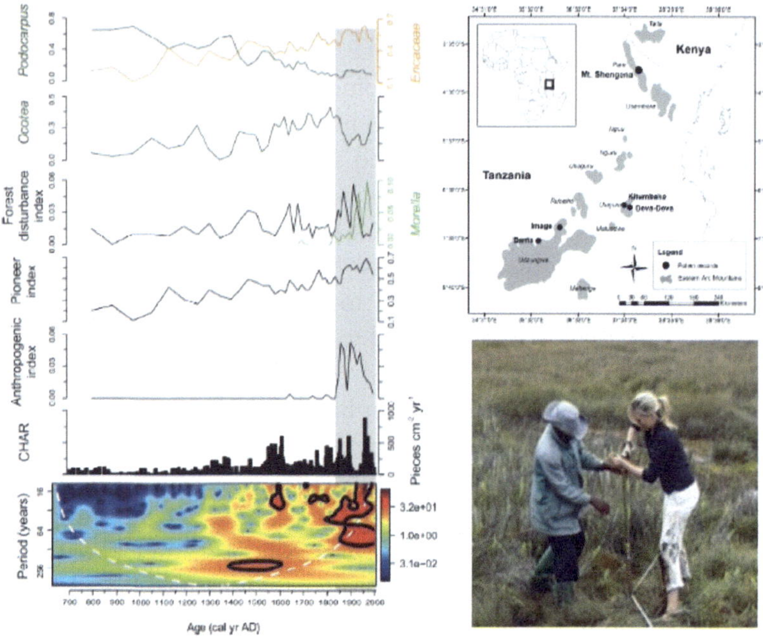

Fig. 3.2 A palaeoenvironmental record generated from a swamp on Mount Shengena, Pare Mountains northern Tanzania that records the past 1300 years of environmental and human history within the catchment. There are increasing signs of human modification of the Montane forest with recent pervasive change detected by an increase in fire activity from 1300 CE and again with the arrival of *Zea mays* at 1810 CE. The arrival of colonial forest officers who were interested East African camphor (*Ocotea usambarensis*) and the planting of fast-growing exotics such as *Eucalyptus*, *Pinus* and *Acacia* (Modified from Finch et al. [2017])

the wettest episodes of the last millennium, as can also be seen in Western Uganda (Colombaroli et al., 2014). However, conditions in the Lake Simbi catchment were never wet enough to allow wooded savannah to outcompete grasslands, implying that fire was sufficiently frequent to exert a positive feedback, impeding the encroachment of woody plants (Cochrane et al., 1999). At Lake Emakat, Tanzania, an increase in charcoal fragments c. 700 yr BP is contemporaneous with enhanced sedimentation suggesting in-wash associated with increased erosion (Ryner et al., 2008). The increase in sediment may also be linked to the increasing number of charcoal fragments c. 700–500 yr BP, in that the surrounding slopes become subject to increased erosion after stabilising vegetation had been burnt away. Frame et al. (1975) also propose that the present vegetation mosaic of woodland and bush in the Lake Emakat catchment is a result of extensive forest clearings by early agropastoral communities in the Crater Highlands. This sedimentary and vegetation evidence, when combined with evidence from linguistic and archaeological sources (Ehret, 1999, 2001; Fosbrooke, 1972; Sutton, 1993), may point to an expansion of human population levels from c. 700 yr BP in Northern Tanzania, however this link is currently unproven. At Lake Duluti (Öberg et al., 2012, 2013) burning was generally limited during both the wettest (ca. 600–400 yr BP) and driest episodes (ca. 800–730 and 200–150 yr BP) (Colombaroli et al., 2014; Stager et al., 2005) of the past millennium. Coastal areas of East Africa experienced a low sea level between 1400 and 100 yr BP (Punwong et al., 2017). However, over the past 100 years or so there have been moderate sea level rises until the present day that likely reflect global sea level rise during the last century (Stocker et al., 2013).

3.2 Developing and Managing Mixed Agricultural Complexes

Distinct changes occur in the archaeological record to the west of Lake Victoria from around 1000 yr BP where the iron-working communities became both numerous and intensive (Fig. 3.3). Ultimately, in some areas these developed into increasingly complex societies and eventually states, as seen in Buganda and Buhaya (Berger, 1981; Reid, 1994, 1996, 2013; Robertshaw, 1997; Robertshaw & Taylor, 2000; Schoenbrun, 1998). These include the appearance of new ceramic styles, characterised by a preference for various kinds of roulette decoration along with changes in vessel form and size (Ashley, 2010), the rapid spread of which may even be indicative of the expansion and intensification of trade and exchange around Lake Victoria (Reid, 2013). Historical linguistic data point to a virtual 'explosion' of new terms associated with banana cultivation around the same times as these material developments, and also

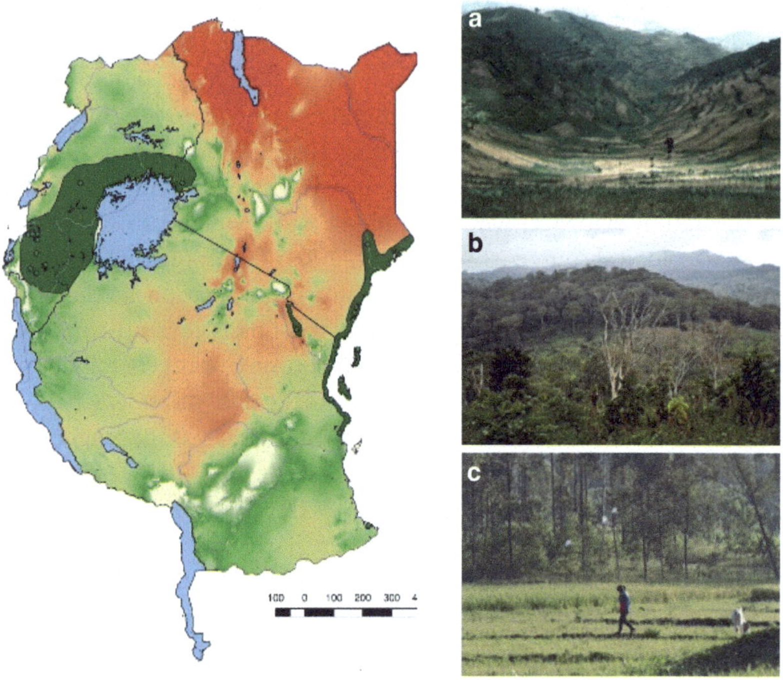

Fig. 3.3 The extent of agricultural systems at 1000 CE across East Africa. It is likely a mix of crops, particular sorghum, millet, yams, cassava, beans and more recently rice (**c**) would have been cultivated. In fertile mountain areas such as the Rukiga Highlands (**a**) intensive intercropping was likely (Map produced by Oliver Boles and modified by the author) (All photographs: Rob Marchant)

in the terminologies of herding and pastoralism (Schoenbrun, 1998), pointing to further expansion and sophistication of food production strategies in the Great Lakes region. There is also material evidence for growing specialisation, including salt production at Kibiro on the eastern shores of Lake Albert (Connah, 1996), and in cattle herding at sites such as Ntusi, occupied c. 950–450 yr BP and situated in Mawogola in South-Western Uganda (Reid, 1994–1995), that allowed for food processing and storage in highly organised and settle locations. One such settlement that extends over c. 100 ha is Ntusi; this was the focal point in the

Mawogola landscape for almost five hundred years linked to an extensive network of smaller sites in the vicinity (Reid, 1994–1995). Faunal remains from Ntusi indicate a focus on cattle herding from the initial establishment of the site onwards. Reconstructed mortality profiles indicate that most livestock were slaughtered at an extremely young age—a practice that would have been possible only with the existence of large herds (Reid, 1996, 2013). Storage pits, numerous grinding stones, some reaping knives and two clusters of carbonised sorghum attest to the importance of agriculture, which likely also included banana plantations, alongside cattle. The presence of a combination of working debris and finished ivory objects also highlights the significance of elephant hunting locally (Reid, 2015). Based on the material evidence from Ntusi, establishment of the site appears to mark the beginnings of a ranked society and socio-political complexity in this part of East Africa, laying a base from which various kingdoms developed. A marked feature of some of the immediate political successors of Ntusi was the construction of a series of deep encircling trenches around the central part of the occupation area, extending over several kilometres at some sites, as illustrated at Bigo bya Mugenyi (13 km north of Ntusi on the Katonga River), and at the site of Munsa in the more forested areas of Western Uganda. Although massive, the configuration of these earthworks does not suggest they were principally defensive in nature, and they may well have served more political and symbolic roles associated with marking centres of power and ritual authority (Reid, 2013; Robertshaw, 2010). Though cattle dominate the faunal remains recovered from excavations at both Munsa and Bigo, the presence of several large grain storage pits at Munsa emphasises the importance of crop cultivation for at least some of these capital sites (Robertshaw, 1988), alongside the maintenance of banana plantations. The scale of the earthworks also points to an ability to pool the labour of large groups and sufficient surplus production to feed this workforce, likely with environmental consequences. As noted above, for example, the pollen record and other environmental proxies for the period point to widespread forest disturbance and increased burning associated with this settlement phase (Lejju et al., 2005).

The earthwork sites appear to have been abandoned around 300–250 yr BP. Archaeological survey data also suggest a change from nucleated towards more dispersed settlement patterns at the village and homestead levels of the settlement hierarchy (Robertshaw, 1994). New capitals for the kingdoms of Bunyoro, Nkore, and Buganda emerged; reports compiled by the first Europeans to visit the Great Lakes kingdoms in the nineteenth century indicate that their capitals were sizeable settlements, structured by distinct spatial divisions based on class/occupation and housing several thousand inhabitants (Reid, 2013). Coupled with the preference for using wood, mud, and thatch for house construction, and the damp, warm soils of the region, archaeological preservation, especially of the animal and plant remains that can provide information about insights into food production strategies and land use, is often poor. At present, only limited archaeological data are available concerning these capitals, with just two, the Nkore capital at Bweyorere and the Mpororo capital of Ryamurari, both being occupied c. 350–250 yr BP having been subjected to detailed archaeological investigation (Reid, 2013). These archaeological data are supplemented by a larger body of ethnohistorical information from linguistic, oral, and documentary sources attesting to food-producing economies based on banana plantations and cattle herding (Schoenbrun, 1998) supplemented by arable cultivation of African cereal crops (Reid & Young, 2000). These same sources indicate continuing economic diversification, including specialised iron (Humphris et al., 2009), salt (Connah, 1996) production and kaolinite mining (Reid, 2003), and a flourishing regional exchange system, much of it by boat across and around Lake Victoria (Kenny, 1979). This period also witnessed growing political complexity, although not necessarily centralisation (Reid, 2003; Robertshaw, 2010), and by c. 200 yr BP aggressive military expansion of the more powerful states, especially Buganda (Reid, 2007), alongside elaboration of kingship (Reid & MacLean, 1995; Sassoon, 1983). Various forms of enslavement, initially mostly for either internal domestic or agricultural purposes, but later also for external trade against food stuff and tools, are also practiced across the interlacustrine region in both the oral and documentary historical sources (Chirikure, 2018; Médard & Doyle, 2007).

On the eastern side of Lake Victoria, the archaeological characteristics of farming and herding communities are more poorly documented. Scattered traces are known from archaeological surveys around the lake margins and along the major rivers draining into the lake (e.g. Lane et al., 2006; Robertshaw, 1991), and from material recovered from the upper horizons of excavated Pastoral Neolithic and Early Iron Age sites, such as Gogo Falls and Wadh Lang'o. Oral histories of the Luo, who are the main inhabitants of this area today and speakers of a Western Nilotic language, suggest that relatively small groups of clan-based, mixed farming-herding-fishing communities established themselves around the lake margins from around 560 yr BP in a series of phases, moving in from areas farther north and west (Cohen, 1983; Ogot, 1967). According to these histories, settlement was initially around Got Ramogi Hill and Yala Swamp, before expanding across the Uyoma peninsula and southwards around the Winam Gulf and into what is now Migori County, Southwest Kenya by c. 300–200 yr BP (Herring, 1979; Ochieng', 1974; Ogot, 1967). A single archaeological site identified in these oral histories and containing roulette-decorated ceramics similar in form and design to ethnographically documented Luo pottery has been excavated. Known as Usare 1, the site comprises a series of three low occupation mounds with shell-midden deposits close to the current lake edge at the southern end of Asembo Bay. Dated to c. 430–300 yr BP, faunal remains recovered from the site suggest a mixed fishing and cattle-based herding subsistence economy (Lane et al., 2006). Subsequently, the practice of enclosing individual homesteads and larger villages within earthen bank-and-ditch enclosures (*gunda buche* in the Dholuo language) became common in the areas north of the Winam Gulf, with over 60 examples now known (Odede, 2008) and likely many more that have been destroyed by farming since the early twentieth century. Only one of these, Lwak, has been excavated. Finds from the site suggest a mixed agropastoral economy, with a similar range of crops (sorghum, millet, beans, maize) and livestock (cattle, sheep, goats, chicken) as described by the first European visitors to this area of Western Kenya (e.g. Johnston, 1902). Intriguingly, the ceramics from the lowest levels of the ditches exhibit typological similarities to MIA pottery from the area, lending support to certain clan histories that suggest that some enclosures were founded by the Bantu-language speakers gradually encountered by the Luo as they moved into the area.

South of the Winam Gulf, enclosures were typically built using drystone walling (*ohinga*, pl. *ohingini*, in Dholuo). The larger and better preserved of these contain one or more stone-walled livestock enclosure with traces of house-platforms and grain-bin foundations arranged concentrically around the inside wall (Lofgren, 1967). Over 135 of these stone-walled enclosures are known; some have droveways arranged in a spoke-like fashion around the main enclosure, and stone terracing is also evident at some sites. Analyses of their distribution and topographic setting indicate a preference for hill-top and upper slope locations close (typically $\leq$ 3 km) to a permanent water source (Onjala, 2003). They also tend to occur in distinct clusters, with the highest density around the site of Thimlich Ohinga (Migori County) which is the largest and best-preserved example dating from c. 280 to 110 yr BP (Wandibba, 1986). Faunal remains from this site attest to the herding of cattle and keeping of small stock, with fish and wild game supplementing diets. Grain-bin foundations and upper and lower grinding stones indicate the importance of quite a diverse range of crops being cultivated. Johnston (1902), for example, noted that the Luo cultivated sorghum, sweet potatoes, peas, eleusine, pumpkins, tobacco, and hemp; there are also references to sesame, green gram, beans, and speckled maize in other oral sources.

In the central Eastern Rift Valley, traces of activity are widespread around the southern end of Lake Baringo. In contrast, diagnostic traces of pastoral iron age communities are relatively scarce, limited to inconclusive remains of settlement activity, a few scatters of Lanet pottery (c. 1050–250 yr BP) and a single occurrence of Kisima ceramic (c. 500–250 yr BP; Petek, 2015). Changing environmental conditions could have had a significant impact on local livelihoods, possibly encouraging settlement abandonment and a shift to hunting and gathering (Petek & Lane, 2017). This was followed by another major drought of a sub-continental scale around 150 yr BP (Bessems et al., 2008), resulting in major socio-ethnic disruptions and the formation of new ethnic groupings and both a faunal and human depopulation of the Baringo basin until c. 120 yr BP (Anderson, 2016). One of the new ethnic communities were the Il Chamus who, after the conditions ameliorated, practiced large-scale irrigation farming on the lowlands south of Lake Baringo, though stock-keeping and hunting remained part of the economy (Anderson, 2002;

Petek & Lane, 2017). They lived in densely settled villages with activities concentrated around nearby well-watered areas, expanding, and intensifying their irrigation system between 130 and 70 yr BP due to the increasing population resulting from immigrant pastoralists from the Central Kenyan Rift and Laikipia and due to the arrival and intensification of the caravan trade. The surrounding Baringo-Bogoria lowlands were seasonally utilised by various pastoralist groups such as Il Doijo and several Maasai groups (Anderson, 2002; Petek & Lane, 2017).

Human occupation of highlands like Pare Mountains, Mount Meru, and Kilimanjaro have a long history, and although archaeological data is patchy it would appear, as one might expect, that farming communities favoured the comparatively wet southern and eastern slopes of these mountains. A variety of sources attest to farming on the southern and eastern slopes from at least 1000 yr BP. These include multiple finds of Iron Age pottery types (e.g. Odner, 1971b) that are commonly found elsewhere in association with agriculture (first Kwale ware and then Maore Ware from c. 1300 yr BP), as well as historical linguistic evidence suggesting occupation in the early first millennium BP by the Maa-speaking Ongamo, who grew finger millet and sorghum, and raised livestock (Ehret, 1984; Nurse, 1979). However, on the north-western slopes of the mountain finds of grinding stones, stone rings (potentially used as digging stick weights), and Kwale and Maore ceramics, in lesser amounts, also provide indirect evidence for farming and iron working communities going back as far as two millennia (Fosbrooke & Sassoon, 1965; Mturi, 1986; Odner, 1971b), though the area may have been primarily used by pastoral communities for grazing (Fig. 3.4).

The date by which bananas were introduced into these highlands is still unknown, but De Langhe et al. (1995) used the number of species variants now farmed on Kilimanjaro and the adjacent Pare Mountains to estimate local use of the crop since at least 1000 yr BP, while Montlahuc and Philippson (2006, citing Rossel, 1998) argue that the crops could not have been introduced until the area became incorporated into long-distance trade networks with the Indian Ocean and beyond after c. 1250 yr BP. The onset of soil erosion in the North Pare highlands from the early second millennium BP can, however, be plausibly linked to clearance of forested land for agriculture (Heckmann, 2011), and there

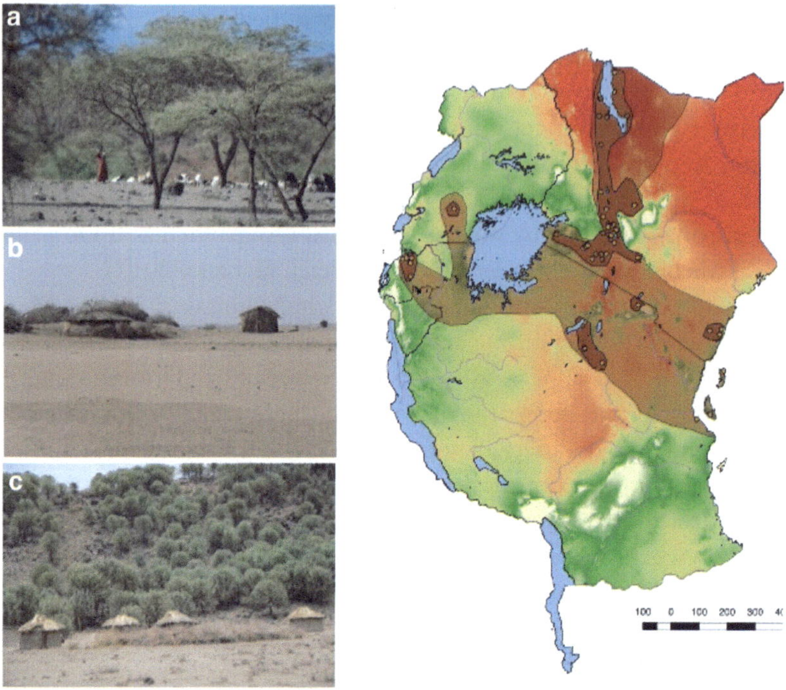

Fig. 3.4 Spread of pastoralism at 1000 CE in East Africa. This was particularly focused on cattle and was augmented by the arrival of new pastoral groups and expansion through the 'Rangelands'. This land use allowed for the seasonal migration and transhumance to access grazing in highland areas (Map produced by Oliver Boles and modified by the author) (All photographs: Rob Marchant)

is no reason to believe that similar small-scale clearances for shifting agriculture were not also taking place in the remaining highland areas in the Eastern Arc Mountains. Increases in the scale and intensity of soil erosion in North Pare in the early first millennium BP (Heckmann, 2011) correlate well with radiocarbon dates on some iron working furnaces on the lower foot slopes of the mountains (Odner, 1971a), while the commencement of more severe soil erosion from c. 350 yr BP coincides with radiocarbon dates from a larger number of iron smelting and smithing furnaces from Pare, and with oral historical evidence from both Pare and Kilimanjaro suggesting that settlements and irrigation structures were already established at this time (see Stump & Tagseth,

2009 and references therein). Oral historical sources, in some cases with support from archaeological data, strongly suggest that similar processes were taking place at about the same time in South Pare (Sheridan, 2002), Usambara (Feierman, 1990), and Meru (Spear, 1993). Land use on the well-watered slopes of the highlands of Northern Tanzania can be characterised as being based on the cultivation of grains and bananas since c. 350 yr BP, but it is perhaps not until the establishment of some degree of local political centralisation and the intensification of the caravan trade from c. 110 yr BP that the landscape started to resemble that of open fields and tree-shaded banana plantations evidenced today.

Climatic change, human population dynamics, and technological development are commonly invoked as causes for the origins of agriculture or precursors of migrations (Bar-Yosef, 1998). Certainly, there is strong evidence for drastic fluctuations between more humid and arid phases during this period of transitions to structure, settled, and highly organised societies that must have had an enormous impact on human populations and their subsistence patterns (Neumann, 2005). However, this is an interconnected process with multiple feedbacks between the environment and people and we should avoid simple correlation and the consequent assumption of cause and effect. For example, episodes of lake level fluctuations from Lake Naivasha have been linked to oral traditions of droughts (cf. Verschuren et al., 2000; Webster, 1979). Based on the oral history, drought-induced famine and political unrest have been suggested as the reasons for large-scale migrations occurring in the three drought periods around 560–580 yr BP (Wamara drought), 390–325 yr BP (Nyarubanga drought), and 190–160 yr BP (Lapanarat-Mahlatule drought) (Verschuren et al., 2000). However, as the temporal and spatial precision of the drought reconstructions from the oral history are questionable, separating cause and effect is highly challenging and unlikely to be so simple. Conversely, the contributions of anthropogenic versus natural forcing on the climate system before the Industrial Revolution remains unknown and charactering such feedbacks remains a major science goal (Battistel et al., 2017).

3.3 Managing Water and the Growth of Industrial Agriculture

The findings from both the archaeological and palaeoecological records (with the addition of historical evidence for the later part of this time period) are better resolved than for any other past period considered here. This increased body of evidence suggests that the environment of the past 1000 years is characterised by a highly variably hydroclimatic regime (Nash et al., 2016). In the western sectors of the East African plateau, the driest conditions of the entire Holocene seem to have already been registered around 2000 years ago, followed by a modest long-term wetting trend (Nash et al., 2016). Lake levels fluctuated by tens of metres over decadal timescales (Fig. 3.1). In addition to inundating or exposing low-lying areas around the numerous East African lakes, this hydrological change also resulted in contraction and expansion of swamp forest (Rucina et al., 2010). As well as a mean change in hydrology, it is most likely that this period of climate variability was characterised by shifts in seasonality of rainfall that would have, as today, impacted on farming communities.

Irrigation in East Africa has a complex and dynamic history (Fig. 3.5). Indigenous practices appear to have been widespread in the precolonial period and some have persisted to the present day, documenting a history of innovation, and attempted induced change (Adams & Anderson, 1988). In response to the high climate variability (Sect. 3.1) and steep slopes, farming systems developed in a way to build and sustain productive soils (Lang & Stump, 2017; Widgren & Sutton, 1999) (Fig. 3.5). For example, intercropping in fertile and wet areas where a permanent and extensive cover with multiple crop types protected the soil from erosion and regenerated the productivity, as well as providing farmers a more diverse output secured from crop failures. Farmers also improved and conserved the soil base and water availability by investing labour to build terraces, cut-off drains, contour plough, apply manure, mulch, rotate crop types, and select crop types suitable for the specific location (Reij et al., 2013). In some cases, where local conditions allowed or necessitated, the irrigation of a spatially more extensive and organisationally more complex system, based on stream diversion and the transport

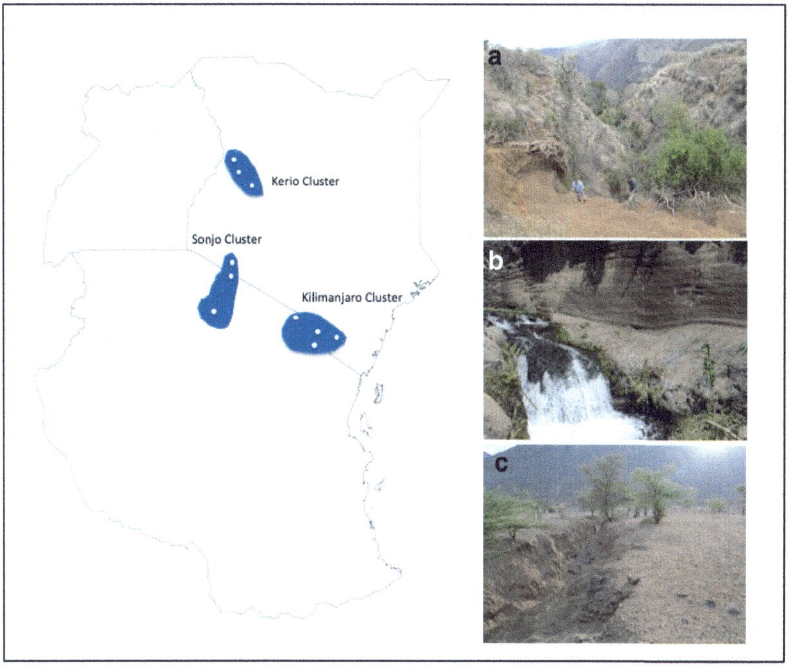

Fig. 3.5 Management of water has always been key to allowing populations to grow crops in a highly seasonal environment as seen at three large irrigation clusters. Elaborate terraces were established to allow water from adjacent highlands (**a**), that flowed down a series of incised valleys (**b**), to be channelled across a lowland plain via a complex series of walls and revetments (**c**). This photographic sequence are taken from Engaruka that was part of the Sonjo cluster where an elaborate and highly organised population farmed an area of some 20 km^2 until their rapid demise around 1500 CE, possibly due to rapid erosion of the headwater sediments (**a**) combine with a phase of climate change (All photographs: Rob Marchant)

of water by canals or furrows, was practiced. These irrigation systems can be found at an astonishing diversity of locations in East Africa (Adams & Anderson, 1988), with such water management still practiced today. However, water management is not just about applying hard infrastructure, it also must include the selection of crops of known water requirements and tolerances, and the adaptation of cropping patterns and strategies to suit environmental conditions (Adams & Anderson,

1988). A major cluster of furrow-based irrigation systems occurs in the hills to the south and east of Mount Kilimanjaro, and on Kilimanjaro across to Kenya's Taita Hills around Taveta, in the South Pare Hills and the Usumbara (Fig. 3.5). The Chagga of Kilimanjaro developed an extensive furrow system fed from mountain streams that, together with water storage ponds, was used to irrigate a wide variety of crops. Several of these sites were important supply points for the trading caravans that crossed the East African interior in the nineteenth century; so, here again travellers' accounts can be combined with more recent anthropological and historical research to give us a picture of past and present agricultural practice (Adams & Anderson, 1988).

Farther to the west in Tanzania, elaborate terrace and irrigation systems are observed in the Manyara Basin and at the base of the Crater Highlands, most notably at Engaruka (Fig. 3.5) (Stump, 2006; Sutton, 1984; Westerberg et al., 2010). Radiocarbon determinations indicate that parts of the 2000 hectares of agricultural fields and associated settlements at Engaruka were in place by 650 yr BP (Westerberg et al., 2010), while historical sources demonstrate that the site was entirely abandoned—likely pulsed rather than by mass migration c. 200–150 yr BP (Sutton, 2004). The construction of the field system was elaborate, as were its attendant network of irrigation channels and canals, with large parts built by capturing sediments eroded off adjacent high ground (Lang & Stump, 2017; Stump, 2006). This suggests the occurrence of severe soil erosion in highlands following the removal of vegetation being a plausible trigger, though the extent of this removal and whether it had a human or climatic origin is the subject of ongoing work. Attempts have been made to relate the development of the field and irrigation system to palaeoclimatic variability interpreted from proxy evidence in the sediments from the nearby Lake Emakat (Westerberg et al., 2010). Taken together the combined evidence from the site indicates an emphasis on cereals (Sorghum and Pearl millet), but the quantities and age profiles of animal bones recovered from excavations of the settlement areas (e.g. Robertshaw, 1986) suggests that some cattle and small stock were kept on or near the site, which in turn implies the economy required access to grasslands outside the more intensively cultivated area.

In many cases not well dated, the use of irrigation and/or agricultural terracing was not restricted to the communities at Engaruka and those in the highlands of Northern Tanzania. Another cluster of furrow systems, the Kerio cluster, is found on the Marakwet Escarpment and the Wei Wei Valley below the Cherangani Plateau in Kenya (Adams & Anderson, 1988) (Fig. 3.5). The furrows show considerable engineering skill and represent what is probably the most extensive and complex indigenous water management in Africa, south of the Sahara. In places, these cultivations are supported by elaborate systems of terracing. The picture of flexibility in the indigenous irrigation system of the II Chamus, contrasts sharply with the model of irrigation development which evolved in Baringo and elsewhere during the colonial period, and has been carried over into the era of modern development planning (Adams & Anderson, 1988). Irrigation and the construction of terraces for homesteads and agricultural fields is a feature of crop production among the Marakwet and agricultural Pokot, but prior to c. 100 yr BP both communities had very dispersed settlement patterns based on individual households, and agricultural plots associated with these households are likely to have been relatively short-lived (Davies, 2008); a point that is essential in reconstructions of historical land cover since even 'intensive' systems of agricultural production need not lead to extensive or permanent land clearances. These agricultural communities in the Rift Valley occupied relatively niche locations, growing mainly finger millet, sorghum, leafy greens, and legumes (Davies & Moore, 2016; Petek & Lane, 2017). As the agricultural Pokot were moving south, the faction that would become the pastoral Pokot received a significant influx of immigrants from Western Turkana and Karamoja in Uganda, likely due to the previously mentioned sub-continental drought. Despite oral history recording that most of the newcomers and the Pokot ancestors practiced mainly mixed subsistence economies, the community reorganised itself around a specialised pastoral lifestyle and a society centred around cattle, and rapidly expanded to exploit the patchily spread resources from Karamoja to the Leroghi Plateau and north of Lake Baringo by about 100 yr BP (Anderson & Bollig, 2016; Bollig, 2016; Bollig & Österle, 2013).

Stimulated by the demands of the caravan trade (Chapter 4), significant expansion in cultivation took place at Baringo from the 1840s to the 1870s after which the irrigation system went into a slow decline. The irrigated fields produced a surplus of grain (sorghum and millets), which could be bartered with the coastal merchants who visited Baringo in search of ivory. Modern developers have pursued an archetype of the modern irrigation scheme for more than three decades with conspicuous lack of success yet have ignored the history of a wide range of indigenous irrigation practices, many of which are still extant, and which have been sustained for centuries. Developers should explicitly recognise the reality and importance of community history and experience; this does not only involve awareness of 'indigenous knowledge', but also implies recognition of the operation of the prevailing social systems and economic networks that sustain local production (Adams & Anderson, 1988).

3.4 The Development of Modern Pastoralist Societies

Pastoralists are defined as groups who depend primarily on the products of their hoofed domestic animals, and who organise their settlement and mobility strategies to suit the needs of their livestock. They include the Maasai of East Africa, who place ideological emphasis on consuming only the products of their herds and flocks. East African pastoralists bleed cattle (both male and female) with the blood consumed alone in liquid or clotted form or mixed with milk. Like milking, this permits people to draw nourishment from an animal without killing it. It is predominantly used when other sources of animal and plant food fail, during dry seasons and droughts (Dahl & Hjort, 1976). Though such a diet was characteristic of the warrior class, Maasai exchanged animal products for agrarian and forest produce with symbiotic farmers and foragers (Kjekshus, 1977). Pastoral groups meet the needs of their herds by moving seasonally as the foraging range is depleted in one area, and others open for grazing. Where local forage and watering conditions permit, pastoral groups may keep their livestock at one settlement year-round, only shifting after several years. Where forage is not located near potable water or cultivable land, pastoralists may organise into labour groups according to age and gender, some carrying on activities at a residential encampment while others take livestock to find areas of good forage (Gifford-Gonzales, 2005). Pastoralism emerged in what is now the Sahara Desert and adjacent Sahelian grasslands that has undergone major pulses of greening and desertification from the last glacial period. Environmental rhythms in these climatically sensitive zones would likely play a major role both in the development of early pastoralism and movements of herding peoples into East Africa (Gifford-Gonzales, 2005). Novel animal disease challenges associated with the sub-Saharan grasslands may have stalled pastoral expansion into the region (Gifford-Gonzalez, 2000). Even away from the tsetse-infested bush, sub-Saharan savanna environments present livestock with new disease challenges. These diseases massively impact on cattle and could have led to a serious reduction, or even local extirpation, of cattle from immigrating pastoralist herds (Fig. 3.6). Two diseases are likely to have challenged the sustainability of cattle herding in East Africa: Rift Valley fever (RVF), a tick-borne protozoan parasite infesting African buffalo, kills up to 15% of calves a year; and Malignant catarrhal fever (MCF), a herpes virus transmitted to cattle from very young wildebeest, is 99% fatal to modern cattle within two weeks of exposure (Gifford-Gonzales, 2005).

If the first pastoralists to enter Eastern Africa were practicing the now nearly vanished economy of animal husbandry plus wild plant-gathering rather than agropastoralism, they would not have been able to fall back on farming to make up nutritional shortfalls (Gifford-Gonzales, 2005). The safety net of far-flung livestock loans, marriage alliances, and other obligations for mutual aid would have been thin along such an expanding front. Thus, in contrast to recent pastoralists, early livestock owners who lost their cattle would have been disadvantaged relative to local foragers. Completely communal usage of land was only beneficial in areas unsuitable for cultivation due to environmental unpredictability, such as the savanna grasslands, as communities needed large areas for grazing and mobility to adapt to variable climatic conditions. In these pastoral systems, communities often adapted a nomadic existence following the rains with their herds (Warren, 1995). Evidence

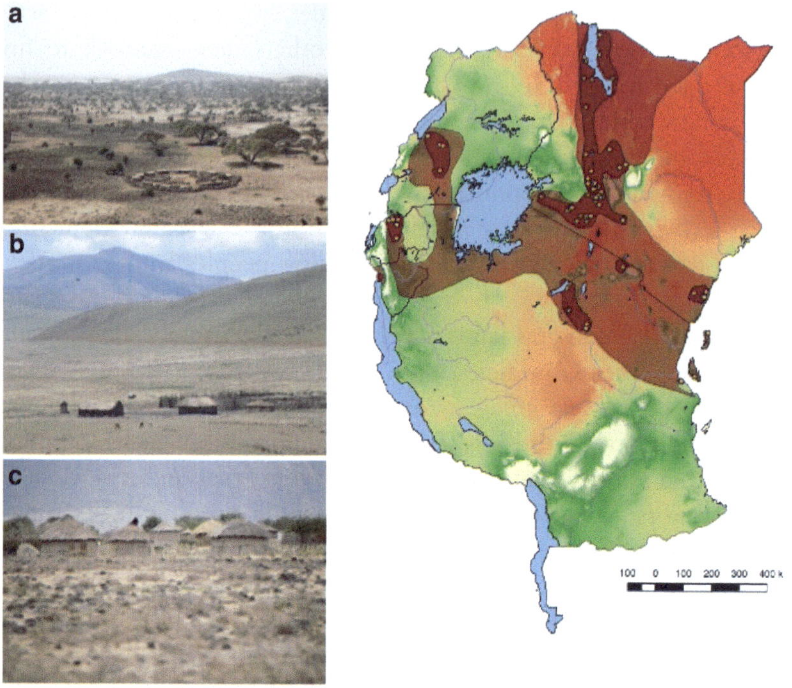

Fig. 3.6 Spread of pastoralism at 1500 CE in East Africa. This was particularly focused on cattle with new pastoral groups, such as the Maasai arriving into the north of the region and expanding rapidly through the 'Rangelands'. This migration into East Africa was likely driven, in part, by droughts and transhumance would continue to be important to buffer seasonal variation in grazing resource via access to highland areas (Map produced by Oliver Boles and modified by the author) (All photographs: Rob Marchant)

suggests that these pastoral communities nonetheless developed elaborate management strategies for the communal lands, which they enforced through strict social customs and cultural traditions. In all cases, social networks were a vital part of indigenous communities to buffer for droughts, labour, or capital shortages (Wynants et al., 2019).

Neolithic herding communities arrived via South Sudan into Northern Uganda and Kenya where they form distinct tribes today such as the Maasai, the Turkana, and the Kalenjin. One of the dominant pastoral groups in East Africa today are the Maasai (Fig. 3.6) who, although inhabiting large areas and being famous for their fearsome reputations as warriors and cattle-rustlers, they are a relatively new addition to the landscape. According to their oral history, the Maasai originated from the lower Nile valley north (Northwest Kenya) and began migrating south around the fifteenth century, arriving in a long trunk of land stretching from Northern Kenya to central Tanzania between the seventeenth and late eighteenth centuries (Galaty, 1993). Many ethnic groups that had already formed settlements in the region were forcibly displaced by the incoming Maasai while other, mainly Southern Cushitic groups, were assimilated into Maasai society. After a period of aggressive expansion throughout the Rift Valley, the Maasai began to suffer a series of setbacks in the second half of the nineteenth century. These began with the Iloikop Wars (which ended in the 1870s) between Maasai and other closely related pastoralist groups, which left the Maasai in control of large areas of East Africa and perhaps overextended (Waller, 1976).

The Maasai developed an extensive nomadic pastoral system regulated by the availability of water, pastures, and sometimes the existence or absence of diseases. The upland areas of the Maasai range provided dry season pasture and water while the lower plains were used during wet seasons. The Maasai avoided the areas that they knew harboured tsetse, east coast fever, and bovine pneumonia. The traditional Maasai life is spatially and ecologically designed to provide a stable system of grazing that insured the availability of pasture, water, and salt licks all year round. They practiced a system of selective breeding of their livestock by crossing Zebu cows with the hardier Boran bulls from Northern Kenya; the resulting crossbreeds could withstand dry conditions and diseases more common in the Rift Valley (van Zwanenberg & King, 1975). Through careful husbandry and extensive territory (Fig. 3.6), the quality of pastures on Maasai range improved (Western, 1994).

Although there is much written about the transition from pastoral societies to sedentary agriculture, like many cultural transitions past and present, the lines would have been blurred. Cultivation appears to have been an integral part of the primarily pastoral subsistence strategy of early pastoral societies around Lake Turkana (Adams & Anderson, 1988). Indeed, there is a strong historical pattern of temporary and permanent occupation of many irrigation sites along the Rift Valley by Maasai-speakers and consequently a close connection between irrigation cultivation and pastoralism (Adams & Anderson, 1988). To the west of Baringo the end of the Sirikwa society appears to have led to the formation of what are now two distinct communities—the Pokot and Marakwet centred around the Kerio Valley and Northern Cherangani Hills. Both communities appear to have commenced land clearance and begun using irrigation around 250 yr BP, and both formerly used pottery styles suggesting a cultural affiliation with Sirikwa traditions (Davies & Moore, 2016). Between c. 200 and 150 yr BP the ancestors of the modern-day Pokot split into an agricultural and a pastoral faction, thereafter the agricultural Pokot expanded southwards from the drier lowlands to the well-watered highlands in the Cherangani Hills (Davies, 2008), with the pastoral sections becoming more mobile and eventually employing rangelands that extended eastwards into Baringo and Eastern Uganda (Davies & Moore, 2016). Pastoral systems in East Africa thus must be characterised as dynamic points along a continuum depending on the local environmental, social, and economic conditions influencing the extent and patchwork constitution of settlements with permanent inter-cropping and/or shifting cultivation areas, natural or human-created grasslands, primary forest and recovering secondary forest.

3.5 Globalised and Commodified World: The Rise of the Swahili Coast

The Swahili coast (Fig. 3.7) encompasses the coastline and offshore islands stretching some 2500 km from Somalia to Mozambique, including Northern Madagascar, and the Comoro and Zanzibar archipelagos. The Indian Ocean, particularly north of the equator, is characterised by a system of monsoon winds and currents that facilitates maritime trade and, through trade, wider global interaction (Pouwels, 2002). The East African coast is a zone of cultural interaction between East Africa and Middle East/Asia, from which emerged a synthesis—the Swahili. The timing of the Swahili, like any foundation, is difficult to pinpoint with any degree of accuracy but it would have developed over

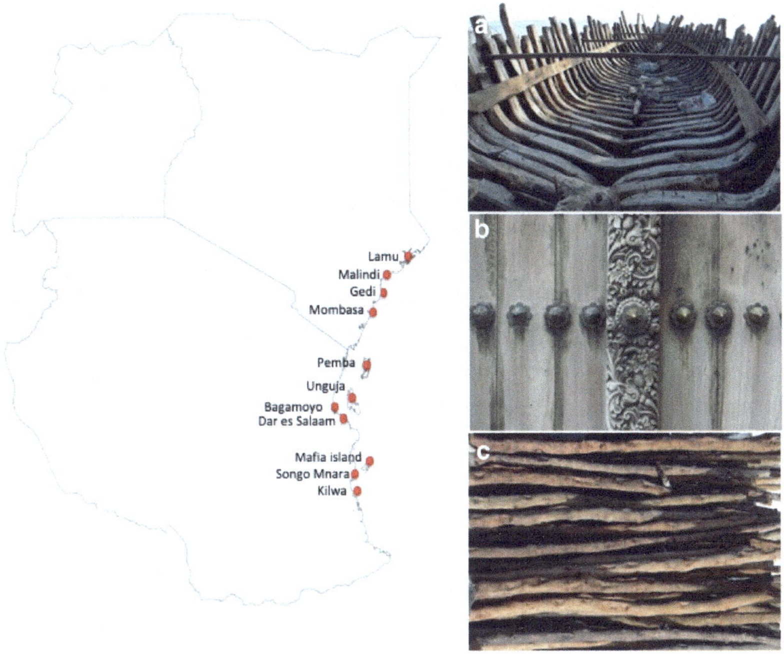

Fig. 3.7 Swahili Towns along coast East Africa and islands communities—these would have formed trading nodes the for the annual trading missions along the coast that were associated with the monsoonal winds. Given the coastal nature there was a particular impact and extensive exploitation of mangrove timber for boat construction (**a**), buildings (**b**) and poles for export to the Middle East (**c**)

the past 2000 years. Early Greek and Roman demand for ivory induced South Arabian traders to the East African coast. One of the earliest descriptions of the Kenya Coast is in the writings of the Greek explorer Diogenes who undertook a voyage of 'exploration' before returning to Egypt around AD 110. Another written account documents trade at the beginning of the 1st millennium CE comes in the form of the Periplus Maris Erythraei (mid–1st c. AD) and Ptolemy's Geographia that charted the Kenyan coastline in the maps of the world c. AD 150 (Chirikure, 2017). Trade to East Africa from Siraf, an ancient port city located in the Persian Gulf, played an important role in facilitating maritime trade and connecting the Indian Ocean and Chinese routes. During the tenth century trade usually consisted of spices, cotton cloth, perfumes, and kitchenware in exchange for ivory, gold, and slaves (Chapter 4). It is from these locations where some of the earlier text document the nature of the trade and provide a little insight into the character of the time. For example, Ibn Hawqal, in 961 CE, met a merchant in Basra by the name of Ahmad ibn 'Umar al-Sirafi and reported that the merchant was very wealthy and that his ships traded along the East African littoral, particularly at Zanzibar, carrying spices, precious stones, and perfumes. Foodstuffs such as rice and wine as well as timber (mangrove poles [Fig. 3.7] and teakwood) were also exchanged, the rice and wine coming from the Persian Gulf and the timber from East Africa (Ricks, 1970). The thirteenth-century Chinese chronicle, Chu-fan-chi, mentions Zanzibar as a place of trade (Ricks, 1970).

Clearly the Swahili built on the previous occupation of the coastal area (Fig. 3.7). Along the coast, and in the coastal hinterland, many sites date to the middle iron Age, though some of these—such as the Juani Primary School site (Crowther, Faulkner, et al., 2016), Mgombani in the hinterland (Helm, 2000), Fukuchani, and Unguja Ukuu on Zanzibar—also had substantial earlier occupations. At these sites there is clear archaeobotanical evidence for farming, mainly of African cereals such as sorghum, pearl millet, and finger millet, as well as the cowpea (Crowther et al., 2017). Faunal exploitation is highly variable from one site to the next (Quintana Morales & Prendergast, 2017). For example, faunal remains from Zanzibar sites point to a marine source, with fish forming a key part of the diet, followed by hunting of terrestrial game, especially at Fukuchani; this is contrasted at Unguja Ukuu where caprines and chickens are a key part of the economy, alongside wild terrestrial, and marine fauna (Prendergast et al., 2017). At Shanga, in the Lamu archipelago, domestic animals form the bulk of the diet (Mudida & Horton, 1996). Systematic archaeological surveys of several areas, including the area between Mombasa northward to Mida Creek, on the central Kenya coast (Helm, 2000), along the lower Pangani valley, Northeast Tanzania (Walz, 2010), around Kilwa (Wynne-Jones, 2010) and Mikindani (Pawlowicz, 2011) in Southern Tanzania, and on Pemba (Kessey, 2003) all indicate a trend towards increasing density of rural settlements and the filling out of the landscape, likely due to demographic growth.

It should be emphasised that East African trade would not have started solely with the arrival of maritime traders. Throughout the Holocene, communities were involved in trade and exchange (Chirikure, 2017): consequently crops, technologies and land use strategies quickly spread across the region. Certainly, the Swahili coastal developments opened another conduit for increased trade that is documented by artefacts with glass beads arriving in substantial numbers from the eleventh century (Horton, 1996). Evidence of trade between Greek and Roman empires and East Africa, including that of glass beads, have been uncovered at several East African sites (Chami, 1998, 2004, 2019). Excavations at Kaole near Bagamoyo (Fig. 3.7) in Tanzania produced two gold/silver glass beads along with some imported ceramics: Sasanian-Islamic and Turquoise glaze pottery in the upper layers and white earthenware/creamware at the base originating from the seventeenth to eleventh century AD (Pollard, 2007). During the tenth century the use of the monsoon winds and star navigation techniques were supplemented by other navigational aids such as primitive astrolabes, shore charts, and sounding devices, as well as identification of types of birds, fish, and the currents, all of which became extremely detailed in later centuries (Ricks, 1970). These international connections would have led to social formations along the East African coast induced by international trade to produce surplus products for exchange (Middleton, 1961, 1992; Prins, 1971); mangrove poles and ivory being exported to the Persian Gulf.

After the collapse of the Roman Empire, the main export was to India and China (Horton & Middleton, 2001; Kusimba, 1999). In the mid to late thirteenth century, there was a shift in East Africa from Islamic to Chinese ceramics (Patel, 2004). This clearly marks the globally connected place East Africa had at this time and documents that much of the trade was coming from the east rather than the north (Wood, 2017). Among the various types of beads that are found in East African assemblages one sort stands out. It consists of small wound beads, most of which are transparent-translucent ruby red and a few that are amber-coloured. They are distinctive not only because small wound beads are rare, but the colours are also unusual. This type of bead was made in

China based on the high lead levels, but also because the Chinese were the only beadmakers known to have fashioned small wound beads at this time. The largest number of these beads was found at Kilwa's Husuni Kubwa where 33 ruby red examples were recovered (Chittick, 1974). Chinese wound beads, which are time consuming and thus expensive to make, would have been able to compete with the small drawn Indian beads in the market. It is most likely that these unusual beads were gifts from the Chinese visitors (Wood, 2017). Indicative with these prestige goods comes information, a set of values and social procedures which are more readily adopted because of the sophistication of the source society's products and prestige in which they are held (Wood, 2017). During this period Kilwa became the main trading port linking trade and traders from the south with markets to the north. It was the height of the gold trade from the Southern African interior, spurred on by growing demand for gold in Europe that, in the context of East Africa, was largely through Kilwa and by the thirteenth to fourteenth centuries Kilwa completely controlled the gold trade coming out of the south (Wood, 2017). Its prominent position may also have been due to direct trade arriving from Southern India via the Maldives (Pouwels, 2002). After the late fourteenth century the focus of wider commercial activity began to shift northward to Mombasa, Gedi, Malindi, and Pate (Pouwels, 2002). The 'Golden age' of the Swahili coast followed the arrival of Islam around 800 CE where a number of towns, such as Mombasa evolved. By the fourteenth century the distinctive Swahili culture was becoming quite wealthy and extensive with its glorious architectural heritage consisting of palatial homes, mosques, and monuments in some key ports like Mombasa, Watamu, Malini, Lamu, and Kilwa. Some of the previous cities are ruin sites such as Gedi (Fig. 3.7) that has its foundation in the thirteenth century until becoming abandoned in the seventeenth century, possibly as a result of sea level recession or incursions from the Oromo.

The relationship between these coastal city states and their hinterlands is poorly understood (Gilbert, 2002). Although it is fairly certain that the Swahili did not politically dominate the people of the interior, as trade was central to the Swahili towns, presumably they had trade connections into the interior. The extent and range of those trade connections are debated, and the wider relationship between the coastal urban development and interior regions of East Africa remains poorly understood, particularly with regard to economic intensification or changes in land use and labour relations (Lane, 2010, 2014). As most of the trade goods that the Swahili sold (ivory, rhino horn, timber, slaves, turtle shell, and ambergris) could have been obtained quite near the coast the initial sphere of influence was quite restricted. The one exception to this trade portfolio was gold and the richest Swahili towns were those that controlled the gold trade (Gilbert, 2002). A handful of studies have explored the inland networks to which coastal towns were connected and a variety of relationships of trade and production have been documented. The Tana Valley of Northern Kenya (Abungu & Muturo, 1993), the Pangani Valley of Northern Tanzania (Walz, 2010), and the deep hinterland of Mikindani in Southern Tanzania (Pawlowicz, 2012), were all occupied over the past 2000 years or so by village communities practicing small-scale agriculture; each related to the urban development of the coast periodically or continuously. In Southern Kenya research has found evidence for foragers and agriculturalists connected with the coast in a 'mosaic' of peoples and practices (Kusimba et al., 2005). But again, there seems to have been no empire building by the Swahili and no significant migration between the coast and the interior or vice versa (Gilbert, 2002). The expansion of trading routes between the interior and the coast, starting around 1300 years ago and intensifying in the eighteenth and nineteenth centuries forming a series of 'caravan routes', will be explored in Chapter 4. In addition to being viewed as trade routes, these acted as conduits for spreading New World crops such as maize (*Zea mays*), tobacco (*Nicotiana* spp.), and tomatoes (*Solanum lycopersicum*), although the processes and timings of their introductions remains poorly documented. The introduction of Southeast Asian domesticates, especially banana (*Musa* spp.), rice (*Oryza* spp.), taro (*Colocasia esculenta*), and chicken (*Gallus gallus*), via transoceanic biological transfers around and across the Indian Ocean, from at least around 1300 yr BP and potentially significantly earlier, also had profound social and ecological consequences across parts of the region.

From c. 900 yr BP, urban centres (commonly known as stone towns due to coral-rag architecture) develop along the coast and islands

(Fig. 3.7), where they served as important entry points for the exchange of goods from the interior to the coast, and from the coast and islands to the broader Indian Ocean maritime world (Boivin et al., 2013; Horton & Middleton, 2001; Wynne-Jones, 2016). This trade is evident in the arrival of new crops, including Asian rice, mung bean (*Vigna radiata*), and coconut (*Cocos nucifera*). While Asian crops appear at sites such as Unguja Ukuu, Zanzibar (Crowther et al., 2017; Crowther, Lucas, et al., 2016), Mafia (Crowther et al., 2014), and Tumbe on Pemba (Walshaw, 2010), they (especially rice) become more abundant after c. 900 yr BP, as evidenced, for example, at Chwaka on Pemba (Walshaw, 2015). Apart from this localised shift towards rice agriculture, sites remain mostly dependent on local crops and on small-scale agriculture and household-level production. This suggests a relatively circumscribed area of farmed land around the towns, which would account for the regionality of subsistence practices in coastal towns, drawing on local crop regimes and practices (Walshaw, 2015). One of the key legacies today of the trade has been a series of monumental towns in various stages of ruin today such as Gedis (Fig. 3.7). Swahili stone towns populations regularly exceeded 5000, with several centres such as Kilwa and Mombasa larger than 15,000 (LaViolette & Fleisher, 2005). Many northern stone towns were surrounded by town walls (Takwa, Pate, Gede) that may represent protective features, or rather be symbolic markers defining 'cultured' spaces (LaViolette & Fleisher, 2005). All stone towns on the Northern Swahili coast contained elite architecture and mosques, as well as unique pillar tombs commemorating political and religious figures. In contrast, most towns on the Southern coast contained few stone buildings that were scattered among much larger earth-and-thatch neighbourhoods (LaViolette & Fleisher, 2005). Kusimba (1999) argued that Swahili stone towns managed economic relationships between town and country, and that agricultural goods, raw materials, and finished products came from the surrounding countryside to support stone town dwellers. Stone town elites thus sat atop a settlement hierarchy, managing a system of small towns and villages to create an efficient network for the production and movement of goods. Imported goods, including Far Eastern and Middle Eastern glazed ceramics, glass vessels, stone and glass beads, and small copper-alloy objects are found in a variety of rural settings including villages with no evidence of stone building (Fleisher, 2003).

Agricultural land beyond the town centres population would have been drawn into cultivation as agriculture expanded across East Africa (Fig. 3.8). Although archaeological evidence for this is slight (Vernet, 2015), it would have likely contributed to a transformation of the coastal strip, with the large-scale clearance of coastal forests and particularly

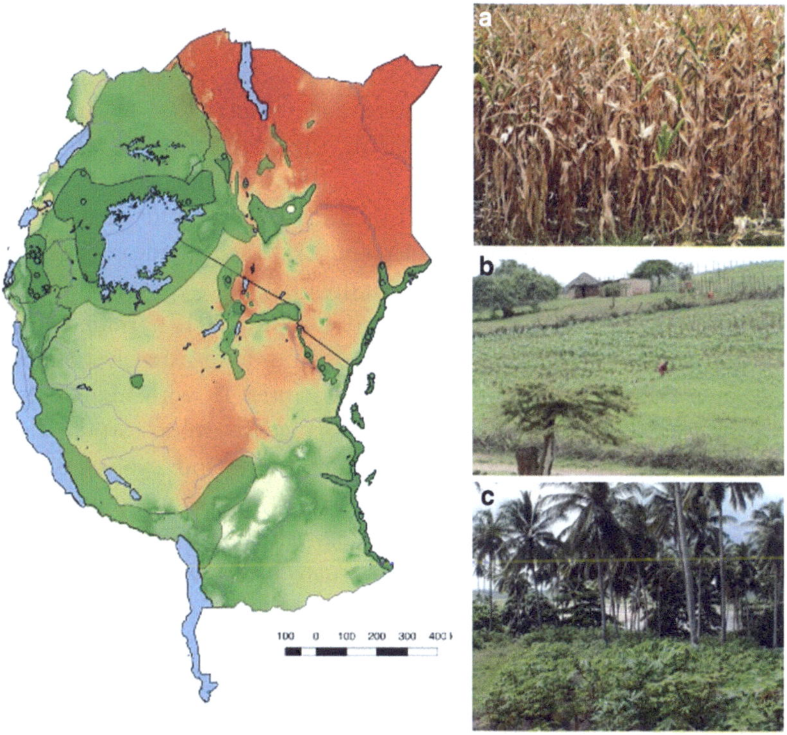

Fig. 3.8 The extent of agricultural systems at 1500 CE in East Africa. New crops started to arrive into East Africa, particularly from the Americas after 1608 CE when Maize (**a**) was first recorded on Pemba—this was quickly followed by potatoes, tomatoes, and squash. Coconut groves (**c**) would have been important on the coast to supply the rapidly developing Swahili trade (Maps produced by Oliver Boles and modified by the author) (All photographs: Rob Marchant)

mangrove forest for the assembly of ocean-going ships (Fig. 3.7). Extensive and pervasive land cover change has been associated with population growth, immigration, and movement of people (Fig. 3.8). These initial centres became increasingly dense settlements with faunal evidence for increasing ratios of domestic stock and chickens, pointing to the intensification of food production within the towns (Quintana Morales & Prendergast, 2017), but also the development of regional procurement networks. Specialised fishing technologies are employed to catch larger prey, including sharks, in open waters (Quintana Morales, 2013) as a more explicitly maritime socio-cultural orientation also emerged (Fleisher & Sulas, 2015).

The revisionist work of the last several decades has all recognised that external elements are present in the culture; Islam is clearly central to Swahili identity (Wood, 2017). The Swahili language, though structurally Bantu, includes much Arabic, Persian, and South Asian vocabulary. Trade promoted contact and exchange of ideas, language, commodities, and services between individuals and communities and variously transformed coastal societies (Chirikure, 2017). Trade became a core element of coastal economies although did not extensively impact hinterland communities until later on (Chapter 4), with many export goods coming from the coast, such as mangrove poles, tortoise shell, and ambergris. As the trade linkages developed and evolved so did the focus in terms of goods and sphere of influence: indeed, this would not have been restricted to the coast and there would have been numerous connections inland. Writers such as al Masudi (c. 856–956 CE) comment on the trade between East Africa and the hinterland believed to be the Zimbabwe Plateau. The hinterland supplied iron, gold, ivory, and possibly slaves in exchange for glass beads, cowries, and cloth (Chapter 4) with regions like Taita-Tsavo area's abundance of highly lucrative resources such as ivory, rhinoceros' horns, rock crystals, skins, and beeswax, making it an important trading partner for the coastal cities (Kusimba & Kusimba, 2005).

Arab market towns developed at Kismayu, Lamu, Mombasa, Gedi, Kilwa, and Zanzibar becoming centres of barter, trade, and exchange (Fig. 3.7). For 10 centuries from the 6th to the 16th, Arab sea power commanded the Indian Ocean and Arabs of Oman exercised their political and tight commercial hold on the East African coast. Trade was good and extensive as ships came from the Persian Gulf, India, China, and Malaysia. Many of the Arabs who came to trade settled and married becoming part of the culture. Following the Arab influence came the Portuguese. Before the arrival of the Portuguese in Indian Ocean waters, trade between India and East Africa was based primarily on the exchange of gold from Southern Zambesi and ivory from the coastal hinterland of East Africa for cotton cloths from India and glass beads from both India and Venice (Alpers, 1976). The first unambiguous references to Gujarati traders in East Africa are provided by the earliest accounts of the Portuguese on that coast during the 1500s (Alpers, 1976). Throughout the sixteenth century the Portuguese declared a monopoly of the Portuguese captains of the coast (Alpers, 1976). By the end of the century the Portuguese must have recognised how very dependent on the support of Gujarati merchants they had become, for in 1595 they were forbidden to trade beyond the ports of Western India, and Hindus were prohibited from acting as agents for Portuguese officials and from holding royal contracts (Alpers, 1976). The first wave of Europeans was led by Vasco da Gama in 1498 with the Portuguese occupying Mombasa from AD 1593 through to AD 1720 when the Arabian people took control of the areas until 1886 when the ivory and slave trade flourished, the latter until 1845 when the British forced the Omani sultan Sayyid Said to stop. The Swahili culture is not just characterised by the Kiswahili language but also by its cuisine and dress; for example, the simple wrap around cloth, the Kanga, has its origins in the mid-nineteenth century and originally was a black cloth with white spots that resemble the Guinea fowl or Kanga in Kiswahili. With such a growing trade and source of riches, thus started the international scramble for control of East Africa (Alpers, 1976); and trade rapidly transitioned from development of coastal societies, urban centred and interlinked trade to a much more exploitative footing that will be explored in Chapter 4.

References

Abungu, G. H. O., & Muturo, H. W. (1993). Coast-Interior Settlements and Social Relations in the Kenya Hinterland. In T. Shaw, P. J. J. Sinclair, B. Andah, & A. Okpoko (Ed.), *The Archaeology of Africa: Food, Metals and Towns* (pp. 694–704). Routledge.

Adams, W. M., & Anderson, D. M. (1988). Irrigation Before Development: Indigenous and Induced Change in Agricultural Water Management in East Africa. *African Affairs, 87*, 519–535.

Alin, S. R., & Cohen, A. S. (2003). Lake-Level History of Lake Tanganyika, East Africa, for the Past 2500 Years Based on Ostracod-Inferred Water-Depth Reconstruction. *Palaeogeography, Palaeoclimatology, Palaeoecology, 199*, 31–49.

Alpers, E.A. (1976). Gujarat and the Trade of East Africa, c. 1500–1800. *The International Journal of African Historical Studies, 9*(1), 22–44.

Anderson, D. M. (2002). *Eroding the Commons: The Politics of Ecology in Baringo, Kenya 1890–1963*. Ohio University Press.

Anderson, D. M. (2016). The Beginning of Time? Evidence for Catastrophic Drought in Baringo in the Early Nineteenth Century. *Journal of Eastern African Studies, 10*(1), 45–66.

Anderson, D. M., & Bollig, M. (2016). Resilience and Collapse: Histories, Ecologies, Conflicts and Identities in the Baringo-Bogoria Basin, Kenya. *Journal of Eastern African Studies, 10*, 1–20.

Ashley, C. Z. (2010). Towards a Socialised Archaeology of Ceramics in Great Lakes Africa. *African Archaeological Review, 27*(2), 135–163.

Ashley, G. M., Maitima Mworia, J., Muasya, A. M., Owen, R. B., Driese, S. G., Hover, V. C., Renaut, R. W., Goman, M. F., Mathai, S., & Blatt, S. H. (2004). Sedimentation and Recent History of a Freshwater Wetland in a Semi-Arid Environment: Loboi Swamp, Kenya, East Africa. *Sedimentology, 51*, 1301–1321.

Barker, P. A., Telford, R., Merdaci, O., Williamson, D., Taieb, M., Vincens, A., & Gibert, E. (2000). The Sensitivity of a Tanzania Crater Lake to Catastrophic Tephra Input and Four Millennia of Climate Change. *The Holocene, 10*, 303–310.

Bar-Yosef, O. (1998). Introduction: Some Comments on the History of Research. In O. Bar-Yosef (Ed.), The Transition to Agriculture in the Old World. *The Review of Archaeology, 19*(2), 1–11.

Battistel, D., Argiriadis, E., Kehrwald, N., Spigariol, M., Russell, J. M., & Barbante, C. (2017). Fire and Human Record at Lake Victoria, East Africa, During the Early Iron Age: Did Humans or Climate Cause Massive Ecosystem Changes? *The Holocene, 27*(7), 997–1007.

Berger, I. (1981). *Religion and Resistance: East African Kingdoms in the Precolonial Period*. Musee Royal de l'Afrique Centrale.

Bessems, I., Verschuren, D., Russell, J. M., Hus, J., Mees, F., & Cumming, B. F. (2008). Palaeolimnological Evidence for Widespread Late 18th Century Drought Across Equatorial East Africa. *Palaeogeography, Palaeoclimatology, Palaeoecology, 259*(2), 107–120.

Boivin, N., Crowther, A., Helm, R., & Fuller, D. Q. (2013). East Africa and Madagascar in the Indian Ocean World. *Journal of World Prehistory, 26*(3), 213–281.

Bollig, M. (2016). Adaptive Cycles in the Savannah: Pastoral Specialization and Diversification in Northern Kenya. *Journal of Eastern African Studies, 10*(1), 21–44.

Bollig, M., & Österle, M. (2013). The Political Ecology of Specialisation and Diversification: Long-Term Dynamics of Pastoralism in East Pokot District, Kenya. In M. Bollig, M. Schnegg, & H. P. Wotzka (Eds.), *Pastoralism in Africa: Past, Present and Future* (pp. 289–315). Berghahn.

Buckles, L. K., Weijers, J. W. H., Verschuren, D., Cocquyt, C., & Sinninghe Damste, J. S. (2016). Short-Term Variability in the Sedimentary BIT Index of Lake Challa, East Africa Over the Past 2200 Years: Validating the Precipitation Proxy. *Climates of the Past, 12*, 1243–1262.

Burton, R. F. (1860). *The Lake Regions of Central Africa, a Picture of Exploration.* Longman, Green, Longman and Roberts.

Chami, F. A. (1998). A Review of Swahili Archaeology. *African Archaeological Review, 15*(3), 199–218.

Chami, F. A. (2004). The Egypto-Graeco-Romans and Panchaea/Azania: Sailing in the Erythraean Sea. In P. Lunde & A. Porter (Ed.), *Trade and Travel in the Red Sea Region* (pp. 93–104). BAR International Series. Archaeopress.

Chami, F. A. (2019). The Archaeology of the Mafia Archipelago, Tanzania: New Evidence for Neolithic Trade Links. *Studies in the African Past, 4*, 73–101.

Chirikure, S. (2017/2020). Documenting Precolonial Trade in Africa. In *Oxford Research Encyclopaedia, African History*. Oxford University Press.

Chirikure, S. (2018). *Precolonial Metallurgy and Mining Across Africa* (p. 2020). Oxford University Press.

Chittick, N. (1974). *Kilwa: An Islamic Trading City on the East African Coast: The Finds 2*. The British Institute in East Africa.
Cochrane, M. A., Alencar, A., Schulze, M. D., Souza, C. M., Nepstad, D. C., Lefebvre, P., & Davidson, E. A. (1999). Positive Feedbacks in the Fire Dynamic of Closed Canopy Tropical Forests. *Science, 284*(5421), 1832–1835.
Cohen, A. S., Talbot, M. R., Awramik, S. M., Dettman, D. L., & Abell, P. (1997). Lake-Level and Palaeoenvironmental History of Lake Tanganyika, Africa, as Inferred from Late Holocene and Modern Stromatolites. *Geological Society of America Bulletin, 109*, 444–460.
Cohen, D. W. (1983). Luo Camps in the Seventeenth Century in Eastern Uganda: The Use of Migration Tradition in the Reconstruction of Culture. *Sprache Und Geschichte in Afrika, 5*, 145–177.
Colombaroli, D., Ssemmanda, I., Gelorini, V., & Verschuren, D. (2014). Contrasting Long-Term Records of Biomass Burning in Wet and Dry Savannas of Equatorial East Africa. *Global Change Biology, 20*, 2903–2914.
Connah, G. (1996). *Kibiro: The Salt of Bunyoro, Past and Present*. British Institute in Eastern Africa, Memoir 13.
Crossley, R., Davison-Hirschmann, S., Owen, R. B., & Shaw, P. (1984). Lake Level Fluctuations During the Last 2000 Years in Malawi. In J. C. Vogel (Ed.), *Late Cainozoic Palaeoclimates of the Southern Hemisphere* (pp. 305–316). Balkema.
Crowther, A., Faulkner, P., Prendergast, M. E., Quintana Morales, E. M., Horton, M., Wilmsen, E., Kotarba-Morley, A. M., Christie, A., Petek, N., Tibesasa, R., & Douka, K. (2016). Coastal Subsistence, Maritime Trade, and the Colonization of Small Offshore Islands in Eastern African Prehistory. *The Journal of Island and Coastal Archaeology, 11*(2), 211–237.
Crowther, A., Horton, M., Kotarba-Morley, A., Prendergast, M., Quintana Morales, E., Wood, M., Shipton, C., Fuller, D. Q., Tibesasa, R., Mills, W., & Boivin, N. (2014). Iron Age Agriculture, Fishing and Trade in the Mafia Archipelago, Tanzania: New Evidence from Ukunju Cave. *Azania: Archaeological Research in Africa, 49*(1), 21–44.
Crowther, A., Lucas, L., Helm, R., Horton, M., Shipton, C., Wright, H. T., Walshaw, S., Pawlowicz, M., Radimilahy, C., Douka, K., & Picornell-Gelabert, L. (2016). Ancient Crops Provide First Archaeological Signature of the Westward Austronesian Expansion. *Proceedings of the National Academy of Sciences, 113*(24), 6635–6640.
Crowther, A., Prendergast, M. E., Fuller, D. Q., & Boivin, N. (2017). Subsistence Mosaics, Forager-Farmer Interactions, and the Transition to

Food Production in Eastern Africa. *Quaternary International* [online], *489*, 101–120. https://doi.org/10.1016/j.quaint.2017.01.014

Dahl, G., & Anders, H. (1976). *Having Herds: Pastoral Herd Growth and Household Economy* (Vol. 2). Department of Anthropology, University of Stockholm.

Davies, M. I. J. (2008). The Irrigation System of the Pokot, Northwest Kenya. *Azania: Archaeological Research in Africa, 43*, 50–76.

Davies, M. I. J., & Moore, H. L. (2016). Landscape, Time and Cultural Resilience: A Brief History of Agriculture in Pokot and Marakwet, Kenya. *Journal of Eastern African Studies, 10*, 67–87.

De Cort, G., Bessems, I., Keppens, E., Mees, F., Cumming, B., & Verschuren, D. (2013). Late-Holocene and Recent Hydroclimatic Variability in the Central Kenya Rift Valley: The Sediment Record of Hypersaline Lakes Bogoria, Nakuru and Elementeita. *Palaeogeography, Palaeoclimatology, Palaeoecology, 388*, 69–80.

De Cort, G., Verschuren, D., Ryken, E., Wolff, C., Renaut, R. W., Creutz, M., Van der Meeren, T., Haug, G., Olago, D. O., & Mees, F. (2018). Multi-Basin Depositional Framework for Moisture-Balance Reconstruction During the Last 1300 Years at Lake Bogoria, Central Kenya Rift Valley. *Sedimentology, 65*(5), 1667–1696.

de Langhe, E., Swennen, R., & Vuylsteke, D. (1995). Plantain in the Early Bantu World. *Azania, 29*(30), 147–160.

Driese, S. G., Ashley, G. M., Li, Z. H., Hover, V. C., & Owen, R. B. (2004). Possible Late Holocene Equatorial Palaeoclimate Record Based Upon Soils Spanning the Medieval Warm Period and Little Ice Age, Loboi Plain, Kenya. *Palaeogeography, Palaeoclimatology, Palaeoecology, 213*, 231–250.

Ehret, C. (1984). Between the Coast and the Great Lakes. In D. J. Niane (Ed.), *The UNESCO General History of Africa, 12th–16th Centuries* (pp. 481–497). University of California Press.

Ehret, C. (1999). *An African Classical Age: Eastern and Southern Africa in World History, 1000BC to AD400*. University of Virginia Press.

Ehret, C. (2001). Bantu Expansion: Re-Envisioning a Central Problem of Early African History. *International Journal of African Historical Studies, 34*, 5–41.

Feierman, S. (1990). *Peasant Intellectuals. Anthropology and History in Tanzania*. University of Wisconsin Press.

Finch, J., Marchant, R., & Courtney Mustaphi, C. J. (2017). Ecosystem Change in the South Pare Mountain Bloc, Eastern Arc Mountains of Tanzania. *The Holocene, 27*(6), 796–810.

Fleisher, J. (2003). *Viewing Stonetowns from the Countryside: An Archaeological Approach to Swahili Regional Systems, AD800–1500*. Unpublished Ph.D. dissertation, University of Virginia.
Fleisher, J., & Sulas, F. (2015). Deciphering Public Spaces in Urban Contexts: Geophysical Survey, Multi-Element Soil Analysis, and Artifact Distributions at the 15th–16th-Century AD Swahili Settlement of Songo Mnara, Tanzania. *Journal of Archaeological Science, 55*, 55–70.
Fosbrooke, H. (1972). *Ngorongoro: The Eighth Wonder*. Deutsch.
Fosbrooke, H., & Sassoon, H. (1965). Archaeological Remains on Kilimanjaro. *Tanganyika Notes and Records, 64*, 62–64.
Frame, G., Frame, L. H., & Spillett, J. J. (1975). An Ecological Survey and Development Plan for the Empakaai Crater Ecosystem (Ngorongoro Conservation Area). *Serengeti Research Contribution, 212*, 472.
Galaty, J. G. (1993). Maasai Expansion and the New East African Pastoralism. In T. Spear & R. Waller (Ed.), *Being Maasai: Ethnicity and Identity in East Africa*. James Currey.
Garcin, Y., Melnick, D., Strecker, M. R., Olago, D., & Tiercelin, J. J. (2012). East African Mid-Holocene Wet–Dry Transition Recorded in Palaeo-Shorelines of Lake Turkana, Northern Kenya Rift. *Earth and Planetary Science Letters*, 331–332 and 322–334.
Gilbert, E. (2002). Coastal East Africa and the Western Indian Ocean: Long-Distance Trade, Empire, Migration, and Regional Unity, 1750–1970. *The History Teacher, 36*(1), 7–34.
Gibert, E., Bergonzini, L., Massault, M., & Williamson, D. (2002). AMS ^{14}C Chronology of 40 cal ka BP Continuous Deposits from a Crater Lake (Lake Masoko, Tanzania): Modern Water Balance and Environmental Implications. *Palaeogeography, Palaeoclimatology, Palaeoecology, 188*, 1–16.
Gifford-Gonzalez, D. (2000). Animal Disease Challenges to the Emergence of Pastoralism in Sub-Saharan Africa. *African Archaeological Review, 17*, 95–139.
Gifford-Gonzalez, D. (2005). Pastoralism and Its Consequences. In A. B. Stahl (Ed.), *African Archaeology: A Critical Introduction* (pp. 187–224). Blackwell.
Hartwig, G. W. (1979). Demographic Considerations in East Africa During the Late Nineteenth Century. *International Journal of African Historical Studies, 4*, 653–672.
Heckmann, M. (2011). *Soil Erosion History and Past Human Land Use in the North Pare Mountains: A Geoarchaeological Study of Slope Deposits in NE Tanzania*. Unpublished Ph.D. dissertation, University of York.

Helm, R. (2000). Recent Archaeological Research on the Iron-Working, Farming Communities of Coastal Kenya. *Azania: Archaeological Research in Africa, 35*(1), 183–189.

Herring, R. S. (1979). The Influence of Climate on the Migrations of the Central and Southern Luo. In B. A. Ogot (Ed.), *Ecology and History in East Africa (Hadith 7)* (pp. 77–107). Kenya Literature Bureau.

Horton, M. (1996). *Shanga: The Archaeology of a Muslim Trading Community on the Coast of East Africa*. The British Institute in Eastern Africa.

Horton, M., & Middleton, J. (2001). *The Swahili: The Social Landscape of a Mercantile Society*. Blackwell.

Humphris, J., Martinon-Torres, M., Rehren, T. H., & Reid, A. (2009). Variability in Single Smelting Episodes: A Pilot Study Using Iron Slag from Uganda. *Journal of Archaeological Science, 36*, 359–369.

Johnsen, S. J., Dahl-Jensen, D., Gundestrup, N., Steffensen, J. P., Clausen, H. B., Miller, H., Masson-Delmotte, V., Sveinbjornsdottir, A. E., & White, J. (2001). Oxygen Isotope and Palaeotemperature Records from Six Greenland Ice-Core Stations: Camp Century, Dye-3, GRIP, GISP2, Greenland and NorthGRIP. *Journal of Quaternary Science, 16*, 299–307.

Johnston, H. (1902). *The Uganda Protectorate* (Vol. 2). Hutchinson and Company.

Kaplan, J. O., & Krumhardt, K. M. (2011). The KK10 Anthropogenic Land Cover Change Scenario for the Preindustrial Holocene, Link to Data in NetCDF Format. *PANGAEA, 21*, 775–791. https://doi.org/10.1594/PANGAEA.871369

Kaplan, J. O., Krumhardt, K. M., Ellis, E. C., Ruddiman, W. F., Lemmen, C., & Goldewijk, K. K. (2011). Holocene Carbon Emissions as a Result of Anthropogenic Land Cover Change. *The Holocene, 21*(5), 775–791.

Kenny, M. G. (1979). Pre-Colonial Trade in Eastern Lake Victoria. *Azania: Archaeological Research in Africa, 14*, 97–107.

Kessey, E. (2003). Iron Age Settlement Patterns and Economic Change on Zanzibar and Pemba Islands. In C. M. Kusimba & S. B. Kusimba (Eds.), *East African Archaeology: Foragers, Potters, Smiths, and Traders* (pp. 117–131). University of Pennsylvania Museum Press.

Kiage, L. M., & Liu, K. B. (2006). Late Quaternary Paleoenvironmental Changes in East Africa: A Review of Multiproxy Evidence from Palynology, Lake Sediments, and Associated Records. *Progress in Physical Geography, 30*(5), 633–658.

Kjekshus, H. (1977). *Ecology Control and Economic Development in East African History*. University of California Press.

Konecky, B., Russell, J., Huang, Y., Vuille, M., Cohen, L., & Street-Perrott, F. A. (2014). Impact of Monsoons, Temperature, and CO_2 on the Rainfall and Ecosystems of Mt. Kenya During the Common Era. *Palaeogeography, Palaeoclimatology, Palaeoecology, 396*, 17–25.

Krapf, J. L. (1860). *Travels, Researches, and Missionary Labors During an Eighteen Years' Residence in Eastern Africa*. Ticknor and Fields.

Kusimba, C. M. (1999). *The Rise and Fall of Swahili States*. Altamira.

Kusimba, C. M., & Kusimba, S. B. (2005). Mosaics and Interactions: East Africa, 2000 bp to the Present. In A. B. Stahl (Ed.), *African Archaeology: A Critical Introduction* (pp. 392–419). Blackwell.

Kusimba, C. M., Kusimba, S. B., & Wright, D. (2005). The Development and Collapse of Precolonial Ethnic Mosaics in Tsavo, Kenya. *Journal of African Archaeology, 3*(2), 243–265.

Lamb, H., Darbyshire, I., & Verschuren, D. (2003). Vegetation Response to Rainfall Variation and Human Impact in Central Kenya During the Past 1100 Years. *The Holocene, 13*(2), 285–292.

Lane, P. J. (2010). Developing Landscape Historical Ecologies in Eastern Africa: An Outline of Current Research and Potential Future Directions. *African Studies, 69*(2), 299–322.

Lane, P. J. (2014). The Archaeological Potential for the History of Labor Relations in East Africa, ca. 1500–1900. *History in Africa, 41*, 227–306.

Lane, P. J., Ashley, C. Z., & Oteyo, G. (2006). New Dates for Kansyore and Urewe Wares from Northern Nyanza, Kenya. *Azania, 41*, 123–138.

Lang, C., & Stump, D. (2017). Geoarchaeological Evidence for the Construction, Irrigation, Cultivation, and Resilience of 15th–18th Century AD Terraced Landscape at Engaruka, Tanzania. *Quat Res* [online], *88*, 382–399. https://doi.org/10.1017/qua.2017.54

LaViolette, A., & Fleisher, F. (2005). The Archaeology of Sub-Saharan Urbanism: Cities and Their Countrysides. In A. B. Stahl (Ed.), *African Archaeology: A Critical Introduction*. Blackwell Studies in Global Archaeology. Blackwell.

Lehmann, C. E., Anderson, T. M., Sankaran, M., Higgins, S. I., Archibald, S., Hoffmann, W. A., Hanan, N. P., Williams, R. J., Fensham, R. J., Felfili, J., & Hutley, L. B. (2014). Savanna Vegetation-Fire-Climate Relationships Differ Among Continents. *Science, 343*(6170), 548–552.

Lejju, B. J., Taylor, D., & Robertshaw, P. (2005). Late-Holocene Environmental Variability at Munsa Archaeological Site, Uganda: A Multicore, Multiproxy Approach. *The Holocene, 15*(7), 1044–1061.

Lofgren, L. (1967). Stone Structures of South Nyanza. *Azania, 2*, 75–88.

Médard, H., & Doyle, S. (2007). *Slavery in the Great Lakes Region of East Africa*. James Currey.

Middleton, J. (1961). Land Tenure in Zanzibar. *Colonial Research Studies* (p. 23). HMSO.

Middleton, J. (1992). *The World of the Swahili*. Yale University Press.

Mills, K., Ryves, D. B., Anderson, N. J., Bryant, C. L., & Tyler, J. J. (2014). Expressions of Climate Perturbations in Western Ugandan Crater Lake Sediment Records During the Last 1000 Years. *Climates of the Past, 10*, 1581–1601.

Mohammed, M. U., Bonnefille, R., & Johnson, T. C. (1996). Pollen and Isotopic Records in Late Holocene Sediments from Lake Turkana, Kenya. *Palaeogeography, Palaeoclimatology, Palaeoecology, 119*(3–4), 371–383.

Montlahuc, M. L., & Philippson, G. (2006). Before Coffee, Pre-Colonial Agricultural Glossary. In F. Bart, M. J. Mbonile, & F. Devenne (Eds.), *Kilimanjaro: Mountain, Memory, Modernity* (pp. 59–69). Mkuki na Nyota Publishers.

Mudida, N., & Horton, M. (1996). Subsistence at Shanga: The Faunal Record. *Shanga: The Archaeology of a Muslim Trading Community on the Coast of East Africa*. British Institute in Eastern Africa.

Mturi, A. A. (1986). The Pastoral Neolithic of West Kilimanjaro. *Azania, 21*(1), 53–63.

Nash, D. J., De Cort, G., Chase, B. M., Verschuren, D., Nicholson, S. E., Shanahan, T. M., Asrat, A., Lézine, A.-M., & Grab, S. W. (2016). African Hydroclimatic Variability During the Last 2000 Years. *Quaternary Science Reviews, 154*, 1–22.

Neumann, K. (2005). The Romance of Farming: Plant Cultivation and Domestication in Africa. In A. B. Stahl (Ed.), *African Archaeology: A Critical Introduction* (pp. 249–275). Blackwell.

Nicholson, S. E. (1996). A Review of Climate Dynamics and Climate Variability in Eastern Africa. In T. C. Johnson & E. Odada (Eds.), *The Limnology, Climatology and Paleoclimatology of the East African Lakes* (pp. 25–56). Gordon and Breack.

Nicholson, S. E. (2000). The Nature of Rainfall Variability Over Africa on Time Scales of Decades to Millennia. *Global and Planetary Change, 26*, 137–158.

Nicholson, S. E. (2001). Climatic and Environmental Change in Africa During the Last Two Centuries. *Climate Research, 17*(2), 123–144.

Nicholson, S. E., & Yin, X. (2001). Rainfall Conditions in Equatorial East Africa During the Nineteenth Century as Inferred from the Record of Lake Victoria. *Climatic Change, 48*(2), 387–398.

Nurse, D. (1979). *Classification of the Chaga Dialects: Language and History on Kilimanjaro, the Taita Hills, and the Pare Mountains.* Buske.

Öberg, H., Andersen, T. J., Westerberg, L. O., Risberg, J., & Holmgren, K. (2012). A Diatom Record of Recent Environmental Change in Lake Duluti, Northern Tanzania. *Journal of Paleolimnology, 48*(2), 401–416.

Öberg, H., Norström, E., Ryner, M. M., Holmgren, K., Westerberg, L. O., Risberg, J., Eddudóttir, S. D., Andersen, T. J., & Muzuka, A. (2013). Environmental Variability in Northern Tanzania from AD 1000 to 1800, as Inferred from Diatoms and Pollen in Lake Duluti. *Palaeogeography, Palaeoclimatology, Palaeoecology, 374*, 230–241.

Ochieng, W. R. (1974). *An Outline History of Nyanza Up to 1914*. East African Literature Bureau.

Odede, F. (2008). Gunda-Buche: The Bank-and-Ditch Fortified Settlement Enclosures of Western Kenya, Lake Victoria Basin. *Azania: Archaeological Research in Africa, 43*, 36–49.

Odner, K. (1971a). Usangi Hospital and Other Archaeological Sites in the North Pare Mountains, North-Eastern Tanzania. *Azania, 6*, 89–130.

Odner, K. (1971b). A Preliminary Report of an Archaeological Survey on the Slopes of Kilimanjaro. *Azania, 6*, 131–149.

Ogot, B. A. (1967). *History of the Southern Luo* (Vol. 1). East Africa Publishing House.

Onjala, I. O. (2003). Spatial Distribution and Settlement System of the Stone Structures of South-Western Kenya. *Azania: Archaeological Research in Africa, 38*, 99–120.

Owen, R. B., Crossley, R., Johnson, T. C., Tweddle, D., Kornfield, I., Davison, S., Eccles, D. H., & Engstrom, D. E. (1990). Major Low Levels of Lake Malawi and Their Implications for Speciation Rates in Cichlid Fishes. *Proceedings of the Royal Society of London B: Biological Sciences, 240*(1299), 519–553.

Patel, A. (2004). Communities and Commodities: Western India and the Indian Ocean, Eleventh-Fifteenth Centuries. *Ars Orientalis, 34*, 7–18.

Pawlowicz, M. (2012). Models of the Swahili Past: The Archaeology of Mikindani, Southern Coastal Tanzania. *Azania: Archaeological Research in Africa, 47*(4), 486–506.

Pawlowicz, M. C. (2011). *Finding Their Place in the Swahili World: An Archaeological Exploration of Southern Tanzania*. Unpublished Ph.D. dissertation, University of Virginia.

Petek, N. (2015). An Archaeological Survey of the Lake Baringo Lowlands 2014: Preliminary Results. *Nyame Akuma, 83*, 100–111.

Petek, N., & Lane, P. J. (2017). Ethnogenesis and Surplus Food Production: Aommunitas and Identity Building Among Nineteenth- and Early Twentieth-Century Ilchamus, Lake Baringo, Kenya. *World Archaeology* [online], *49*(1), 40–60. http://dx.doi.org/10.1080/00438243.2016.1259583

Pollard, E. J. D. (2007). *An Archaeology of Coastal Tanzanian Landscapes in the Middle Iron Age (6th to 15th centuries AD)*. Ph.D. thesis. University of Ulster.

Pouwels, R. L. (2002). Eastern Africa and the Indian Ocean to 1800: Reviewing Relations in Historical Perspective. *The International Journal of African Historical Studies, 35*(2/3), 385–425.

Prendergast, M. E., Quintana Morales, E. M., Crowther, A., Horton, M. C., & Boivin, N. L. (2017). Dietary Diversity on the Swahili Coast: The Fauna from Two Zanzibar Trading Locales. *International Journal of Osteoarchaeology* [online], *27*(4), 621–637. https://doi.org/10.1002/oa.2585

Prins, A. H. J. (1971). *Didemic Lamu: Social Stratification and Spatial Structure in a Muslim Maritime Town*. Instituut voor Culturele Antropologie der Rijksuniversiteit.

Punwong, P., Sritrairat, S., Selby, K., Marchant, R., Pumijumnong, N., & Traiperm, P. (2017). An 800 Year Record of Mangrove Dynamics and Human Activities in the Upper Gulf of Thailand. *Vegetation History and Archaeobotany*. https://doi.org/10.1007/s00334-017-0651-x

Quintana Morales, E. (2013). *Reconstructing Swahili Foodways: The Archaeology of Fishing and Fish Consumption in Coastal East Africa, AD 500–1500.* Unpublished Ph.D. dissertation, University of Bristol.

Quintana Morales, E., & Prendergast, M. E. (2017). Animals in the Swahili World. In A. LaViolette & S. Wynne-Jones (Eds.), *The Swahili World* (pp. 335–350). Routledge.

Reid, D. A. M. (1994–1995). Ntusi and the Development of Social Complexity in Southern Uganda. In G. Pwiti & R. Soper (Ed.), *Aspects of African Archaeology* (pp. 621–628). University of Zimbabwe Press.

Reid, D. A. M. (1996). Cattle Herds and the Redistribution of Cattle Resources. *World Archaeology, 28*(1), 43–57.

Reid, D. A. M. (2003). Buganda: Unearthing an African Kingdom. *Archaeology International, 7*, 40–43.

Reid, D. A. M. (2013). The Emergence of States in Great Lakes Africa. In P. J. Lane & P. Mitchell (Ed.), *The Oxford Handbook of African Archaeology*. Oxford University Press. https://doi.org/10.1093/oxfordhb/9780199569885.013.0061

Reid, D. A. M. (2015). Archaeological Ivory and the Impact of the Elephant in Mawogola. *World Archaeology, 47*(3), 467–485.

Reid, D. A. M., & MacLean, R. (1995). Symbolism and the Social Contexts of Iron Production in Karagwe. *World Archaeology, 27*, 144–161.

Reid, D. A. M., & Young, R. (2000). Pottery Abrasion and the Preparation of African Grains. *Antiquity, 74*, 101–111.

Reid, R. (2007). *War in Pre-Colonial Eastern Africa: The Patterns and Meanings of State-Level Conflict in the 19th Century*. James Currey.

Reij, C., Scoones, I., & Toulmin, C. (2013). *Sustaining the Soil: Indigenous Soil and Water Conservation in Africa*. Routledge.

Ricks, T. M. (1970). Persian Gulf Seafaring and East Africa: Ninth-Twelfth Centuries. *African Historical Studies, 3*(2), 339–357.

Robertshaw, P. T. (1986). Engaruka Revisited: Excavations of 1982. *Azania, 21*, 1–20.

Robertshaw, P. T. (1988). Environment and Culture in the Late Quaternary of Eastern Africa: A Critique of Some Correlations. In J. Bower & D. Lubell (Eds.), *Prehistoric Cultures and Environments in the Late Quaternary of Africa* (pp. 115–126). Cambridge University Press.

Robertshaw, P. T. (1991). Gogo Falls: Excavations at a Complex Archaeological Site East of Lake Victoria. *Azania, 26*, 63–195.

Robertshaw, P. T. (1994). Archaeological Survey, Ceramic Analysis, and State Formation in Western Uganda. *African Archaeological Review, 12*, 105–131.

Robertshaw, P. T. (1997). Munsa Earthworks. *Azania: Archaeological Research in Africa, 32*, 1–20.

Robertshaw, P. T. (2010). Beyond the Segmentary State: Creative and Instrumental Power in Western Uganda. *Journal of World Prehistory, 23*, 255–269.

Robertshaw, P. T., & Taylor, D. (2000). Climate Change and the Rise of Political Complexity in Western Uganda. *Journal of African History, 41*(1), 1–28.

Robertshaw, P. T., Taylor, D., Doyle, S., & Marchant, R. (2004). Famine, Climate and Crisis in Western Uganda. In R. W. Battarbee, F. Gasse, & C. E. Stickley (Eds.), *Past Climate Variability Through Europe and Africa* (pp. 535–549). Kluwer.

Rossel, G. (1998). *Taxonomic-Linguistic Study of Plantain in Africa*. Leiden University.

Rucina, S. M., Muiruri, V. M., Downton, L., & Marchant, R. (2010). Late Holocene Savanna Dynamics in the Amboseli Basin, Kenya. *Holocene, 20*, 667–677.

Russell, J. M., & Johnson, T. C. (2005). A High-Resolution Geochemical Record from Lake Edward, Uganda Congo and the Timing and Causes of Tropical African Drought During the Late Holocene. *Quaternary Science Reviews, 24*(12), 1375–1389.

Russell, J. M., & Johnson, T. C. (2007). Little Ice Age Drought in Equatorial Africa: Intertropical Convergence Zone Migrations and El Nino-Southern Oscillation Variability. *Geology, 35*, 21–24.

Russell, J. M., Verschuren, D., & Eggermont, H. (2007). Spatial Complexity of 'Little Ice Age' Climate in East Africa: Sedimentary Records from Two Crater Lake Basins in Western Uganda. *Holocene, 17*, 183–193.

Ryner, M., Holmgren, K., & Taylor, D. (2008). A Record of Vegetation Dynamics and Lake Level Changes from Lake Emakat, Northern Tanzania, During the Last c. 1200 Years. *Journal of Paleolimnology, 40*(2), 583–601.

Sankaran, M., Hanan, N. P., Scholes, R. J., Ratnam, J., Augustine, D. J., Cade, B. S., & Ardo, J. (2005). Determinants of Woody Cover in African Savannas. *Nature, 438*(7069), 846–849.

Sassoon, H. (1983). Kings, Cattle and Blacksmiths: Royal Insignia and Religious Symbolism in the Interlacustrine States. *Azania, 18*, 93–106.

Schoenbrun, D. L. (1998). *A Green Place, a Good Place: Agrarian Change, Gender, and Social Identity in the Great Lakes Region to the 15th Century*. James Currey.

Sheridan, M. J. (2002). An Irrigation Intake Is Like a Uterus: Culture and Agriculture in Precolonial North Pare, Tanzania. *American Anthropologist, 104*(1), 79–92.

Spear, T. (1993). Being "Maasai", but Not "People of Cattle": Arusha Agricultural Maasai in the Nineteenth Century. In T. Spear & R. Waller (Eds.), *Being Maasai: Ethnicity and Identity in East Africa* (pp. 120–136). James Currey.

Ssemmanda, I., Ryves, D. B., Bennike, O., & Appleby, P. G. (2005). Vegetation History in Western Uganda During the Last 1200 Years: A Sediment Based Reconstruction from Two Crater Lakes. *The Holocene, 15*(1), 119–132.

Stager, J. C., Ryves, D., Cumming, B. F., Meeker, L. D., & Beer, J. (2005). Solar Variability and the Levels of Lake Victoria, East Africa, During the Last Millennium. *Journal of Paleolimnology, 33*, 243–251.

Staver, A. C., Archibald, S., & Levin, S. A. (2011). The Global Extent and Determinants of Savanna and Forest as Alternative Biome States. *Science, 334*(6053), 230–232.

Stocker, T. F., Qin, D., Plattner, G. K., Tignor, M., Allen, S. K., Boschung, J., Nauels, A., Xia, Y., Bex, B., & Midgley, B. M. (2013). *IPCC Climate Change 2013: The Physical Science Basis.* Contribution of Working Group I to the Fifth Assessment Report of the Intergovernmental Panel on Climate Change.

Stump, D. (2006). The Development and Expansion of the Field and Irrigation System at Engaruka, Tanzania. *Azania, 41*, 69–94.

Stump, D., & Tagseth, M. (2009). The History of Precolonial and Early Colonial Agriculture on Kilimanjaro: A Review. In T. Clack (Ed.), *Culture, History and Identity: Human–Environmental Relations in the Mount Kilimanjaro Area, Tanzania* (pp. 107–124). Cambridge Monographs in African Archaeology. Archaeopress.

Sutton, J. E. G. (1984). Irrigation and Soil-Conservation in African Agricultural History: With a Reconsideration of the Inyanga Terracing Zimbabwe and Engaruka Irrigation Works Tanzania. *The Journal of African History, 251*, 25–41.

Sutton, J. E. G. (1993). Becoming Maasailand. In T. Spear & R. Waller (Eds.), *Being Maasai: Ethnicity and Identity in East Africa* (pp. 38–60). James Currey.

Sutton, J. E. G. (2004). Engaruka: The Success and Abandonment of an Integrated Irrigation System in an Arid part of the Rift Valley, c. 15th to 17th Centuries. In M. Widgren & J. E. G. Sutton (Eds.), *Islands of Intensive Agriculture in Eastern Africa* (pp. 114–132). James Currey.

Taylor, D. (1993). Environmental Change in Montane South West Uganda: A Pollen Record for The Holocene from Ahakagyezi Swamp. *The Holocene, 3*, 324–332.

Taylor, D. M., Marchant, R., & Robertshaw, P. (1999). Late Glacial-Holocene History of Lowland Rain Forest in Central Africa: A Record from Kabata Swamp, Ndale Volcanic Field, Uganda. *Journal of Ecology, 87*, 303–315.

Taylor, D. M., & Marchant, R. A. (1996). Human Impact in the Interlacustrine Region: Long-Term Pollen Records from the Rukiga Highlands. *Azania, 29–30*, 283–295.

Thevenon, F., Williamson, D., Vincens, A., Taieb, M., Merdaci, O., Decobert, M., & Buchet, G. (2003). A late-Holocene charcoal record from Lake Masoko, SW Tanzania: climatic and anthropologic implications. *The Holocene, 13*(5), 785–792.

Thompson, L. G., Mosley-Thompson, E., Davis, M. E., Henderson, K. A., Brecher, H. H., Zagorodnov, V. S., ... Beer, J. (2002). Kilimanjaro ice core records: Evidence of Holocene climate change in tropical Africa. *Science, 298*(5593), 589–593.

Tierney, J. E., Smerdon, J. E., Anchukaitis, K. J., & Seager, R. (2013). Multi-decadal Variability in East African Hydroclimate Controlled by the Indian Ocean. *Nature, 493*, 389–392.

van Zwanenberg, R. M. A., and King, A. (1975). Nomadic Pastoralism: The Process of Impoverishment. In *An Economic History of Kenya and Uganda 1800–1970* (pp. 79–109). Palgrave Macmillan.

Vernet, T. (2015). Avant le giroflier. Esclavage et agriculture sur la côte swahili, 1590–1812. In H. Médard, M.-L. Derat, T. Vernet, & M. P. Ballarin (Eds.), *Traites et esclavages en Afrique orientale et dans l'océan Indien* (pp. 244–306). Karthala.

Verschuren, D., & Charman, D. J. (2008). Latitudinal Linkages in Late Holocene Moisture-Balance Variation. In R. W. Battarbee & H. A. Binney (Eds.), *Natural Climate Variability and Global Warming: A Holocene Perspective* (pp. 189–231). Wiley.

Verschuren, D., Damsté, J. S. S., Moernaut, J., Kristen, I., Blaauw, M., Fagot, M., & Haug, G. H. (2009). Half-precessional dynamics of monsoon rainfall near the East African Equator. *Nature, 462*(7273), 637–641.

Verschuren, D., Johnson, T. C., Kling, H. J., Edgington, D. N., Leavitt, P. R., Brown, E. T., & Hecky, R. E. (2002). History and Timing of Human Impact on Lake Victoria, East Africa. *Proceedings of the Royal Society of London B: Biological Sciences, 269*(1488), 289–294.

Verschuren, D., Laird, K. R., & Cumming, B. F. (2000). Rainfall and Drought in Equatorial East Africa During the Past 1,100 Years. *Nature, 403*(6768), 410–414.

Vincens, A., Williamson, D., Thevenon, F., Taieb, M., Buchet, G., Decobert, M., & Thouveny, N. (2003). Pollen-Based Vegetation Changes in Southern Tanzania During the Last 4200 Years: Climate Change and/or Human Impact. *Palaeogeography, Palaeoclimatology, Palaeoecology, 198*(3), 321–334.

Waller, R. (1976). The Maasai and the British 1895–1905 the Origins of an Alliance. *Journal of African History, XVII*(4), 529–553.

Walshaw, S. C. (2010). Converting to rice: Urbanization, Islamization and crops on Pemba Island, Tanzania, AD 700–1500. *World Archaeology, 42*(1), 137–154.

Walshaw, S. C. (2015). *Swahili Trade, Urbanization, and Food Production: Botanical Perspectives from Pemba Island, Tanzania, 600–1500*. Cambridge

Monographs in African Archaeology 90 (British Archaeological Reports). Archaeopress.
Walz, J. R. (2010). *Route to a Regional Past: An Archaeology of the Lower Pangani (Ruvu) Basin, Tanzania, 500–1900 C.E.* Unpublished Ph.D. dissertation, University of Florida.
Wandibba, S. (1986). Thimlich Ohingini. *Azania, 21*, 134.
Warren, A. (1995). Changing Understandings of African Pastoralism and the Nature of Environmental Paradigms. *Transactions of the Institute of British Geographers* [online], *20*, 193–203. https://doi.org/10.2307/622431
Webster, R. J. B. (1979). Noi! Noi! Famines as an Aid to Interlacustrine Chronology. In R. J. B. Webster (Ed.), *Chronology, Migration and Drought in Interlacustrine Africa* (pp. 5–11). Africana.
Westerberg, L. O., Holmgren, K., Börjeson, L., Håkansson, N. T., Laulumaa, V., Ryner, M., & Öberg, H. (2010). The Development of the Ancient Irrigation System at Engaruka, Northern Tanzania: Physical and Societal Factors. *The Geographical Journal, 176*, 304–318.
Western, D. (1994). Ecosystem Conservation and Rural Development: The Case of Amboseli. In D. Western & R. M. Wright (Eds.), *Natural Connections: Perspectives in Community Based Conservation* (pp. 15–52). Island Press.
Widgren, M., & Sutton, J. E. (1999). *'Islands' of Intensive Agriculture in the East African Rift and Highlands: A 500-year Perspective*. University of Stockholm.
Wood, M. (2017). Glass Beads and Indian Ocean Trade. In *The Swahili World* (pp. 458–471). Routledge.
Wynants, M., Kelly, C., Mtei, K., Munishi, L., Patrick, A., Rabinovich, A., Nasseri, M., Gilvear, D., Roberts, N., Boeckx, P., Wilson, G., Blake, W. H., & Ndakidemi, P. (2019). Drivers of Increased Soil Erosion in East Africa's Agro-Pastoral Systems: Changing Interactions Between the Social, Economic and Natural Domains. *Regional Environmental Change, 19*, 1909–1921.
Wynne-Jones, S. (2010). Lines of Desire: Power and Materiality Along a Tanzanian Caravan Route. *Journal of World Prehistory, 23*(4), 219–237.
Wynne-Jones, S. (2016). *A Material Culture: Consumption and Materiality on the Coast of Precolonial East Africa*. Oxford University Press.

4

Elephants, Maize, and Pervasive Societal Environmental Transformations

4.1 Introduction

The arrival of Bantu agriculturalists, pastoralists, terracing technologies, pot making, metal working, international trade, and growing populations, as outlined in the previous chapter, led to large-scale ecosystem transformations and altered relationships between people, ecosystems, and their environments through the expansion of global trade and exploitation. Although the earliest confirmed trade from East Africa was during the time of the Roman empire, the expansion of overseas trade was explosive in the last 500 or so years as increasing amounts of ivory, slaves, and other commodities such as timber, mangrove poles and copal were exported to China, India, the Middle East, America, and Europe, and goods flowed into the region on return (Alpers, 2020). Particularly during the fifteenth to nineteenth centuries, the European arrival in sub-Saharan Africa established economic and political foundations that would ultimately lead to the colonial expansion and imposition of an industrialised, capitalist society. These trade connections and the rapid expansion of exploitation would have impacted on wider societies and

R. Marchant, *East Africa's Human Environment interactions*,
https://doi.org/10.1007/978-3-030-88987-6_4

ecosystems in the region, particularly along the coasts where coastal and mangrove forests would have continued to bear the brunt of the impacts. However, these impacts would have been throughout the region with legacies that continue through to the current day.

4.2 The Arterial Caravan Routes—Founding the Transport Network

The East African ivory trade is an ancient one (Chapter 3) and is mentioned in the first accounts of geographers and travellers into East Africa. It may have been the search for ivory that brought the first ships around Cape Guardafui, and then southwards along the East African coast (Beachey, 1967). The Periplus (1–18) provides a detailed account of the routes to East Africa. Vessels were loaded up with trade goods such as clothing, metals for making into ornaments or utensils, weapons and, for the tribal chieftains, luxury garments and objects of gold and silver. As the traders proceeded along the eastern shore of the Red Sea to Ras Hafun, still picking up myrrh and incense, and then continued south along the eastern shore of Africa to Menouthias Island, which is either Pemba or Zanzibar, and finally, Rhapta, which is either Dar es Salaam or at the mouth of the Rufiji River (Casson, 1980). The bidirectional monsoon winds would only allow for a single return visit each year (the return journey could not take place before October) with multiple stops along the coast when the early northeast monsoon provided favourable winds for traversing the Gulf of Aden. Once on the coast there would be ensuing trade into and from the hinterland. Reference to the export of ivory from the East African coast continues throughout the Middle Ages. Al Masudi, writing in the early tenth century, details that elephants were extremely common in the land of Zinj and that it was from this country that large elephant tusks were obtained. Most of the ivory is transport to Oman whence it is sent to India and China as part of the ancient silk route. Marco Polo states, in reference to the East African coast, that they have elephants in plenty and drive a brisk trade in tusks (Beachey, 1967). During the Portuguese domination of the coast from the sixteenth to the eighteenth century, ivory continued to be a principal export, receiving more mentions in Portuguese records than the slave trade. Indeed, the main goods to be traded would have been ivory due to its global demand and ability to really connect East Africa to the rest of the world.

Demand for East African ivory from around the world resulted from Asian elephants, unlike East Africa elephants (Fig. 4.1) being 'poorly provided with ivory' alongside Asian ivory being 'hard and brittle', more difficult to polish, and discolouring to yellow with time. East African ivory is soft, ideal for carving and can be used in a wide number of commodities. Today, many are concerned about the demand for ivory driven by the Asian market but this pales into insignificance compared with this initial wave of ivory trade. Increasingly driven by Europe, the large ivory carving centres which had emerged in Southern Germany and in the Low Countries during the Middle Ages, supplied religious

Fig. 4.1 East African elephants. The numbers of elephants across the East African landscape is massively reduced compared to historical estimates. In addition to overall depopulation, the number of large elephants is relatively low as the large 'tusker' elephants would have been selectively killed—impacting on the gene pool and current population structure (All photographs: Rob Marchant)

reliquaries and carvings for Christian Europe (Beachey, 1967). These bespoke users were the forerunner for the more industrial demand that accompanied the transition through the Industrial Revolution of Europe and America and rapidly expanding use of ivory in goods such as cutlery, handles, musical instruments, and furniture inlays that saw the ivory demand expand exponentially. One good example of this expansion is the ivory-carving industry in Connecticut centred around settlements such as the towns of Deep River or Ivoryton that grew massively through the seventeenth, eighteenth, and nineteenth centuries. Indeed, whole towns were built on the ivory trade and ensuing processing, such as Ivoryton close to what is modern-day Philadelphia. Ivory trading became a major transformative issue in East Africa to demand by affluent capitalist societies of the west as it was fuelled by the rise of Victoriana and industrial revolution. Ivory was the foundation upon which some of the merchant houses in various parts of the world were built, as well as an important element in the development and profitable operation of various transport route to the East African coast. In addition to being a source of profit for these firms, ivory was a source of income for multiple people along the chain from ships crews to carvers, from agricultural supply to the caravans to skilled industrial workers (Shayt, 1992).

East African elephants have been hunted for their ivory for millennia, but the nineteenth century witnessed strongly escalating demand from Europe and North America. The direct consequence of this trade was that, by the eighteenth century, elephant herds along the coast had become so scarce (Fig. 4.4) and impacted, that to meet demand, trade caravans trekked farther into interior regions of East Africa, extending the extraction frontier (Coutu et al., 2016). Writing in the 1840s the missionary Krapf observed that, 'although the elephant was still found in some areas near the coast, ivory caravans were now making regular trips into Usagara, Masailand and the Kikuyu countries' where the elephants (and their tusks) were larger—Krapf was surprised to see an elephant tusk from Kikuyuland so large that it required three Akamba tribesmen to carry it. An increased demand for ivory in America and Europe coincided with the opening up of East Africa by Arab traders and European explorers, and this led to the intensive exploitation of the ivory resources of the interior. To access the elephants there were a series of 'caravans' set up—these would essentially be a group of porters to carry the ivory out to the coastal ports and beyond and reversely the trade goods from the coast to the interior of the continent. The nature of these is largely undocumented but it is likely there would have been a moving frontier of Elephant exploitation and an established series of arterial routes (Fig. 4.5). The travel was largely linear with increased production spreading out spatially as the frontier zone of hunting increased with exchange expanding out along these linear routes. This does not mean that the ivory trade had the same consequences everywhere, or that it always moved at an equal rate. Rather, the assumption is that ivory production necessarily moved through space, continually driven by the need to find more elephants to kill to supply the demands. These new sites of 'production' are highly differentiated in political terms as the territories varied from centralised states, where they were open to trade, versus some societies that were not accommodating to the elephant hunters.

The term 'caravan' suggests a small-scale bespoke set up, but these caravans were highly organised numbering up to several 1000 people strong (Lane, 2010; Rockel, 2006) with the need for dedicated supply chains and food. It is difficult to generalise about the size, or indeed other characteristics, of caravans. The early accounts on the central route of caravans suggest mass movements were common, though away from the 'main highways' (Fig. 4.2), it would be likely that comparatively small parties comprising 30 or 40 members would have been more common. In terms of the composition, the caravans consisted largely of professional porters hired at the coast and in Unyamwezi, or slaves hired out by their masters. Loads carried by porters were staggering in their weight, apart from their bulk and awkwardness in handling. Felkin and Wilson, writing in 1879, stated that the loads were usually 70 lb. Grant, in his *A Walk Across Africa*, notes porters in the Southern Sudan carrying 50–60 lb on their heads (Beachey, 1967). To place this in context, an average 2-m-long tusk would be a single load whereas the large tusks would have taken two people to carry. Within the long-distance caravans, carriers of ivory had a higher status than carriers of other trade goods (Cummings, 1973). Porters who could carry large tusks single-handedly (up to double

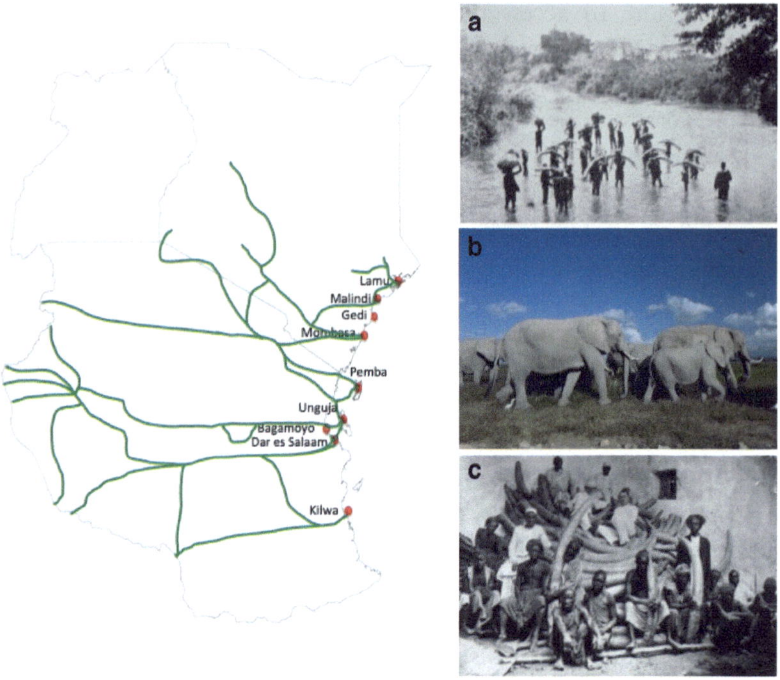

Fig. 4.2 Nineteenth century Caravan routes across East Africa redrawn from Coutu et al. (2016), Beachey (1967), Rockel (2006), Lane (2010), and Cummings (1973). The caravans varied in size some industrial ventures comprised some 2000 porters! Photograph (**a**) shows a small caravan crossing a river in the Congo with (**c**) a collection of ivory at Zanzibar ready for export (edited from the ED Moore Collection, Ivoryton Library Association and Treasure of Connecticut Libraries)

the standard load of 60 lbs.) were given special status and substantially larger food rations (Lamden, 1963).

As we will see the impact was much more than 'just' the decimation of Elephant populations. The increasing scale of ivory capture and export in the nineteenth century would have had significant consequences for humans, elephants, and the wider landscape; these impacts resonate with the landscape we see today. The ivory trade initiated a distinctive yet predictable chain of consequences; with changes brought by the trade, whether negative or positive, being irreversible. It is imperative to know where this ivory was being extracted so that these impacts can be placed in the context of current planning of elephant herds. For example, coastal populations would have been extirpolated early on, possibly several millennial ago (Prendergast et al., 2016). As the trade developed, some of the biggest and most valuable tusks came down from the area around Lake Tanganyika and into Uganda. There would have been four main arterial routes (Fig. 4.2) with numerous tributaries that are still lacking in detail, the location of these is often reconstructed based on isotopic data on ivory that can be used to uncover the provenance of historic ivory trade (Coutu et al., 2016). However, better geographic provenance of ivory could inform which habitats in Africa were most depleted of elephants by the ivory trade through times when there was a surge in extraction across the continent. The isotopic data sets do suggest that a range of habitats were exploited for elephants from the late nineteenth to the mid-twentieth centuries, but particularly interior regions of East Africa (Coutu et al., 2016). The passage of these routes seems to have started at the coastal towns of Kilwa, Bagamoyo, Tanga, and Mombasa (Fig. 4.2) then later going towards Tsavo, an area that was an important source of trade goods, including ivory, and persons bound for coastal and international slavery (Kusimba, 2004). The most northerly route, and those still preferred by some missionaries in the latter part of the century, followed the Pangani river from Tanga to Moshi and Arusha, then westward to the south-west corner of Lake Victoria. A southerly route ran from Kilwa on the coast to Lake Nyasa; a variation of this route was along the Rovuma River to Lake Nyasa. One of the most frequently travelled routes into the interior, and that chosen by Burton, Speke, and other European travellers, left the coast at Bagamoyo and, after crossing the dry, barren strip immediately behind the coast known as the 'Nyika', ran through the savannah and scrub bush country of what is now central Tanzania, the route of the present central railway from Dar es Salaam to Tabora. From Tabora it continued westward to Ujiji on Lake Tanganyika. The total travelling distance by this route from the coast to Ujiji, 'prolonged by the sinuosities of the road', was 955 miles; it took Burton and Speke some seven months to travel. Caravans seldom arrive to Lake Tanganyika in under six months, and even those lightly laden took four months. Two large inland markets for ivory were Unyanyembe (Tabora) in what is now central Tanzania, and Ujiji on the east coast of

Lake Tanganyika (Cameron, 1872). Indeed, commercial networks spread rapidly along these corridors between lakes Tanganyika and Nyasa and north to the interlacustrine states, notably Buganda.

Through the nineteenth century the number of caravans travelling into the interior grew until, by 1885, it was not unusual to have over 2000 porters in a single caravan. The ivory caravans developed a life of their own, and the supply of their needs led to a system somewhat similar to that of ship chandlering (Beachey, 1967). There is a record of one caravan in the 1880s carrying 27,000 yards of *merikani* made up in loads of 30–40 yards each (Beachey, 1967)—this would require in itself more than 700 porters to carry the load. It was not just cloth, beads, and wire—guns and powder were also important items of trade. But the Arabs took care that effective arms did not fall into African hands, and usually disposed only of flintlock muskets that were surplus from the Crimean War, and almost useless. European traders were less careful. Stokes, an ex-missionary turned ivory trader, sold Martini-Henry and Winchester repeaters to the ruler of Buganda at the rate of one firearm for two *frasilah* of ivory (about 75 lb.). A huge quantity of gunpowder passed into the interior, much of it American gunpowder and the first reference to American trade with the East African coast mentions this article. On the coast, gunpowder sold for $5–10 a keg whereas inland, in Unyamwezi, it was worth six times that amount. The British East Africa Company purchased ivory in Buganda at the rate of 35 lb. of ivory for two kegs of powder. The Berlin Act of 1885, the Brussels Act of 1890, and the Anglo-German agreements of 1886 and 1890, all attempted to deal with the arms and ammunition trade but with small success (Beachey, 1967). Other trade articles included scissors, looking glasses, picture books, jointed jumping dolls, rings, daggers, naval and cavalry sabres, and cooking pots. In Buganda there was a special taste for parasols, and Lugard found here a demand for 'white donkeys and opera glasses' (Beachey, 1967). Imported goods were clearly an integral part of the trade, this can be seen in evidence that archaeologists recovered a significant quantity of imported ceramics from Swahili households such as Shanga, Gedi, and Kilwa (Chirikure, 2017). Indian Ocean based traders placed a very high value on gold whereas many hinterland communities did not. Across the majority of Africa, copper and its alloys bronze and brass were more valuable than gold (Chirikure, 2017). Beads, shells, and copper dating to the fifteenth century have been found in the ruined irrigation community of Engaruka (Sutton, 1984). Glass beads, cowrie shells, and worked ivory from the thirteenth and fourteenth centuries were excavated in Bunyoro in Eastern Uganda (Robertshaw, 1999), indicating an early wide extension of the ivory trade.

This rise in demand for ivory was not just exported to Europe and North America, and this rejuvenated East African trade with India and opened the region to the world. Zanzibar was at the centre of the ivory market, supplying some 75% of world demand in 1891. There had been a substantial ivory export for many years from the dhow ports on the mainland, such as Malindi, Kipini, Lamu, and Kismayu. In the late 1890s, Mombasa and Dar es Salaam began to hold their own ivory auction sales. Mr. Marsden, Chief Customs Officer at Mombasa in 1901, endeavoured to make that port the centre of the ivory trade. He circulated printed catalogues, advertised periodical sales, and persuaded the government to provide free passages on government steamers so that both Zanzibar and coast merchants could attend. These efforts were more than successful in attracting buyers, chiefly Germans, and Italians, to the ivory sales at Mombasa. Mombasa's position as the centre of the ivory trade was maintained well into the twentieth century: in 1960–1961 the entire export of East African ivory (75,000 kgs) passed through this port.

Of great commercial significance to Zanzibar was the first entry of American traders, especially those from the small port of Salem. The first recorded American visit to East Africa was that of Captain Johnson in 1823, trading at Zanzibar and Mombasa for gum copal (resin) and ivory. American traders also began to penetrate the East African market from India under a treaty that regulated American commercial trade with Zanzibar. However, America did not wait until the treaty of 1835 was in place. As external demand for African commodities grew there was 'cut-throat competition' between foreign traders that was advantageous to Zanzibar merchants due to lowered price of imports and raised price of exports. Said's ambition was to create an economic empire built on trade between East Africa and Europe, Arabia, India, and America. The chief product was again Ivory and about £70,000 worth (equivalent to over £7 million today) was bought annually in the 1840s in exchange for American and British cottons, glass beads and brass wire. By the 1860s large and well-armed Arab expeditions were common. However, large fortunes from the ivory trade were not common—an investment of £750 in trade goods rarely brought in more than 70 *frasilah* (2450 lbs) of ivory, and assuming an average of £30 per *frasilah* at Zanzibar prices, this was worth £2000, giving a gross profit of £1500. But against this potential profit there was the cost of porterage and rations, at least £3 per *frasilah*; and the enormous interest on borrowed capital, sometimes as high as 100%, in addition, the possibility of loss made this a risky business (Beachey, 1967). The quest for ivory was never-ending as the price on the world market was remarkably free from fluctuations through the nineteenth century. Fresh supplies kept the price on the world market from rising, although in the interior the price might fluctuate in terms of the value of trade goods; for example, it rose from 10 lb of ivory for 1 lb of beads in 1848, to almost weight for weight in 1859 (Beachey, 1967). Americans complained of the custom masters 'hard grasping character' and the way he encouraged all foreign traders to visit Zanzibar. Americans from Salem and New York had offices in Zanzibar where they bought ivory, as well as hides, dried chillies, and gum copal. Like the ivory, the hides and gum copal were raw materials sought by industry in North America. Indian merchants also bought ivory and transported it to India where this was cut and carved and re-exported to Britain. As trade developed different nations started playing roles: the first German ship to call at Zanzibar was in 1844 and was followed by German firms setting up business, something that a Great British consul quickly emulated—laying for the foundations for the ensuing colonial power struggle (Chapter 5). On 16th of April 1854, Said ibn Sultan left Zanzibar for Muscat for the last time; after his death, his son Said Majid ibn Said became ruler of Zanzibar and, in 1859, a treaty of amity and commerce was concluded between him and the Hanseatic Republic of Lubec, Bremen, and Hamburg. In 1862 the attention of Said focused on Dar es Salaam as an outlet for trade with the interior and in 1866 he started to build on the banks of the harbour, however, he died before his grand plans had reached completion. By 1870 German trade with Zanzibar had outstripped that of both America and France and was second only to that of the British—the economy continued to thrive, and connections were growing between Zanzibar and Europe as well as establishing some wealthy and powerful merchants. The assumption that traders like Tippu Tip were merely the mechanism by which external demand for ivory was transmitted from the coast to the interior is an over-simplification; these were successfully transferring or translating goods from one kind of production system into another (Bennett, 1986). The Sultan and his regime depended on a variety of sources of income related to trade: direct production for trade as in the case of cloves, control of shipping, shares in the wages of slave porters in caravans, and a system of taxation designed to squeeze the maximum amount of surplus out of the trade passing through Zanzibar (Sheriff, 1987).

There would have also been export of agricultural products, particularly cloves from the islands of Zanzibar and Pemba, and grain and coconuts from the mainland; the coastal towns of Tanga, Pangani, and Mombasa all produced millet and sorghum for export to South Arabia and the Persian Gulf. In addition to direct export of mangrove poles there would have been a flourishing ship building/repair market with high demand on mangrove and coastal forest for timber (Fig. 4.3). The labourers working on the growing number of plantations producing agricultural export products were slaves, most of who came from the interior. In Zanzibar, the clove plantations that Seyyid Said had encouraged his followers to plant became more central to the economy and as expanded as trade grew around the world for spice (Gilbert, 2002).

Given the wide range of consequences of the ivory trade as articulated by Håkansson (2004), it is important that efforts are made to refine the geographical and temporal impact. By 1890, elephants had been virtually exterminated from a broad belt along the coast, were no longer common across much of what is now mainland Tanzania and were still very common only across a narrow band of the Kenya highlands from Amboseli northward to Lake Turkana (Fig. 4.4).

At Kamurasi's capital there was a great hoard of ivory and a 'present' of thirty tusks was placed at Baker's door. Baker himself, despite his regret at the destruction of the elephant, was associated with the ivory trade; he left Bunyoro with a tremendous load of ivory as conveyed in

Fig. 4.3 Coastal and mangrove forests were at the forefront of increased deforestation and exploitation, particularly for boat building and export. Large mangrove trees are virtually absent across East Africa although their large stumps attest to their durability as a boat-building timber (All photographs: Rob Marchant)

the following account 'the quantity of ivory in camp was so large that we required 700 porters to carry both tusks and provisions'. Ivory from Bunyoro had very little outlet until almost the end of the century as ivory traders from the north did not penetrate this far south, and Bunyoro's trade south and eastwards was strictly controlled by Buganda. Thus, cut off from the outside markets, ivory piled up in Bunyoro. The introduction of game regulations in Uganda from 1897 ensured the preservation of the great ivory resources in the Murchison Falls area, which today has perhaps the greatest concentration of elephant in the world. Lieutenant-Colonel Martyr and Major Delme Radcliffe, administrative officers in this area at the turn of the century, commented on the sheer numbers of animals where a single herd might consist of 700 elephants. Their migratory field extended from North-central Uganda across the Kafu River and north-westwards to the White Nile. There was a remarkable elephant road, apparently used during the seasonal migrations only, running from the east through Lira and on into the Acholi country. It was perfectly defined, smooth and hard, and seen by early colonial visitors (Chapter 5) as 'by far the best road in the whole country'.

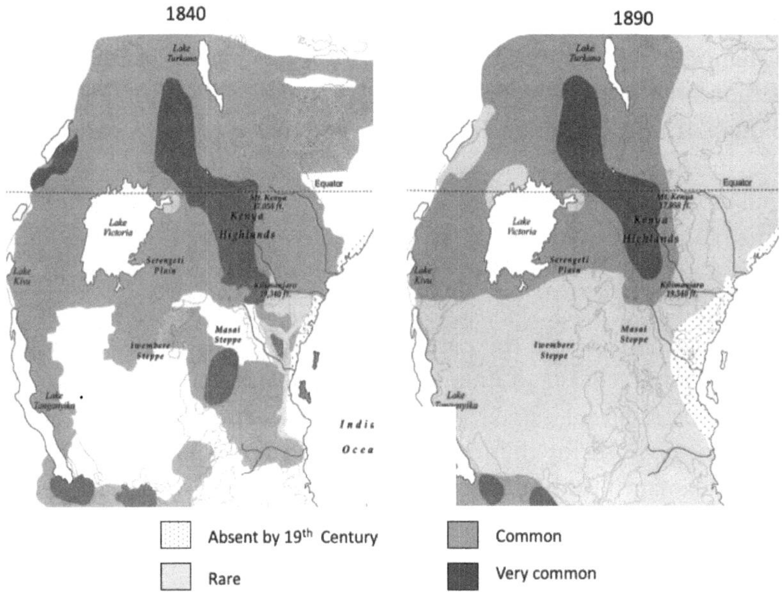

Fig. 4.4 Elephant population shifts occurred as a moving front—as populations became locally extinct from the coastal zone caravan trades developed to access the large herds inland. Elephant population range and change in this at 1840 and 1890 across East Africa. Redrawn from Alpers (1977), Coutu et al. (2016), Milner-Gulland and Beddington (1993), Thorbahn (1979) and Håkansson (2004)

What was the ultimate destination of the thousands of tusks of ivory shipped every year from East Africa? A vast quantity went to England, where the Victorian love of ornate furnishing and decor was expressed in ivory inlay work in myriad forms, ranging from ivory-handled umbrellas to ivory snuffboxes and chess pieces. There were also large imports for the great cutlery works of the Midlands; William Rodgers of Sheffield used up to 20 tons of ivory a year in making handles for cutlery. In the Latin countries' ivory was used in many articles such as delicate ivory fans, ivory diamond-shaped inlays for the fingerboards of Spanish guitars, the keys of Italian accordions, and finely carved boudoir articles. Many tusks went to ivory-carving centres in Europe: Dieppe was famous for its miniature ornaments, statuettes, crucifixes, mathematical instruments, little book-covers, paper-cutters, combs, serviette-rings, and the more general *articles de Paris*. St Claude in the Jura, Geislingen in Wurtemberg, and Erbach in Hesse, were ivory-carving centres of long tradition. Ivory was used for false teeth, making buttons and clasps, although rivalled more and more by vegetable ivory and casein from milk. In the United States a demand came with the rapid increase in population, and ivory was used for piano and organ keys, musical instruments, and billiard and bagatelle balls, not to mention the ivory inlaid butts of pistols for the American west. America was the market for 80% of the soft ivory exported from Zanzibar in 1894. But a significant market remained in the East. India and China supplied a multitude of toys, models, chess and draughtsmen, puppets, work-box fittings, the tremendous range of carved figures and ornaments, and the thousand artistic forms both for the home market and for export abroad. Additionally, damaged, or broken ivory was ground into powder and sold for love potions and medicines, further perpetuating Eastern demand for this material.

4.3 Impact of Ivory Trade on African Elephant Populations

One of the key transitions felt across East Africa has been the ivory trade that decimated the population of East Africa's elephants (Fig. 4.5). The caravan trades clearly had a massive impact on elephant populations and societies across East Africa, an impact that still resonates today across the wider landscape where the East African elephant population is currently around 200,000 individuals. As prices and demand rose so did a constant expansion of the hinterland resulting in a cyclical rapid rise in demand and growing impact on elephants. By 1891, 75% of the entire world's supply of ivory was shipped from Zanzibar, with estimates of East African exports ranging from 8000 to 30,000 tusks per year for the latter half of the nineteenth century (Coutu et al., 2016). The scale of the extraction was enormous, between 1840 and 1875 British demand alone rose from 200,000 kg to over 800,000 kg per annum that equates to between 4000 and 16,000 elephants per year being killed for this single

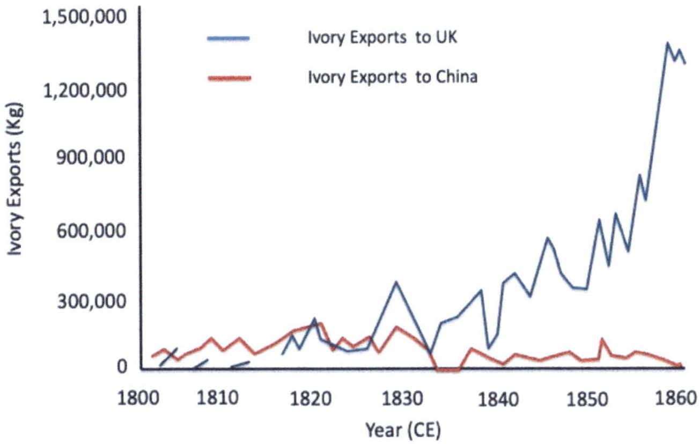

Fig. 4.5 Plot of historical ivory exports from Zanzibar from 1800 to 1860. This is just from one export point at Zanzibar and clearly shows the rise in ivory exports to the UK while those to China were stagnant and decreasing. Redrawn from Sheriff (1987)

market (Fig. 4.5). One of the main uses for ivory was in the construction of pianos—in terms of the numbers it is estimated that an 'average tusk' would be needed to cover the keys of 45 pianos. Just to provide a context how this would translate to elephants from 1900 to 1910, some 350,000 pianos were made in Ivoryton, Connecticut alone! Various figures have been put forth to document the number of elephants killed to supply ivory exports. With annual exports of about 175,000 kg and an average pair of tusks weighing 50 kg (this may be on the high side), would mean that at least some 3500 elephants had died to provide this ivory. Livingstone estimated that 44,000 elephants were killed annually to supply the ivory imported into England alone in 1870, much of this exported through Zanzibar. In 1894 the annual mortality of elephants for all Africa was estimated at 65,000 elephants (Gazette of Zanzibar and East Africa, 1894). Figures of ivory exports from East Africa during the early nineteenth century are not easy to obtain and as these relate to tax are best viewed as guesstimates. Various estimates range as low as 20,000 kg a year to as high as 100,000 kg but no indication is given as to how these figures were arrived at. Following the arrival of Colonel Rigby as British consul at Zanzibar in 1858, customs returns become available: in 1859 some 244,300 kg of ivory (worth £146,666) were exported annually—this equates to around 5000 elephants per year. Ivory exports remained around 200,000 kg, despite the price rise, and continued at this level almost until the end of the century, except for 1885, when they dropped to 130,000 kg in 1890–1891 before rising to 470,000 kg, the result of an accumulation of ivory in the interior during the blockade of the German East African coast in the previous year. By 1894 the figure of exports was back to 211,460 kg, and it remained around this mark until the late 1890s when game laws and a closer administrative control (Chapter 5) began to be felt. In 1899 the figure was down to just over 50,000 kg per year.

It was not just about the total number of tusks that can be used to reconstruct the impact on elephant populations, although this does provide some insight into the numbers of elephants. Ivory tusks ranged in weight, from the small tusks destined for the Indian market weighing no more than a few pounds, to the huge tusks of 100 kg or more. Female tusks, being softer and more malleable, were highly prized for billiard balls for the American market (Beachey, 1967). As large tusks were obviously more prized there was selectivity in the hunters in favour of larger tusks and tusk size distributions was found to reduce through the trade (Parker, 1989). Using analyses of stable (^{13}C, ^{15}N, and $^{18}O/^{16}O$) and radiogenic ($^{87}Sr/^{86}Sr$) isotopes from elephant hair and ivory, it is possible to determine more precisely the geographical locations from which ivory was extracted that broadly is a moving frontier of exploitation and local extirpolation (Fig. 4.4).

Today, we know that trade in wildlife products, such as ivory, can rapidly decimate species on a continental wide scale (Wasser et al., 2004) and that this has extensive impacts throughout the wider ecosystem, certainly this was the case for East Africa. It is not possible to estimate the original population, but the carrying capacity of East African savanna has been estimated to extend to some 10 million plus animals (Milner-Gulland & Beddington, 1993). To provide a little bit of context to these estimates of past populations, the present population of elephants across East Africa is c. 200,000 elephants (Hauenstein et al., 2019). Coastal herds that used to migrate freely have all but become extinct, with only a few small resident populations remaining within protected areas such as the Shimba Hills. The consequences of this trade and the effects on elephant populations and ecologies have been widely discussed although it remains largely unknown how this impact itself manifests on the present day. Thorbahn (1979) and Håkansson (2004), for example, charted how the increase in ivory trade would have transformed interior landscapes as elephant stocks on the coast dwindled and traders travelled farther inland. Although it is difficult to discern the scale and ecological impact of these extractions clearly elephants have massively reduced in numbers and range (Figs. 4.4 and 4.5), especially as they were pushed into less hospitable environments. Elephants play a crucially important ecological role in the transformation of wooded areas into grassland (Fig. 4.6), affecting a wide variety of species. Elephants function as 'ecosystem engineers' (Hayes & Schradin, 2017) and have profound effects on structure, composition, and distribution of ecosystems (Fig. 4.6) across wide swaths of East Africa; removing millions of elephants would have led to an expansion of woodland across East African savanna. In turn, this might have led to changes in agricultural practice as disease threats posed by the spread of tsetse fly intensified (Giblin, 1990; Lane, 2010) and the availability of grazing resources declined.

4.4 The Slave Trade and the Caravan Routes

Slavery and the slave trade are ancient practices that, like Elephant hunting, can be traced back more than two millennia in East Africa. For centuries, humans were part of the cargo in trade conducted between Africa and Eurasia, along with ivory, gold, and other commodities (Kusimba, 2004). Like the historical ivory trade, this would have far-reaching ramifications in the underdevelopment of Africa through depopulation and warfare and the destruction of indigenous African technologies and economies (Kusimba, 2004; Nunn, 2007; Shitta-Bey, 2014). Early Chinese sources indicate the exporting of enslaved East Africans: *Yu- yang-tsa-tsu* (ca AD 860) and *Chu-fan-chih* (ca AD 1266) specify that the main products of the East Coast were ivory but do also

Fig. 4.6 Elephants as ecosystem engineers and the rapid reduction in numbers would have profound effects on structure, composition, and distribution of ecosystems over a very short period of time. Demonstrated clearly by this animal exclosure from Amboseli National Park (Photograph: Rob Marchant)

mention the slave trade (Kusimba, 2004). Parallel to the trade in ivory, it was not until the expansion of European trade, and wider global trade in the seventeenth century that the Slave trade expanded as the industrial revolution impacted every corner of the word.

The slave trade grew in the late eighteenth century from Kilwa, Bagamoyo, and Zanzibar to provide labour in the plantations founded by the French on the islands of Mauritius and Réunion in the Indian Ocean. Plantation slavery fuelled most of the traffic in humans (Fig. 4.7) from the mid-nineteenth century, sending East Africans to Reunion and the Seychelles (Nwulia, 1975). Between 1820 and 1830, some 15,000 slaves per year were exported from East Africa, rising to 17,000 per year during

Fig. 4.7 Slavery and the slave trade can be traced back more than two millennia in East Africa. The slave trade grew rapidly in the late eighteenth century to provide labour in clove plantations, and growing trade in sisal (**a**, **b**). In addition to forced labour within the region there was also slave export from coastal towns such as Zanzibar Town (**c**) (Photograph: Rob Marchant)

the 1830s (Nwulia, 1975). During the same decade, 15–18 Brazilian slave ships would arrive into Mozambique every year (Alpers, 1975a). Although documented evidence is scanty, it is likely that the shipping of enslaved East African to the Americas also began in parallel with the much more extensive and widely documented West Atlantic slave trade (Alpers, 1975b). Though these numbers are clearly harrowing, perhaps more harrowing is the estimation that for every person that was sold into slavery from Zanzibar, three others would have died in transit (Lodhi, 1973).

The European demand for enslaved East Africans is symbolised by Monsieur Morice's treaty with the King of Kilwa in AD 1776, in which he promised to provide 1000 slaves annually (Freeman-Grenville, 1965).

Although slave population figures in the eighteenth century are extremely haphazard (Alpers, 1970) during the late 1770s French traders were taking away about 4500 slaves each year from the ports of Ibo (Mozambique), Kilwa, and Zanzibar (Alpers, 1970). During the Seven Years' War, France had lost the West African colonies of Senegal to the British resulting in the increased tempo of the French slave trade at Kilwa, particularly exporting slaves to the French West Indies (Alpers, 1970). French domination of the slave trade in East Africa reached its zenith between 1785 and 1794. During a brief period of peace following the American War for Independence up to the beginning of the Napoleonic Wars, many French merchantmen called at East African ports (Alpers, 1970) where along with the Omani Arabs, they played a particularly crucial role in the rise of the slave trade during the eighteenth century. This alliance was not just commercial but a conscious entry into the local politico-economic struggle along the East African coast between the various Swahili polities attempting to maintain their independence, and the expanding Omani hegemony. Although Morice was a monopolist who sought to 'exclude all European rivals' he did not want to alienate the Omanis. On Morice's death in 1781 the agreement lapsed and intense competition between French traders ensued in the first half of the 1780s that greatly increased prices. Trade continued to shift throughout late 1700s; and by the early 1800s Kilwa had become an outpost of trade through Zanzibar. By 1812, Kilwa Kisiwani was described as merely 'a pretty village'. The Napoleonic wars (between 1793 and 1810) were a 'catastrophe' for Omani and Swahili merchant classes in East Africa with the consequential reduction in slavery. At the beginning of the nineteenth century, the estimated volume of East African slave trade was between approximately 6000 people per year, reaching 13,000 by 1840, between 14,000 and 15,000 people per year in 1850s, and about 20,000 in 1860s coinciding with 'clove mania'. The spread of clove cultivation resulted in an increase of labour as new commodities needed to meet the demand for rapidly industrialising nations in Europe and America. Coconut production increased to meet vegetable oil demands in France and Germany and large quantities of sesame began to be produced along the northern coast of Kenya (more than half went to France and about 1/3 to Germany) much of this from plantations fuelled by slave labour.

It should be emphasised that, like the ivory trade, this impact was not restricted to coastal areas and would have likely affected the entire region in quite alarming and finite ways. Hence, after 1800 most of the East African interior was 'opened up' by the expansion of slave and ivory traders. Since human porterage provided the only transport an increasing traffic in slaves required larger numbers of retainers to manage the caravan, met by paid carriers and soldiers. As both slaves and hired porters adopted elements of the Zanzibari and coastal culture, a new class of waungwana (lit., freemen) emerged among whom distinctions of slave and free became blurred. John Hanning Speke observed many 'waungwana' and noted Arabs, on circumcising them, teach them to repeat the words Allah and Mohammed. Throughout c. 19th Zanzibar was a beneficiary of extremely favourable barter terms of trade, which can be attributed to industrialisation in Europe and North America.

As early as 1822, the British signed the first of a series of treaties with Sultan Said to curb the Slave trade, but not until 1876 was the sale of slaves finally prohibited and Barghash signed a treaty which prohibited the export of slaves from the continent, with the slave market in Zanzibar closing in 1873, largely due to British pressures. The slave trade was officially abolished in 1890 following the Brussels anti-slavery conference in 1889–1890, however slavery itself remained legal in Zanzibaruntil 1897. Although this earmarked distinct progress and a fundamental transformation of the slave sector, due to restrictions placed on export of slaves, slavery continued in several forms. The various plantations absorbed more and more of the slaves, many agricultural industries went through boom and bust but generally expanded (Gilbert, 2002). Like the grain plantations, clove plantations needed lots of labour and even after the abolition of slavery illegal slave trade continued to feed plantations with labour (Gilbert, 2002). Once the British stamped out this illegal slave trade they were left in an awkward position, as the British rule was dependent on revenue from the clove industry. Thus, in their zeal to end the slave trade, the Zanzibar clove industry had 'recruited' slaves a few years before (Gilbert, 2002). Similarly, there was a large recruitment following the 15th of April 1872 when a hurricane devastated the island and essential rebuild called for an increased demand for labour and slaves. Hence, the region was transformed from one based on slave export to one based on production by slave labour for export of commodities such as spices, coconuts, and copal (Hopper, 2007). Many dhow crews of the Western Indian Ocean during the nineteenth century were slaves, as were many pearl divers who were estimated to number 27,000–30,000 that continued well into the twentieth century (Hopper, 2007).

4.5 The Arrival of Maize, Potatoes, and Tobacco

One of the largest transformations at this time, although not exploitative like the ivory and slavery trade, was additive and transformed East African use of land and agriculture, but was the arrival of Maize. In 1498 CE, the Portuguese explorer Vasco da Gama's small flotilla sailed up the East African coast before crossing to India (Russell-Wood, 1998). This process also ushered in a period of coastal domination led by the Portuguese. On Zanzibar, trading posts (*feitorias*) that were simultaneously warehouses, markets, and customs offices, were established during the sixteenth century at Mvuleni, Fukuchani, and at the Gereza in Stone Town, where a cruciform church was built. Another church survives almost intact in Malindi, which originated from the Portuguese who also established an unfortified *feitoria* here and erected a stone cross to commemorate their presence (Finneran, 2002). Their main settlement, however, was on Mombasa Island, which became their centre of political authority and commercial operations after 1590 AD (Kirkman, 1974; McConkey & McErlean, 2007). The arrival of Maize is a clear 'Anthropocene marker' for East Africa (Odada et al., 2020) and can be traced to its arrival on Zanzibar in 1608—and one that spread across the region to transform agricultural practice and associated forest clearances (Finch et al., 2017). Maize was locally widespread by 1643 on Zanzibar and Pemba where it was grown, among another crops, to supply the Portuguese garrison (Miracle, 1965).

Arab and early modern European sources indicate that surplus food production for regional exchange was already established prior to the arrival of the Portuguese. Perhaps the most significant consequence of the establishment of a Portuguese presence on the East African coast, was the introduction of North American domesticates, particularly maize and tobacco, but also other crops such as potatoes (*Solanum tuberosum*) and sweet potatoes (*Ipomoea batatas*), although the timing, rate, and directions of their spread in East Africa remains under-researched. The speed of spread of these new crops was quite amazing expedited by the caravan trade routes (Fig. 4.2); for example in 1715 six coffee plants were introduced to Bourbon from Moka and in 1718 the plants reached harvestable size. By 1723 the Compagnie, which had again been reorganised in 1717, established a monopoly on the coffee trade, and the growth of rapidly expanding plantations ensued (Alpers, 1970). The establishment of small Portuguese enclaves encouraged food production in coastal areas to supply growing urban communities, trading ships and the caravans. The caravans, composed mostly of African porters, required a food that was calorific, storable, and transportable: sorghums, maize, and rice satisfy all these requirements. There is an account from Andrew Dick, ex-chief accountant of The Imperial British Company, who by 1895 had a chain of 12 stores established from the coast to the Lake Tanganyika, principally to facilitate and supply transport (Miracle, 1965). By the end of the nineteenth century Maize was present through the interior of East Africa, with the major exception of western Uganda, where Matoke (Plantains) were the main source of carbohydrate. Thus, maize spread rapidly across East Africa, facilitating the energy-demanding Caravan trades, and resulting in a large wave of associated forest clearance (Finch et al., 2017). Sissons (1984) estimates that 80,000 persons traversed Ugogo each year between the 1850s and the 1880s. The caravans obtained food in exchange for cloth, beads, iron hoes, and other goods. In the second half of the nineteenth century approximately 1,090,909 kg of grain was produced per year as surplus and sold to caravans in exchange for approximately 80,000 *doti* of cloth (Sissons, 1984). There would have been highly varied population changes; some areas would have experienced growth throughout the region as new trade opened opportunities for the absorption of caravan porters and slaves from the caravans, conversely other areas would have been decimated by slavery. For example, Sissons (1984) claims that the population of Ugogo, one of the key trading centres in 1860 was about 200,000 and closer to 360,000 in 1890. Conversely, Near Lake Nyasa (Malawi), Livingstone upon meeting a

Waiyau chief who supplied Arab caravans with slaves recorded a scene of decimation: 'They almost depopulated the broad fertile tract, of some three or four miles between the mountain range and the Lake, along which our course lay - It was wearisome to see the skulls and bones scattered about everywhere' (Livingstone, 1880).

The combined consequences of an increased demand for supplies of elephant ivory and slaves (Beachey, 1967; Sheriff, 1987) clearly created a profound shift in interior landscape food production systems (Pradines, 2004). A significant change came with the development of plantation agriculture, notably for cloves in the Zanzibar archipelago (Croucher, 2014) and sugar on the mainland, with associated transformations in the nature and extent of slavery and slave raiding (Kusimba, 2004, 2014; Lane, 2013) from the middle of the nineteenth century onwards. There are few historical examples where it is possible to directly assess the impact of the ivory trade on agriculture and the environment. However, research is beginning to show that human-induced erosion in parts of Africa extends back centuries before the beginning of colonialism (Håkansson, 2004). The voluminous trade was accompanied by an expansion of cultivation onto more marginal soils without soil conservation methods such as terracing (Håkansson, 2004). Ugogo was covered with open grassland interspersed with woodland and shrubs; by the end of the nineteenth century wooded vegetation was heavily cut around settlements and caravan routes, leaving soils bare and exposed to erosion through wind and water creating denuded flats (Christiansson, 1981). Large areas may have been denuded of trees by that time (Ambler, 1988) and descriptions from around 1900 are of landscapes with high grasses on level ground and grass mixed with trees and scrub on the hills. Many hills had been cleared for farming and isolated of gullies and bare ground were also reported as supported by the photographs taken by the ethnographer Lindblom (1914) in 1910–1912, which document both bare hills and mixed grass and woodland mosaics (Fig. 4.9).

Thus, the caravan trade created new livelihood regimes that transformed landscapes like the Lake Baringo catchment that was turned over to crop production to supply the caravan trade by the Il Chamus group who has previously practiced a pastoralist economy (Anderson, 2002;

Petek & Lane, 2017). Historical records also document an intensification of agricultural extraction around the newly created centres built along Tanzanian caravan routes, such as Tabora or Ujiji, based largely on a slave economy (Rockel, 2006). Caravan halts in the Pangani Valley have revealed evidence for continuity in local subsistence practices, based on a mixed economy of cattle, hunted fauna, and fishing, despite a growing engagement with coastal trade (Biginagwa, 2012; Wynne-Jones, 2010). The local Zigua communities farmed maize supplying caravans and thousands of people arriving to well documented overnight halts (Biginagwa, 2009), much like a modern-day service station on a major road. Opportunities arose for communities lying along the main caravan routes to change their food production systems to produce a surplus to feed passing caravans, and through these exchanges obtain access to some of the imported trade goods (most commonly, cloth, brass wire, and glass beads). Håkansson (1995) suggested this change strategy was extended to some of the upland areas, such as the South Pare Mountains, by intensifying agricultural production through the introduction of terracing, irrigation, and manuring systems; this is a process that could have resulted in the largely impacted landscape that was recorded by early cartographers and photographs of this area (Fig. 4.9). This impact was not restricted to highland areas that surround the main arterial routes; on the plains it is likely that Maasai herders changed their herd management system to the kind of specialised and exclusive pastoralism (Håkansson, 2004) that could supply the emerging markets of the caravan trade.

While these changes and the access to trade goods they provided may have brought economic prosperity to communities along the main caravan routes, it has long been recognised that there were less beneficial consequences as well (Lane, 2010). These included increased levels of violence and inter-community raiding, partly triggered by the rise in demand for slaves as plantation economies developed. Furthermore, these new crops from the America's and the expansion and imposition of emerging market-based economies were not the only imports. Although even less is known regarding their impact, it is likely there was an introduction of New World diseases, especially syphilis and polio, among local populations; although their effects were far less than the well-documented devastation caused by the introduction of diseases into the Americas, it is likely there would have been locally significant impacts. There are few historical examples where it is possible to directly assess the impact of the ivory trade on agriculture and the environment. However, clearly change land use and intensification combined with novel crops and cropping strategies led to human-induced erosion and most likely ensuing systems of water management that extends back centuries (Håkansson, 2004) and clearly resonate with contemporary issues of agricultural production and contemporary challenges (Chapter 7).

4.6 Social, Land, and Legacies of Ivory and Slave Trade

The expansion of the ivory trade would have triggered significant socio-ecological and political change. Communities along the trade routes and in ivory extraction areas diversified their economic strategies and labour relations to take advantage of the trade opportunities. This resulted in the emergence of specialist hunters, porters, and middlemen, and the founding of new settlements such as Ujiji on Lake Tanganyika, several of which became prosperous trading hubs (Coutu et al., 2016). These transformations were certainly not always positive; slave and ivory business devastated some regions but brought prosperity to others. Markets for cattle and crops expanded and access to trading routes brought great power imbalances between regions and people. There was clear and deep social change associated with the whole of the caravan trade across Eastern Africa that can be grouped in three main areas: the entrance of intrusive cultural elements into a particular region; the disruption, whether slave traders or not, of the indigenous political and social systems; the establishment of relatively permanent settlement. Given these intrusions and disruptions, a new stratified social order begins to develop further fragmenting African society and leading towards the emergence of distinct cultural groups.

The influence of the ivory trade on human ecology in Eastern Africa derived both from the direct effects of hunting, and the indirect effects of trade and exchange (Håkansson, 2004). No doubt these historical interactions had played a fundamental role in shaping the landscape and people's relationship with ecosystems and adjacent populations. East Africa was integrated through a common value system based on cattle that were exchangeable for other products as well as acting as the most valued prestige goods in social transactions. Many agricultural communities on the coast converted their gains from the coastal trade into cattle. During the first half of the nineteenth century, the Giriama north of Mombasa and the Digo to the south were trading ivory and grain to the coastal towns. While little is known of the ecological effects of this expansion of cattle herds among the Digo, the Giriama were practically an agropastoral group in the middle of the century. The social reproduction of families and kinship was dependent on a regional system of exchange in which cattle were ultimately used to build political power and to secure the growth and wealth of kin groups (Håkansson, 1994, 2004). From the beginning of the nineteenth century these goods were traded with the interior towards the Mount Kenya region (Emery, 1833) and in Northern Tanzania, where the Maasai obtained cloth and beads from the Kamba and later Swahili caravans in exchange for tusks and cattle (Krapf, 1968). It is likely that populations practicing this economic and cultural specialisation grew (Galaty, 1993; Lamphear, 1986; Summer & Vossen, 1993) (Fig. 4.9).

According to oral traditions, the Kamba ancestors first settled in the highlands to the east of Mount Kenya, in what is now known as Machakos, where they developed irrigated agriculture in the 1700s (Lindblom, 1920). The Kamba became the dominant ivory traders in the central part of East Africa from the end of the 1700s, expanding their trading networks from Lake Turkana in the north to Kilimanjaro in the south (Lamphear, 1970). From the mid-nineteenth century the Maasai pastoralists lost control on the central Kenyan plains while the Kamba became stronger, as did the neighbouring Kikuyu agriculturalists located to the north of Kamba range around the slopes of Mt. Kenya. The Kamba became increasingly engaged in the trade of products supporting the caravan trade, such as arrows, arrow poison, hides, snuff bottles, digging tools, and iron (Jackson, 1976). Along the immediate coastal hinterland and in Kambaland, cloth was used to barter for several goods and services such as livestock, palm wine, slaves, and porters, construction workers and poison (Cummings, 1984). During the height of the ivory trade, many groups became professional poison makers as well as elephant hunters. Some hunters abandoned subsistence agriculture in favour of a trading occupation in the numerous towns that had emerged on the coast and along the caravan routes (Kusimba & Kusimba, 2005) where new relationships were forged. For example, the Kamba were excellent in maintaining links with the surrounding ethnic groups and were ideal brokers for passage and access to resources for the caravan. Specifically, the Kamba of central Kenya Tsavo had ties with their WaTaita neighbours in the Taita Hills based upon blood brotherhoods and strong inherited kinships that enabled the exchange of ideas and knowledge, eased tensions arising from competition for resources, and allowed the exchange of technical and sacred knowledge. Such alliances and clientship individuals were welcome in several communities as technicians or brokers involved in the ivory trade (David et al., 1988). The ivory trade was lucrative and the Maasai also shared in it as they drove the Waboni from the southern bank of the Sabaki River, so that they could gain access to the port of Malindi. From the beginning of the nineteenth century, goods were traded with the interior towards the Mount Kenya region (Emery, 1833) and in Northern Tanzania, where the Maasai obtained cloth and beads from the Kamba in exchange for tusks (Krapf, 1968).

Slave raiding would have clearly caused widespread insecurity, fear, famine, disease, and the collapse of farming and pastoral societies, chiefdoms, and states in East Africa (Kusimba, 1999). Although there are no direct historical data or accounts available regarding the direct impact, the cost of these highly exploitative trades would have been significant. David Livingstone stated that up to five people died for every tusk that left East Africa. As the number of tusks to leave would have been in the millions this clearly indicates that a massive impact on Elephant overhunting also accompanied a transition of open savanna to woodland and forest, and a return of the tsetse fly, a vector of sleeping sickness among people and of trypanosomiasis in cattle and wildlife. Declining ecosystem quality was one factor that forced people to retreat to the hills and other tsetse fly-free habitats (Kusimba & Kusimba, 2005). Economic activities connected with hunting, transport, and trading affected regional systems of exchange and thereby, indirectly through the political economy, settlements, patterns of resource utilisation, population parameters, and specialisation of production (Håkansson, 2004). Considering current interest in sustainable development, the impacts of long-distance trade, and the present-day ramifications of these clearly transformative activities, are needed. Particularly, a deeper look at the history of trade in East Africa can be used to re-examine relationships between different ethnic groups and their relationship with land, trade, wildlife, and ecosystems.

4.7 Ecological Impacts and Legacies of Caravan Trade

Over the last 300 years, the ecology of East Africa has changed substantially although these changes were not from a state of zero impact and there was already significant transformation through the arrival of people, agriculture, crops, and technologies as previously mentioned (Chapter 3). With the rapidly expanding population and energy-demanding activities across the region there was clearly a rapid transformation in the agricultural systems as a wider proportion of the landscape was turned over to crop production (Fig. 4.10). It was not just the more extensive nature of crop production, but also the transition to new crops that had arrived from the America's and along with these new forms of land management. Most likely these new forms of land management built on the previous knowledge systems such as the applications of terracing and water management infrastructure. As today, in some places this would work, in others there could have been rapid landscape-scale degradation (Fig. 4.8). Ultimately this expansion of agriculture would be to supply the caravan trade and maximise the market-based opportunities, particularly for those communities living along the main caravan routes (Fig. 4.2).

One of the best-documented changes has been the decline of the African elephant (Gulland & Beddington, 1993). There were multiple impacts on the ecosystems ranging from the direct removal of elephants through to changed relationships and use of ecosystems by people. The

Fig. 4.8 Transformed open landscapes of the 1890s in Chome, Pare Mountains Tanzania that was visited by German missionaries. There is clear increase in tree cover with both picture storied documenting a clear increase in tree cover over the past century. The previously highly transformed landscape is likely associated with supply of caravan trades traversing the Pangani Basin. All photographs from Paul Lane and Pauline van Hellermann

African elephant is a major influence on ecosystem species composition and distribution, positively for some, negatively for others. Drawing on a wealth of historical and anthropological data, and savannah and elephant ecology, the expansion of the ivory trade allows us to examine the environmental repercussions of the trade. Removing elephants on such a large scale, (in their millions) meant that local habitats were significantly, and in some places irrevocably, transformed. Elephants are a keystone species and major 'ecological architects' or 'ecosystem engineers' (Fig. 4.6) that can control what grows where. In very simple terms, when present in large numbers, elephants sustain open bushland and mixed scrub savannah habitats; once the elephants are removed habitats

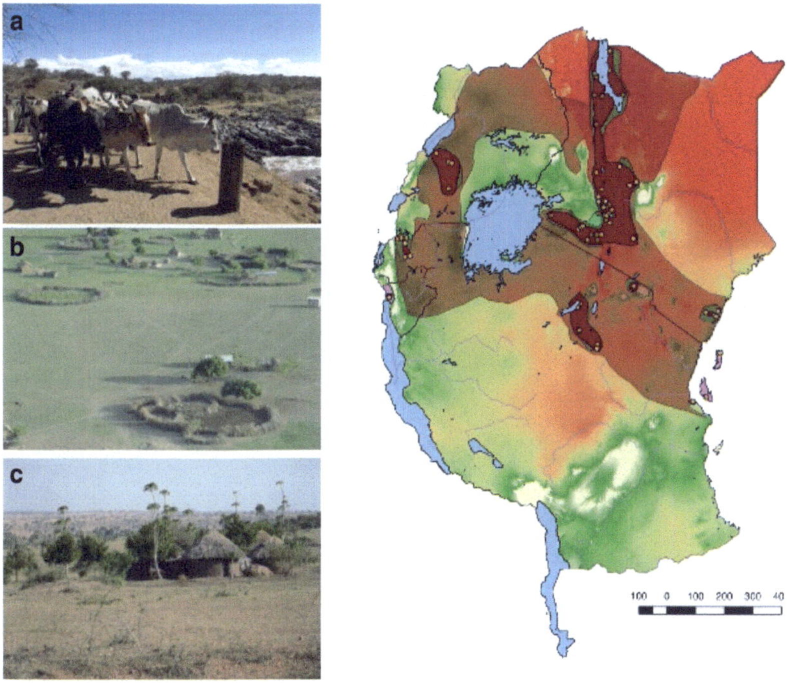

Fig. 4.9 Spread of pastoralism at 1850 CE in East Africa. This was particularly focused on cattle with pastoral groups, such as the Maasai, well established and becoming increasing sedentary. There would likely have been a diversification of livelihoods both in trading and providing services to the caravans. Transhumance would continue to be important to buffer seasonal variation in grazing resource via access to highland areas. Map produced by Oliver Boles and modified by the author (All photographs: Rob Marchant)

become more wooded. Indeed, except for humans, no other mammals have such profound an impact on their environment as elephants. In many areas in Eastern Africa elephants maintain open grasslands with few trees and where they have disappeared, such as in the recent poaching crises (Chapter 6), there has been rapid recovery of woody vegetation (Brockington, 2002). For example, Queen Elizabeth Park in Uganda had a population of 4000 elephants prior to the 1970s when poaching reduced them to a few hundred. Open savanna grassland was rapidly transformed into dense woodland and thickets, which effected

Fig. 4.10 Extent of agriculture at 1850 CE in East Africa. This expanded and diversified to take advantage of new market-based crops such as avocado (**b**) and coffee (**c**) that would be supplied to the caravan trade. Map produced by Oliver Boles and modified by the author (All photographs: Rob Marchant)

the species composition of other animals (Eltringham, 1992). Thus, elephants were partly responsible for maintaining the open grasslands that favoured pastoralist activities by providing grazing and an environment free of sleeping sickness as the tsetse fly (the disease vector) cannot survive without dense woodland and thickets. To provide an indication of the impact and role of elephant; they consume between 100–300 kg of plant food per individual per day (Prins & van der Jeurg, 1993) and, in addition, trample plants, uproot trees, destroy bushes, compact soil, and through de-barking (Fig. 4.6) can destroy trees in large numbers.

Interest in the interactions between fire and herbivory in structuring savanna ecosystems has increased in recent decades. Changes in the mammalian herbivores will have profound effects on ecosystems, and such a rapid change may lead to alternative ecosystem stable states that are very different from those that once existed. Ecosystems that have lost their large mammals are likely to represent a new baseline, with no historic (or prehistoric) analogue (Goheen et al., 2018). A series of relatively large scale, long-term experiments with the dual purpose of revealing how large mammals impact community structure and ecosystem function of rangelands, and of providing insights relevant to conservation of these increasingly impacted on ecosystems (Fig. 4.6). One thing to consider when assessing the impact that the removal of elephant may have had is that although the type of impact is different, a rise in domestic herbivore numbers similarly maintains community structure (Goheen et al., 2018; Veblen et al., 2016). The heterogeneous distribution of large herbivores across savanna landscapes, as well as differences among those species in forage preferences and diet composition (Kartzinel et al., 2015), influences the relative density of woody plant cover (Daskin et al., 2016; Pringle et al., 2007) and spatial patterns in the primary productivity, community composition, and traits of trees. Elephants are particularly potent architects, owing to their ability to topple trees and splinter large branches, which shapes understory plant communities (Coverdale et al., 2016) and creates habitat for small animals (Pringle, 2008). Understanding this response provides an important tool for land managers as changes to community composition correlate with the number of herbivorous changes. Insights from these long-term perspectives are crucial not only to inspire future generations of ecologists and conservation biologists working in these grand ecosystems (Goheen et al., 2018), but to inform and create management interventions that work for the diverse range of social, ecological and economic demands (Chapter 7).

4.8 Social Impacts Late Nineteenth Century Drought

Although not directly a cause of caravan trade, one of the biggest impacts that closed this era was a combined series of intense droughts from the

1880s and the great Rinderpest epidemic of the early 1890s. Data from Lakes Victoria, Chew Bahir, and Naivasha show a sharp fall in lake level in the late 1890s (Nicholson, 1996) as drought impacted across the region. Although droughts and cyclical impacts on livestock and agricultural production (Chapter 3) characterise East Africa, it may have been that the shift towards a more market-based economy, the growing population, and transformation of the land all combined to make the impact that much more severe and underpinned the transition from a 'climate oscillation' to a 'drought'. The Rinderpest epidemic is thought to have killed up to 95% of all cattle in many areas of East Africa (Kjekshus, 1977) and must have impelled pastoralists to engage in the ivory trade more intensively. Epidemics of rinderpest and anthrax would also have decimated populations of African buffalo (*Syncerus caffer*) and impala (*Aepyceros melampus*) (Kiffner et al., 2017). The impact of this catastrophe on the ivory trade in its waning years is unknown but scattered references indicate that selling ivory was one of several strategies employed by pastoralists to rebuild their herds (Barber, 1968; Baumann, 1891). Somali caravans obtained ivory through shorter trade networks in Northern Kenya. The Samburu, the Borana, and the Gabbra were especially active traders when their herds had been reduced by disease or drought. Ivory became a means to generate currency to purchase livestock and re-stock. When Karamojong pastoralists of Northern Uganda lost their stock through the pleuropneumonia and rinderpest pandemics of the 1880s, they recovered their losses by collecting ivory on a massive scale and converting it into cattle, as did the Maasai (Parker & Graham, 1989).

The mass migration and relocation that ensued with the caravan trades created subsistence insecurities and made people vulnerable to famine and disease. The Mwakisenge famine that had occurred in Taita in the 1880s reported by Hobley (1895) is a case in point. The impacts were very widespread and there are reports of Taita, Taveta, Chagga, Pare, and Ukambani starving to death in houses, on roadsides, in gardens and being left unburied for no one had strength to dig graves; the number of bodies was too numerous to be disposed by hyena or other scavengers. The Sagala area in Tsavo was one of the hardest hit areas as people killed one another in competition for food and many

Sagala emigrated to Giriama for relief. Abandoned settlements reverted to wilderness and a depopulated landscape ensued. At the end of the famine, after the rains returned, only 1000 of the estimated 10,000 Taita people survived (Kusimba, 2004; Merritt, 1975). Societal disruptions caused by the slave trade, cattle raiding, and persistent droughts weakened pre-existing regional networks of interaction, exchange, and crisis management. Insecurity confined people within ethnic boundaries further constricting spheres of interaction. The Taita people's response to this crisis was to abandon village settlements in the lowland plains for the Taita Hills, where they remained isolated well into the early twentieth century. Previous droughts had been weathered through the barter of various goods for crops with the highland populations, but the shift in trade routes left the Kitui population in a very dry and drought prone area without means to obtain food from the highlands. From an adaptationist perspective, one could argue that these exchange systems existed to simply maintain subsistence in a highly variable environment (Håkansson, 2004), as the environment got harsher or was open to shocks such as the epidemic then they became prone to significant impact.

4.9 Shifting World Views; Establishing Rather than Hunting Controls and National Park Foundations

With the abolition of slavery towards the end of the nineteenth century (Sect. 4.3) and the expansion of colonial administration in the 1890s, there quickly followed the introduction of game regulations and the demarcation of game reserves—the 'great days' of the ivory trade were over. The British Foreign Office suggested the need for game regulations as early as 1891, but the British East Africa Company never tackled the question. In the German sphere of control, in what is currently Tanzania, game regulations were introduced in early 1896; reserves were set up, and game licences made compulsory for elephant hunting. The killing of cow elephants was prohibited, and all tusks under 7 kg weight were subject to confiscation. In 1898 exclusive rights to elephant hunting in German territories were entrusted to elephant hunters who were required to deliver one tusk of every pair to the district station as payment for their trophy. These privileged hunters were induced to report unlawful killing, and they also learned to hunt in a 'sportsmanlike' manner (Gazette for Zanzibar and East Africa, 1892). Game regulations were promulgated by the British administration in Uganda and the East Africa Protectorate by 1897; these approximated to the German regulations but were slightly more favourable to wealthy sportsmen who brought in money and influence. The shooting of cow elephants was prohibited, and all ivory below 5 kg weight (raised to 15 kg in 1905) was liable to confiscation. Demarcation of reserves followed that were the foundation for the ensuing National Park system (Chapter 5). In North-Western Uganda, where there were still large populations of elephants, demarcation was delayed because it was thought that this area was too remote to attract hunters.

The first international conference for the protection of game met in London in January 1900, largely owing to the initiative of Hermann von Wissmann, former governor of German East Africa. As a result of this conference, a convention was signed in London in May 1900 for the preservation of wild animals in Africa. In Uganda, game regulations were tightened up following this convention; a time limit was laid down for the disposal of hordes of immature ivory, and severe penalties imposed for the violation of the game laws. Game regulations and the consequent setting up of game reserves led to ivory smuggling. Infringements of the game laws and illicit trade in ivory continued to come before the courts throughout the early twentieth century. As common today, corruption was rife and British officials were equally guilty of breaking the laws. Hand in hand with an illegal arms trade there was a substantial illegal ivory trade with destinations to Europe or the far East, a chilling resonance with the poaching crises of the twentieth century (Chapter 6).

References

Alpers, E. A. (1970). The French Slave Trade in East Africa (1721–1810). *Cahiers d'études Africaines, 10*(Cahier 37), 80–124.

Alpers, E. A. (1975a). *Ivory and Slaves: Changing Patterns of International Trade in East Central Africa to the Later Nineteenth Century*. University of California Press.

Alpers, E. A. (1975b). *Ivory and Slaves in East and Central Africa*. Heinemann.

Alpers, E. A. (1977). Madagascar and Mozambique in the nineteenth century: The era of the Sakalava raids (1800–1820). *Omaly sy anio, 5*(6), 36–53.

Alpers, E. A. (2020). *Ivory and Slaves in East Central Africa*. University of California Press.

Ambler, C. H. (1988). *Kenyan Communities in the Age of Imperialism*. Yale University Press.

Anderson, D. M. (2002). *Eroding the Commons: The Politics of Ecology in Baringo, Kenya 1890–1963*. Ohio University Press.

Barber, J. (1968). *The Imperial Frontier*. East African Publishing.

Baumann, O. (1891). *Usambara und seine Nachbargebiete*. D. Reimer.

Beachey, R. W. (1967). The East African Ivory Trade in the Nineteenth Century. *Journal of African History, 8*, 269–290.

Bennett, N. R. (1986). *Arab Versus European: Diplomacy and War in Nineteenth Century East Central Africa*. Africana Publishing Co, Holmes and Meier.

Biginagwa, T. J. (2009). Excavation of 19th Century Caravan Trade Halts in North-Eastern Tanzania: A Preliminary Report. *Nyame Akuma, 72*, 52–60.

Biginagwa, T. J. (2012). *Historical Archaeology of the 19th Century Caravan Trade in North-Eastern Tanzania: A Zooarchaeological Perspective*. Unpublished Ph.D. thesis, University of York.

Brockington, D. (2002). *Fortress Conservation*. Indiana University Press.

Cameron, V. L. (1872, January). Livingstone East Coast Expedition. In *Proceedings of the Royal Geographical Society of London* (Vol. 17, No. 5, pp. 334–337). Royal Geographical Society (with the Institute of British Geographers). Wiley.

Casson, L. (1980). Rome's Trade with the East: The Sea Voyage to Africa and India. *Transactions of the American Philological Association, 110*, 21–36.

Chirikure, S. (2017). Documenting Precolonial Trade in Africa. In *Oxford Research Encyclopedia, African History*. Oxford University Press, 2020.

Christiansson, C. (1981). *Soil Erosion and Sedimentation in Semi-Arid Tanzania*. Ph.D. dissertation, Scandinavian Inst. of African Studies, Uppsala.

Coutu, A. N., Lee-Thorp, J., Collins, M. J., & Lane, P. J. (2016). Mapping the Elephants of the 19th Century East African Ivory Trade with a Multi-Isotope Approach. *PloS One, 11*(10), e0163606.

Coverdale, T. C., Kartzinel, T. R., Grabowski, K. L., et al. (2016). Elephants in the Understory: Opposing Direct and Indirect Effects of Consumption and Ecosystem Engineering. *Ecology, 97,* 3219–3230.

Croucher, S. (2014). *Capitalism and Cloves: An Archaeology of Plantation Life on Nineteenth-Century Zanzibar*. Springer.

Cummings, R. (1973). A Note on the History of Caravan Porters in East Africa. *Kenya Historical Review, I*(2), 109–138.

Cummings, R. J. (1984). Wage Labor in Kenya in the Nineteenth Century. In C. Coquery-Vidrovitch & P. Lovejoy (Eds.), *The Workers of African Trade* (pp. 193–208). Sage.

Daskin, J. H., Stalmans, M., & Pringle, R. M. (2016). Ecological Legacies of Civil War: 35-Year Increase in Savanna Tree Cover Following Wholesale Large-Mammal Declines. *Journal of Ecology, 104,* 79–89.

David, N., Sterner, J., & Gavua, K. (1988). Why Pots Are Decorated. *Current Anthropology, 29,* 365–389.

Eltringham, S. K. (1992). Ecology and Behaviour. In J. Shoshani (Ed.), *Elephants* (pp. 124–127). Rodale Press.

Emery, J. B. (1833). A Short Account of Mombasa and the Neighbouring Coast of Africa. *Journal of the Royal Geographical Society, 3,* 280–283.

Finch, J., Marchant, R., & Courtney Mustaphi, C. J. (2017). Ecosystem Change in the South Pare Mountain Bloc, Eastern Arc Mountains of Tanzania. *The Holocene, 27*(6), 796–810.

Finneran, N. (2002). *The Archaeology of Christianity in Africa.* Tempus.

Freeman-Grenville, G. S. P. (1965). *The French at Kilwa Island: An Episode in Eighteenth Century East African History*. Clarendon.

Galaty, J. G. (1993). Maasai Expansion and the New East African Pastoralism. In T. Spear & R. Waller (Eds.), *Being Maasai: Ethnicity and Identity in East Africa.* James Currey.

Giblin, J. (1990). Trypanosomiasis Control in African History: An Evaded Issue? *The Journal of African History, 31*(1), 59–80.

Gilbert, E. (2002). Coastal East Africa and the Western Indian Ocean: Long-Distance Trade, Empire, Migration, and Regional Unity, 1750–1970. *The History Teacher, 36*(1), 7–34.

Goheen, J. R., Augustine, D. J., Veblen, K. E., Kimuyu, D. M., Palmer, T. M., Porensky, L. M., Pringle, R. M., Ratnam, J., Riginos, C., Sankaran, M., &

Ford, A. T. (2018). Conservation Lessons from Large-Mammal Manipulations in East African Savannas: The KLEE, UHURU, and GLADE Experiments. *Annals of the New York Academy of Sciences, 1429*(1), 31–49.

Håkansson, N. T. (1994). Grain, Cattle, and Power: The Social Process of Intensive Cultivation and Exchange in Precolonial Western Kenya. *Journal of Anthropological Research, 50*, 249–276.

Håkansson, N. T. (1995). Irrigation, Population Pressure, and Exchange in Precolonial Pare, Tanzania. *Research in Economic Anthropology, 16*, 297–323.

Håkansson, N. T. (2004). The Human Ecology of World Systems in East Africa: The Impact of the Ivory Trade. *Human Ecology, 32*(5), 561–591.

Hauenstein, S., Kshatriya, M., Blanc, J., Dormann, C. F., & Beale, C. M. (2019). African Elephant Poaching Rates Correlate with Local Poverty, National Corruption and Global Ivory Price. *Nature Communications, 10*(1), 1–9.

Hayes, L. D., & Schradin, C. (2017). Long-Term Field Studies of Mammals: What the Short-Term Study Cannot Tell Us. *Journal of Mammalogy, 98*(3), 600–602.

Hobley, C. W. (1895). Upon a Visit to Tsavo and the Taita Highland. *Geographical Journal, 5*, 545–561.

Hopper, M. S. (2007, October). The Nineteenth-Century Slave Trade from East Africa to Arabia Reconsidered. In *50th Annual Meeting of the African Studies Association.*

Jackson, K. (1976). The Dimensions of Kamba Pre-Colonial History. In B. A. Ogot (Ed.), *Kenya Before 1900* (pp. 174–261). East African Publishing.

Kartzinel, T. R., Chen, P. A., Coverdale, T. C., et al. (2015). DNA Metabarcoding Illuminates Dietary Niche Partitioning by African Large Herbivores. *Proceedings of the National Academy of Sciences USA, 112*, 8019–8024.

Kiffner, C., Rheault, H., Miller, E., Scheetz, T., Enriquez, V., Swafford, R., Kioko, J., & Prins, H. H. (2017). Long-Term Population Dynamics in a Multi-Species Assemblage of Large Herbivores in East Africa. *Ecosphere, 8*(12), e02027.

Kirkman, J. (1974). *Fort Jesus: A Portuguese Fortress on the East African Coast, BIEA Memoir 4*. Clarendon Press.

Kjekshus, H. (1977). *Ecology Control and Economic Development in East African History*. University of California Press.

Krapf, J. L. (1968). *Travels, Researches, and Missionary Labours During an Eighteen Years' Residence in Eastern Africa*. Frank Cass. (Original work published 1860.)

Kusimba, C. M. (1999). *The Rise and Fall of Swahili States*. Altamira.

Kusimba, C. M. (2004). Archaeology of Slavery in East Africa. *African Archaeological Review, 21*(2), 59–88.
Kusimba, C. M. (2014). The Impact of Slavery on the East African Political Economy and Gender Relationships. In L. W. Marshall (Ed.), *The Archaeology of Slavery: A Comparative Approach to Captivity and Coercion* (pp. 230–254). Southern Illinois University.
Kusimba, C. M., & Kusimba, S. B. (2005). Mosaics and Interactions: East Africa, 2000 bp to the Present. In A. B. Stahl (Ed.), *African Archaeology: A Critical Introduction* (pp. 392–419). Blackwell.
Lamden, S. C. (1963). Some Aspects of Porterage in East Africa. *Tanganyika Notes and Records, 61,* 155–164.
Lamphear, J. (1970). The Kamba and the Northern Mrima Coast. In R. Gray & D. Birmingham (Eds.), *Pre-Colonial African Trade* (pp. 75–102). Oxford University Press.
Lamphear, J. (1986). The Persistence of Hunting and Gathering in a "Pastoral World." *Sprache und Geschichte in Afrika, 7*(2), 227–265.
Lane, P. J. (2010). Developing Landscape Historical Ecologies in Eastern Africa: An Outline of Current Research and Potential Future Directions. *African Studies, 69*(2), 299–322.
Lane, P. J. (2013). Trajectories to Pastoralism in Northern and Central Kenya: An Overview of the Archaeological and Environmental Evidence. In M. Bollig, M. Schnegg, & H. P. Wotzka (Eds.), *Pastoralism in Africa: Past, Present and Future* (pp. 105–143). Berghahn.
Lindblom, G. (1914). *Afrikanska Stro¨vta ˚g.* Bonniers.
Lindblom, G. (1920). *The Akamba.* Lundell.
Livingstone, D. (1880). *Last Journals of David Livingstone in Central Africa From 1865 to His Death.* John Murray.
Lodhi, A. Y. (1973). *The Institution of Slavery in Zanzibar and Pemba.* Nordiska Afrikainstitutet.
McConkey, R., & McErlean, T. (2007). Coastal Survey on Mombasa Island: A Preliminary Report. *International Journal of Historical Archaeology, 11,* 99–121.
Merritt, H. (1975). *A History of the Taita of Kenya to 1900.* Ph.D. thesis, Department of History, Indiana University Press, Bloomington.
Milner-Gulland, E. J., & Beddington, J. R. (1993). The Exploitation of Elephants for the Ivory Trade: An Historical Perspective. *Proceedings of the Royal Society of London: Series B: Biological Sciences, 252*(1333), 29–37.
Miracle, M. P. (1965). The Introduction and Spread of Maize in Africa. *Journal of African History,* 39–55.

Nicholson, S. E. (1996). Environmental Change Within the Historical Period. In W. A. Adams, A. S. Goudie, & A. R. Orme (Eds.), *The Physical Geography of Africa* (pp. 60–87). Oxford University Press.

Nunn, N. (2007). Historical Legacies: A Model Linking Africa's Past to Its Current Underdevelopment. *Journal of Development Economics, 83*(1), 157–175.

Nwulia, D. E. (1975). *Britain and Slavery in East African.* Three Continents Press.

Odada, E. O., Olago, D. O., & Olaka, L. A. (2020). An East African perspective of the anthropocene. *Scientific African*, e00553.

Parker, I. S. C. (1989). *The Raw Ivory Trade 1979–1989*. Report to CITES, Lausanne.

Parker, I. S., & Graham, A. D. (1989). Elephant Decline: Downward Trends in African Elephant Distribution and Numbers (Part II). *International Journal of Environmental Studies, 35*(1–2), 13–26.

Petek, N., & Lane, P. J. (2017). Ethnogenesis and Surplus Food Production: Aommunitas and Identity Building Among Nineteenth- and Early Twentieth-Century Ilchamus, Lake Baringo, Kenya. *World Archaeology* [online]. https://doi.org/10.1080/00438243.2016.1259583

Pradines, S. (2004). *Fortifications et Urbanisation en Afrique Orientale.* BAR International Series 1216.

Prendergast, M. E., Rouby, H., Punnwong, P., Marchant, R., Crowther, A., Kourampas, N., ... Boivin, N. L. (2016). Continental island formation and the archaeology of defaunation on Zanzibar, eastern Africa. *PloS one, 11*(2), e0149565.

Pringle, R. M. (2008). Elephants as Agents of Habitat Creation for Small Vertebrates at the Patch Scale. *Ecology, 89,* 26–33.

Pringle, R. M., Young, T. P., Rubenstein, D. I., et al. (2007). Herbivore-Initiated Interaction Cascades and their Modulation by Productivity in an African Savanna. *Proceedings of the National Academy of Sciences USA, 104,* 193–197.

Prins, H. H., & van der Jeugd, H. P. (1993). Herbivore Population Crashes and Woodland Structure in East Africa. *Journal of Ecology*, 305–314.

Robertshaw, P. T. (1999). Seeking and Keeping Power Bunyoro-Kitara, Uganda. In S. Keech McIntosh (Ed.), *Beyond Chiefdoms: Pathways to Complexity in Africa* (pp. 124–135). Cambridge University Press.

Rockel, S. J. (2006). *Carriers of Culture: Labor on the Road in Nineteenth-Century East Africa.* James Currey.

Russell-Wood, A. J. R. (1998). *The Portuguese Empire, 1415–1808: A World on the Move*. John Hopkins University Press.
Shayt, D. A. (1992). The Material Culture of Ivory Outside Africa. In *Elephant: The Animal and Its Ivory in African Culture*, 367–381.
Sheriff, A. (1987). *Slaves, Spices and Ivory in Zanzibar: Integration of an East African Commercial Empire into the World Economy, 1770–1873*. Ohio University Press.
Shitta-Bey, O. A. (2014). In Defence of the Internalists' Conception of Africa's Under-Development. *International Letters of Social and Humanistic Sciences* (15), 14–21.
Sissons, C. J. (1984). *Economic Prosperity in Ugogo, East Africa, 1860–1890*. Ph.D. dissertation, University of Toronto.
Summer, G., & Vossen, R. (1993). Dialects, Sectiolects, or Simply Lects? The Maa Language in Time Perspective. In T. Spear, & R. Waller (Eds.), *Being Maasai* (pp. 25–37).James Currey.
Sutton, J. E. G. (1984). Irrigation and Soil-Conservation in African Agricultural History: With a Reconsideration of the Inyanga Terracing Zimbabwe and Engaruka Irrigation Works Tanzania. *The Journal of African History, 251*, 25–41.
Thorbahn, P. F. (1979). *The Precolonial Ivory Trade of East Africa: Reconstruction of a Human-Elephant Ecosystem*. Unpublished Ph.D. dissertation, University of Massachusetts.
Veblen, K. E., Porensky, L. M., Riginos, C., & Young, T. P. (2016). Are Cattle Surrogate Wildlife? Savanna Plant Community Composition Explained by Total Herbivory, Not Herbivore Identity. *Ecological Applications, 26*, 1610–1623.
Wasser, S. K., Shedlock, A. M., Comstock, K., Ostrander, E. A., Mutayoba, B., & Stephens, M. (2004). Assigning African Elephant DNA to Geographic Region of Origin: Applications to the Ivory Trade. *Proceedings of the National Academy of Sciences, 101*(41), 14847–14852.
Wynne-Jones, S. (2010). Lines of Desire: Power and Materiality Along a Tanzanian Caravan Route. *Journal of World Prehistory, 23*(4), 219–237.

5

Colonial Transitions

5.1 Colonial Foundations—An Atypical Moment in Time

Historical and palaeoclimatic data record a major drought across Eastern Africa in the late-nineteenth century (Gillson, 2015). The 'Great East African Famine' extended from 1888 to 1892 CE and is thought to have cost the lives of one third of the region's population with local populations extrapolated. The Maasai record this period as 'Emutai' or 'to wipe out' (Saitoti & Beckwith, 1980; Waller, 1988) with mass mortality being documented (Illife, 1987). However, as explored in Sect. 4.7, this was not 'simply a period of drought'. There was a 'perfect storm' of drought, epidemics of bovine pleuropneumonia (BPP) and rinderpest that killed cattle, wildlife, and people in many areas; this was quickly followed by an epidemic of novel diseases, such as smallpox and leishmaniasis (Waller, 1988), that devastated the human populations towards the end of the nineteenth century (Saitoti & Beckwith, 1980). Raiding of stock and crops between different ethnic groups on the back of increasingly in conflict following the caravan trade (Chapter 4) exacerbated

R. Marchant, *East Africa's Human Environment interactions*,
https://doi.org/10.1007/978-3-030-88987-6_5

social fragmentation. As severe famine spread across large areas of East Africa by 1899 CE, there was developing antagonism between European settlers and many East African communities (Ofcansky, 1981; Waller, 1988).

As we have seen through Chapters 2–4, the risks of drought would have been commonplace. Some of the main droughts or climate transitions were coeval with the arrival of populations, development of technologies, or local adaption through trade. Often such risks were 'insured' by the trade relationships with the pastoral populations, such as the Maasai, maintained with their agricultural neighbours, which created a symbiotic relationship between pastoralists and cultivators (Sutton, 1971). However, given the challenges across the landscape it would have been likely that such trade relationships would have been fraught and the drought and epidemics of the 1890s would have meant traditional strategies, trading relationships and medicines could not cope. Many pastoralist survivors sought refuge among neighbours such as the Kikuyu, Luo, Gusii, Nandi, Meru, Chagga, Taveta, Arusha, and Luhya. While the combined effects of these events on pastoralists varied, widespread famine significantly reduced pastoralist populations overall and it is estimated that some two thirds of the Maasai pastoralists population in Kenya and Tanzania perished and the ecological systems they managed collapsed (Waller, 1988).

One of the biggest challenges, that is perpetuated through to the present day, was that as Europeans began to enter East Africa in increasing numbers they arrived at an atypical point in history—East Africa was characterised by a depopulated (it is estimated 80% of all hoofed animals and 1 million people died) and a green landscape verdant landscape. This atypical arrival point meant that many colonial reports; their understanding, and the ensuing policies, many of these still have a legacy today, were not based on a more densely populated landscape. These policies would lay the foundation for what would happen in the next century and beyond, including the impact of the accelerating rate of climate change and human use of fragile ecosystems and managers of biodiversity. This new dynamic, a landscape sparsely populated by people and dominated by rinderpest-resistant wildlife populations, coincided with the post-Berlin Conference (1884–1885) arrival of colonial administrations, assuming this dynamic to be the norm rather than the exception (Lamprey & Waller, 1990). The Berlin Conference, otherwise known as the Congo or West Africa Conference, was set up to partition the continent across the major European Colonial powers and feed into a series of bilateral agreements and wars that resulted in countries passing from the rule of one nation to another. One quite amazing concept was that there was no representation of African rulers, and the conference overrode any form of autonomy!

5.2 The Colonial Explorers and Missionaries

There are numerous people who ventured to 'discover' different regions of East Africa to claim territories (Fig. 5.1), naming lakes, rivers, mountains, towns, and forests in their wake. The group of Victorian British explorers, Livingstone, Speke, Thomson, Burton, and Grant all contributed to communicating what the land was like and setting the foundations for expanding missions. The first European explorers of East Africa describe 'pristine' natural environments, where 'primitive' human societies were 'in the defensive' against forces of nature on which they had little impact (Stanley, 1889; Thompson, 1887; von Höhnel, 1894). Contrary to those reports, precolonial East African social ecological systems, as we have seen in previous chapters, were characterised by millennia of co-evolution between human societies and ecosystems. People had by this time had a transformative influence on their environment, with extensive agropastoral systems (Widgren & Sutton, 1999) in place with ensuing supply and trade links. Additionally, the division of the African population into static tribes or ethnic groups was reenforced as a colonial construct rather than something inherent to African societies. The tribal structures known today are conglomerates of peoples who had previously been carriers of different cultural identities (Klopp, 2001). Instead, the history of the region is characterised by the spread and changing influence of major African groups, Arabic sultanates, trading networks, and slave trade, which influenced land use,

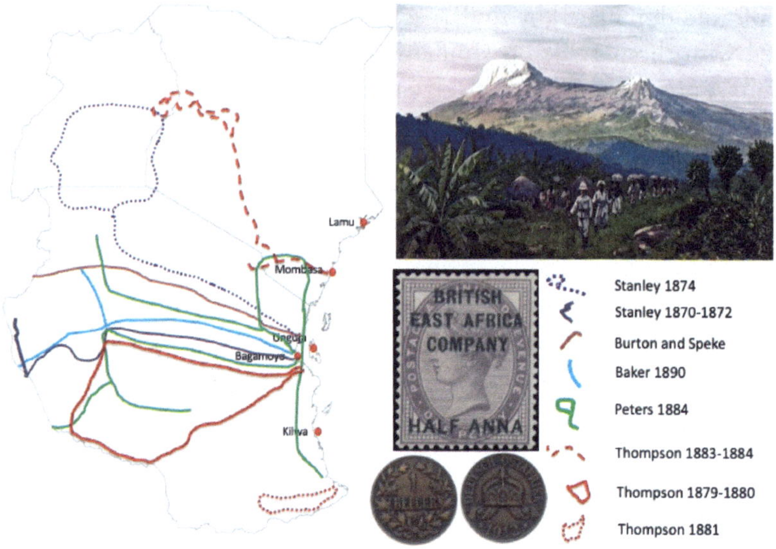

Fig. 5.1 British and German explorer routes; these started from the coastal staging posts of Unguja, Bagamoyo or Mombasa. The numerous people who ventured to 'discover' East Africa were largely driven by exploration, missionary activities, to claim territories or secure natural resources for the developing colonial influence

politics, culture, and economy. Some of these cultures developed into kingdoms and even empires, with dynamic spheres of influence. For example, as Speke entered the kingdom of Buganda, he found it was much more politically developed than the areas he had travelled through to the east. In Buganda the people acknowledged the rule of an autocratic king (the Kabaka), in an area with towns, roads, and markets. As seen through Chapters 3 and 4 population movements and the amalgamation of different peoples have taken place in East Africa since the evolution of our species but particularly since the mass population movements of the past two millennia (Iliffe, 1979; Mamdani, 2018).

Possibly most famous of these early explorers, due to the sponsorship by media and subsequent serialisation of his exploits in the New York Herald and the Daily Telegraph, was Henry M. Stanley. With support from the US Consul, Stanley departed from Zanzibar in March 1871 with some 110 porters reaching the shores of Lake Tanganyika around eight months later when he met the explorer David Livingston who had travelling in the footsteps of John Hanning Speke, Richard Francis Burton, and Samuel Baker to find the source of the Nile. Livingston was a Scottish missionary and a strong anti-slavery voice, indeed some good came of this expedition, and the opening of East Africa to the rest of the world through the media contributed to the subsequent abolition of slavery, particularly by Livingstone's copious accounts. He complained to John Kirk in 1871 that what he observed was 'not slave trade (but) slacking thirst for blood and catching free people'. During the 1870s there was a marked quickening of European interest in East Africa, much of this interest being evangelistic and supported by a powerful and wealthy church. Several British Protestant missionary societies sent volunteers to establish church mission outposts among populations far inland between 1875 and 1879 such as the Scottish missions established near Lake Nyasa. In the early stage, the missionary activity was apolitical, although it was increasingly hard to differentiate between religion and politics when these were both focussed on imposing belief systems and exerting control. The missionary penetration made East Africa known to a wider European public and fuelled a growing commercial and scientific interest across the region.

5.3 The Colonial Partitioning of East Africa

The Berlin conference took place at the end of 1884 when the seal of approval was given on Leopold's claim to the Congo Basin and on Bismarck asserting German rights over a large, although ill-defined, area on the mainland opposite Zanzibar. The colonial powers agreed at the Berlin Conference that they had 'the obligation to ensure the establishment of authority in the regions occupied by them'. Thus, while earlier traders, missionaries, and plantation owners may have concentrated on commercial extractive potential, colonial expansion was much wider and had a mandate to prevent or suppress any defiance of their self-imposed authority. Following the Berlin conference, an international commission with German, British, and French representatives was engaged in reducing the sultans and wider Arabic recognised interests on mainland East Africa until this was reduced to a thin strip along coastline and the offshore islands. In 1886, the London agreement was signed apportioning East Africa into spheres of influence dominated by Germany, Britain, and the Sultan of Zanzibar. In 1890, Germany recognised a northern border for its new East African colony, which acknowledged a British sphere of control in what is now Kenya and Uganda. Largely through the influence and money of Cecil Rhodes and the British South Africa Company the British government asserted control over the region in 1890. British interests were initially focussed inland from the East African coast across two regions; largely what is today Kenya and Uganda. Buganda territory lies on the shores of Lake Victoria where in 1887 the British-sponsored British East Africa Company established itself as a key trading outfit with growing economic and political power. The British East Africa Company opened a trading route from the Kenya coast into the interior, initially with a line of trading stations, ultimately consisting of a railway and steamers on Lake Victoria.

Similarly to the British, early German colonisation efforts in Africa were initially carried out by a number of private explorers motivated by personal gain and commercial ambition. In 1884, Carl Peters founded the Society for German Colonisation, which lobbied for the acquisition of colonies and soon evolved into the German East-Africa Company (Deutsch Ostafrikanische Gesellschaft, DOAG). To exert pressure on Bismarck to have a strong presence in East Africa, and to fulfil his desire for profit and fame, Peters undertook a series of expeditions throughout East Africa during which he signed obscure treaties with whatever local authorities he could; invariably these would have been village chiefs or elders, which transferred their land to the company 'for all times'. In early 1885, Bismarck agreed to Peters' demands (who was threatening to seek support from Leopold II of Belgium) to officially 'legitimized' DOAG's territorial possessions in Africa and the agreements Peters had entered. During the following decade, the borders of German East Africa were defined through interactions with British colonial powers and DOAG. With governmental support, DOAG continuously expanded its territorial presence by establishing several small trading stations at coastal nodes and inland forts. DOAG increasingly intervened in pre-existing cultural, political, and economic structures, predominantly by establishing authority over tariffs and taxes, jurisdiction, and land rights. The principal mode was to seek 'protection contracts' in exchange for land rights. Through these agreements, that seem akin to modern day 'protection rackets', the DOAG was able to control large areas of land towards the Uluguru and the Usamabara Mountains. In 1888 CE, in response to DOAG's heavy interference in political and economic affairs along the coast, local wealthy merchants and landowners began to mobilise resistance against the Germans in the so-called Abushiri Revolt led by Abushiri ibn Salim al-Harthi, a wealthy merchant and plantation owner. This attempt to take over the Pangani area required the first large scale and organised intervention; Bismarck initially hesitated to intervene but ultimately supported the defence of the DOAG's territory. Hermann von Wissmann was charged with recruiting an army comprised of German and Sudanese soldiers and, most importantly, naval support was sent that bombarded coastal towns and quelled the resistance. By 1890 CE, the military campaign had crushed the revolt and executed Abushiri ibn Salim al-Harthi. Violent resistance and repression were endemic to the German colonial rule with the Wehehe rebellion extending from 1891 to 1898 CE and the Maji Maji rebellion in 1905 CE. The later revolt was the greatest uprising against colonial rule in East Africa. It was violently crushed by German military repression in 1907 CE and is estimated to have cost some 250,000 lives (Iliffe, 1967). The reasons behind this are a little opaque but it largely stemmed from the DOAG's wish to grow cotton (Fig. 5.2) that, combined with ensuing famine, appears to have been a deliberate policy to subjugate the indigenous populations, particularly in Southern Tanzania. That policy has resulted in the area remaining relatively divided and underdeveloped right through to the modern era (Gunther, 1955).

From 1886 Kenya came under British colonial rule; Kenya is often seen as the classic British colony that appeared to be comprised of large spaces of 'vacant' land that was only 'sparsely inhabited' by pastoralists (Sect. 5.1). Most early settlers were South African. Immigrant populations from India and relatively few British. To prevent competition from Indian settlers the British government reserved large sections of central Kenya for whites only, creating the infamous 'White Highlands Policy'

Fig. 5.2 Large tracts of land were assigned over to extensive plantations that took different forms, Cotton (**b**) and Sisal (**c**) plantations in the lowland for cloth and fibre, respectively. Much of the highland areas were cleared of forest and used for extensive Tea plantations (**a**, **d**)

focussed on the present-day highland islands from Chyulu through to Kericho. Initially the area for settlement was demarcated as lying between two points, Kiu to the East and Fort Ternan to the west and was a particularly key part in the development of the colony accompanying the development of the railway to open up the country and connect Uganda and Kenya to the coast. Initially, the explorer Frederick Luguard suggested that settlement was not possible in the lowlands and that the highlands should be the focus for European settlers due to the cool environment and relative 'lack of existing settlement' and high agricultural potential (Fig. 5.2). One of the key reasons for the White Highlands being relatively unoccupied is that these comprised the seasonal grazing lands for a number of pastoral communities, particularly the Maasai. The pastoral societies practised a form of transhumance where, during dry seasons in the lowlands, cattle and populations would move to highlands particularly during periods of drought. The timing of the British arrival when the pastoral populations and their cattle were decimated (Sect. 5.1) led to the misconception that these were unoccupied lands. Of the 12,000 square miles of European settled, some 7000 consisted of former Maasai grazing grounds abandoned under agreements between 1904 and 1913 (Morgan, 1963). In the Kenya highlands new settler farmers could raise many temperate climate crops with extensive development of tea, coffee, and sisal plantations (Fig. 5.2). Although covering only some 5% of the land mass these highland areas remain the agricultural driver of the Kenyan economy and of large trading centres such as Nairobi, Eldoret, Nakuru.

The first, and possibly the most ambitious, British enterprise was to construct a railway from the coast to Lake Victoria (Fig. 5.3). This was constructed following the old caravan route from Mombasa to Bussian

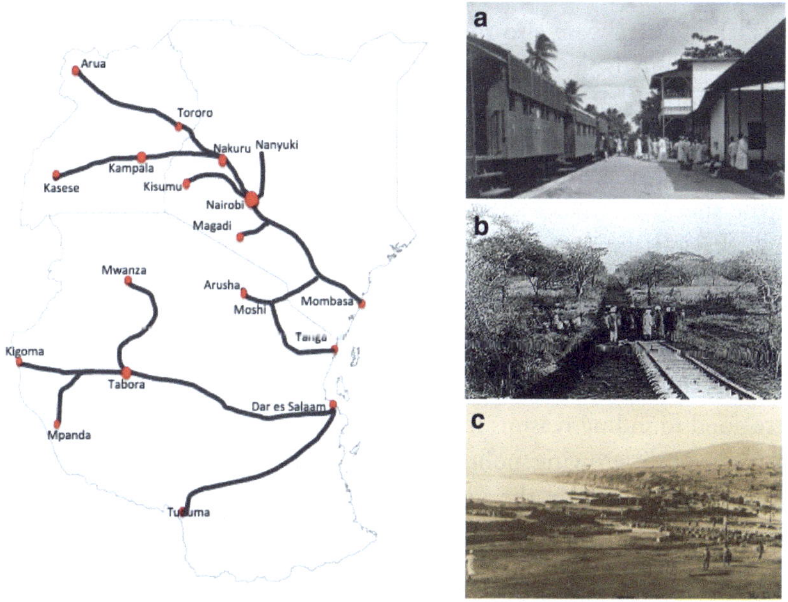

Fig. 5.3 British railway route. One of the key infrastructure developments to take place was the establishment of a rail network that would have replaced the need for large caravans to transport goods and services in and out of East Africa. Dar es Salaam station (**a**), constructing the Uganda Railway through the savannah (**b**) and arrival at Kisumu (**c**)

known as the Makinnion-Sclater road that was an Oxcart route developed by the British East Africa company. At some 800 miles long, the railway had two main aims—to provide transportation via steam power of goods in and out of Uganda; to further establish and expand grip on power. This railway also had a key political aim of curtailing the caravan trade, particularly in slavery: with an alternative form of transport people would no longer be needed to carry goods in and out of East Africa. Captain James Macdonald conducted a year-long survey between 1881 and 1892 CE ahead of construction starting in 1896 CE under the directorship of George Whitehouse. Nearly all the labour for the construction came from India with some 36,000 people directly employed on the project (Whitehouse, 1948). Quite a number stayed in Kenya following the completion and thus the foundation of the strong Asian community present in Kenya today was set. The project was not without challenge; the Nandi people strongly resisted the passage of the railway through their land. There were also the infamous man-eating lions of Tsavo that are thought to be responsible for the death of between 28 and 135 railway workers (Railway Gazette Int, 2009). The line was challenging and expensive to build and it was not until 1901 CE, four years after it was started, that the Uganda railway reached its terminus at Kisumu on the shores of Lake Victoria. The railway was a massive project with all steel goods imported from India that needed the creation of a port facility at Mombasa. Funds came from the British Treasury and even on completion ran at an annual loss, principally as there were many goods and people to be carried into East Africa but there was little to take out during the foundational years of the colony. Although there were good founding reasons for the construction of the railway there was simultaneously growing political resistance to the massive use of treasury funds to support the venture that resulted in this being dubbed the 'Lunatic express'. Although the cost of construction was around £3 million in 1894 CE, equivalent to some £650 million in current times (Knowles, 2016), this is considerably less than the replacement standard gauge railway (Chapter 7) that has recently opened at a cost of c. £2.5 billion. Indeed, there are strong parallels between this first massive engineering project and the contemporary counterpart; today much of the steel, machinery, and labour are imported, this time via China. Certainly, the railway and connecting port at Mombasa opened Kenya to the world and Sir Charles Elliott, first Commissioner of the British East Africa Protectorate, pointed out that after the railways first 300 miles it entered a region where much of the soil was fertile, the altitude high, and the climate healthy and therefore a splendid place for British farmers to take up land and live producing crops for the railway to carry away. As today, there was a period of rapid expansion driven by the policy of Henry Johnson as the colony tried to recoup funds spent on the railway, and numbers of white settlers, with impact on the landscape (Sect. 5.4) rose rapidly from 100 in early twentieth century to more than 1000 by 1914 CE (Ochieng & Maxon, 1992) with active policies to attract and encourage settlers.

The machinery of government in the Kenya colony was a complicated and elaborate affair with many strands to the administrative hierarchy where native councils at the districts level combine to form provinces each with the senior officer in charge reporting to the Country and national governor. Colonial governors, initially centred in Mombasa before shifting the seat of parliament to Nairobi in 1905 CE with the completion of the railways (Fig. 5.3), reported to the colonial office in Downing Street and the British parliament. The indigenous populations continued to be governed through the existing chiefs and institutions rather than directly; native councils were a mixture of the traditional authorities, the Chiefs, and elder councils that acted as an intermediary between the colonial government and the population.

Uganda, named after the central and dominate Buganda kingdom (Fig. 5.4), had a different colonial administration mainly due to the indigenous foundation where it was already divided into a series of well established kingdoms such as Buganda, Ankole, and Toro. As explored through Chapters 2 and 3, like the other countries there was a similarly long history of hunter-gather occupancy, with an increased spread of Bantu agriculture around Lakes and towards the montane areas alongside extensive pastoral communities in the central northern lowland region. Uganda, being farthest from the wave of caravan trade influence with their coastal nodal engines, was relatively late to be exposed to the wave of international influence. Indeed, some of the earlier influence from Sudan was via people like Khedive Ismail Pasha from Egypt

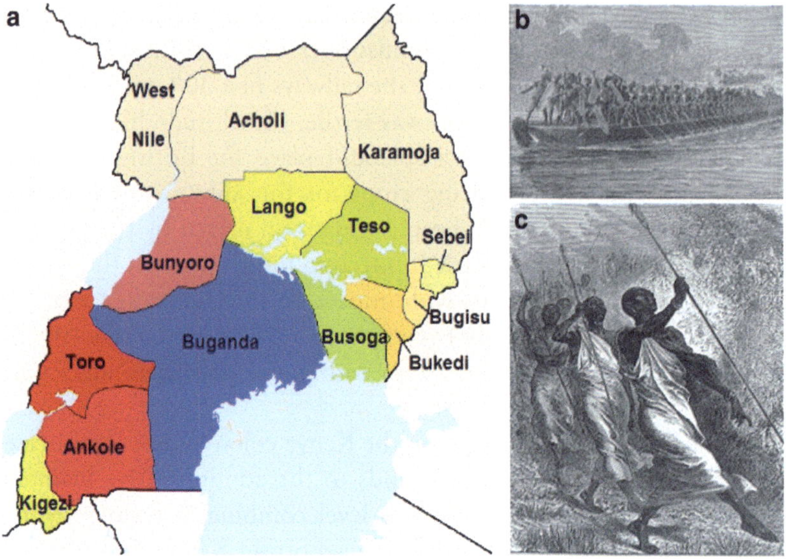

Fig. 5.4 Uganda was divided into a series of administrative kingdoms (**a**), some of them such as Buganda were highly centralised with a king (Kabaka), council and large well trained and extensive army (**c**) with reports of 250 large outrigger canoes that functioned as a navy (**b**)

who supported Samuel Baker to explore the area of Buganda with a multiple remits of tracing the Nile northwards to Sudan, establishing a trading centre in Buganda, engaging in the ivory trade and to document and ultimately stop the slave trade. There soon followed a wave of missionaries, initially by the British Anglican Church in 1877 CE and followed by Roman Catholics from France in 1879 CE. As they encountered the Kabaka (King) of Buganda, Mwanga II, who ruled from 1884 CE, he viewed these foreign ideologies, Muslim, or Christian, as a threat to his power and therefore executed a number of his subjects who had converted to Christianity in 1885 CE along with James Hannington the Archbishop. This interaction between early missionaries with Mwanga II led to quite a large bout of internal conflict towards the end of the nineteenth century as both Muslims and Christians, Catholic and Protestants, fought each other and all faiths pushed back against the traditional outright ruling power of Mwanga II who yielded significant powers in the traditional kingdom. In addition, Buganda was highly organised, socially structured had a navy and extensive army; one report from Livingstone indicated an army of some 125,000 people with a navy of some 250 large outrigger canoes (Fig. 5.4). These uprisings and internal conflicts led to significant instability and the British East Africa Company came under direct threat, ultimately becoming bankrupt and passing 'control' of Buganda to the British Government.

Given the logistical importance of the Nile system as a northerly transport route, the British Government secured a series of treaties, particularly between the kingdoms of Buganda, Ankole, and Toro and some decades later in 1931 CE with Bunyoyo (Apuuli, 1994) to form the Uganda Protectorate. The main differentiation between a protectorate and a colony was that Uganda retained a large degree of self-rule under the chiefs that were largely Protestant. The Uganda agreement of 1900 CE was signed by H.H. Johnston, Apollo Kagwa, Stanislaus Mugwanya, Mbogo Noho, and Zakaria Kizito, and defined the boundaries of the territory along with alliance to Queen Victoria. As well as external boundaries the agreement defined ten internal districts that would form administrative centres across the Kingdom of Uganda. The signing also legitimised the role of the Kabaka with their replacement being a combination of heredity lineage and voting in by the court council—the Lukiko—and subsequent approval by the British state. There was imposition of UK law, but the Kabaka courts were able to function as a lower court system. Each appointed Kabaka and their immediate family received annual allowances and key administrative positions to form an effective hierarchy. To support these, a series of taxes were levied from a hut tax to a gun tax. The chiefs were expected to collect a diverse range of taxes from their populations, largely on houses, guns, and land via cash crops that were exported along the railway through Kenya to Mombasa. The first Governor of the protectorate, Sir Henry Johnson, was charged with making back the considerable money that has been spent on quashing earlier internal battles and religious unrest. Taxes were used to support the protectorate and develop and maintain growing infrastructure. Like most taxation schemes, there was growing resistance to these, and the British began to intervene by trying to instal a system of secondary power brokers. This worked through an extensive system of official appointments that were salaried and came with land allotments. The main architect of these agreement was demarcated land and title such as 1500 square miles of forestland apportioned of land to private landowners in allotments of eight square miles appointed by the Lukiko, the governing council, or parliament.

Following the 1914–1918 war, all the territory between the Indian Ocean and the Great Lakes came under British control as four separate territories (Zanzibar, the Uganda protectorate, the Tanganyika Territory and East African protectorate) governed in different ways. Zanzibar was semi-autonomous with the Sultan of Oman still sitting on the throne. The Uganda protectorate was more complicated as Buganda survived as a single entity with its king, Premiere, Parliament, and system of counties and subchiefs within the Lukiko with the other kingdoms amalgamated. German East Africa transitioned to become the 'Tanganyika Territory' although it was effectively administered under British control from 1916 CE even though the commander of East African German forces (Paul Emil von Lettow-Vorbeck) did not surrender until he received the armistice in November 1918 CE. In 1927 CE Tanganyika joined the wider customs union of the East African protectorate that was governed as a League of Nations until the end of WWII, after which it transitioned to the United Nations mandate under British control until independence in 1961 CE. Following the end of the First World War the number of settlers across East Africa increased dramatically with concomitant impact on East Africa.

5.4 Land Use Transformations

Colonial policies, while not static and differing greatly between the countries, had in all cases major impacts on indigenous farming and pastoral systems (Anderson, 1984; Wynants et al., 2019). Particularly in Tanzania and Kenya, new policies were enforced from a centralised power that replaced the more localised indigenous management (Anderson, 1984; Smith, 1989). While the indigenous land use and land tenure systems were flexible and dynamic, having evolved to work with the environmental systems of East Africa such as the practice of cattle keeping for blood, milk, and meat in a transhumance system that saw cattle migrate across the landscape in response to the rains and ensuing pasture resource this was not the case for newly imposed systems. The new colonial rules imposed rigid legal systems and structures distinguishing between private, public, and government land (Migot-Adholla et al., 1991) in commercial categories such as forest, commercial agriculture, and 'no value'. The most productive land was attributed to the state and given to European settlers and pioneer farmers, particularly in the early twentieth century and following WWI. There was a substantial focus on implementing large-scale plantations, ranches, and monocultures for supporting export. The focus of these varied depending on the soils and environmental regime with sisal plantations for rope and fibre in the lowlands, livestock ranches on the extensive rangelands with highlands preserved for tea and coffee plantation. In the process, local smallholder farmers were forced to move to less productive land or work on the plantations (Conte, 1999; Kjekshus, 1996; Sandford, 1919). Both policies led to a large-scale shift from subsistence crops to cash crops of interest to the colonial powers, such as coffee, cotton, rubber, and tea (Fig. 5.2), replacing the more diverse selection of food crops (Kjekshus, 1996). In the process of this changing land use and land allocation large areas of land were fenced and subdivided (Fig. 5.5)—divisions that continue through to the present day. Indigenous farmers and pastoralists were confined to increasingly marginal areas where they struggled to adapt to the unfamiliar and constricted ecological space, often leading to rapid land degradation, loss of soil fertility, erosion, and exposure to famine (Homewood, 1995; Kjekshus, 1996) as resilience and adaptive capacity was increasingly curtailed. Whether regions have been 'developed' or 'exploited' by colonial rule continue to be argued, although clearly the enormous impacts on agropastoral systems in East Africa cannot be disputed. There were massive shifts in social organisation, political power balances, agricultural production, land tenure, and economic systems, ultimately leading to changed interactions between humans and the environment (Hydén, 1980; Kjekshus, 1996). Given

Fig. 5.5 As land was subdivided fences became an increasingly common feature of the landscape such as these across the Laikipia Plateau (All photographs: Rob Marchant)

the enormity of these impacts, they will be presented in the three themes of Rangelands, Agriculture, and Forestry.

5.4.1 Rangelands

Ever since the Maa-speaking peoples from Southern Sudan migrated through the Rift Valley they have survived through adaptation. At the end of the nineteenth century the combination of the drought and disease (Sect. 5.1) devastated the pastoral economy, and as British and German troops arrived many Maasai survived by taking up agriculture and by moving in with other tribes (Bruner & Kirshenblatt-Gimblett, 1994; Enghoff, 1990). The subsequent recovery and development of pastoral systems have been seen as unnatural, threatening, and ecologically unsound (Marchant, 2010). The degradation of pastoral land in large expanses of Northern Kenya and Uganda did not begin recently, with many suggesting that it can be traced back to European colonisation in the late nineteenth century with the restriction across the transhumance catchment resulting in impacts being concentrated in the most marginal lands that quickly became degraded. Throughout most of the twentieth century the long-term health and sustainability of the rangelands of East Africa were debated and contested; this continues to the present day. From the beginning of the colonial period, indigenous systems of pastoral production had been seen as economically irrational and lawless, based on a system of raiding and incapable of any development other than switching to agropastoralism (Eliot, 1905). The false view that pastoralists do not trade, nor eat agricultural products (Knowles & Collett, 1989), was used to justify taking over large tracts of their pastoral lands, simply because 'they don't use it'. The pastoral production system went against the colonial market-based perspective who viewed East African herders as keeping large numbers of 'non-productive' livestock for 'prestige' or other non-economic purposes (Herskovits, 1926). The 'irrationality' of keeping large herds was challenged by Harold Schneider in the 1950s who argued that a subsistence system based on milk production required large numbers of livestock and that the herding practices of East African pastoral people made perfect economic sense (Schneider, 1957).

Many colonial officials held the view that the Maasai were the richest ethnic group in East Africa because they possessed so many cattle (Sandford, 1919). Colonial intervention in Maasailand led to the formation of reserves in 1904 CE, and again in 1911 CE, and the breakdown of traditional transhumance. The Maasai lost land around Lakes Naivasha, Elmenteita, Nakuru, and Baringo, plus along the rivers flowing into them; these areas formed the dry season range and were crucial to the proper management of their pastoral system. An estimated 10,000 people, 200,000 cattle, and 550,000 sheep and goats were evicted from the annexed land and moved into the Southern Maasai reserve (Sandford, 1919). These numbers added to the populations already living there and brought about physical destruction of land, particularly as European settlers controlled nearly all the permanent streams. The territory within the reserve was either without water, contaminated by disease, or relatively poor in pasture (Lewis, 1934). Officials viewed the problem as overstocking beyond the subsistence requirements of the Maasai; again showing disregard for the value of cattle within a pastoral society. British colonial government intervention particularly impacted on the mobility of pastoral communities—instigating a process of sedentirisation that has continued through to the present day. The creation of definite boundaries and the establishment of the demarcated areas, such as Maasailand, prohibited migration outside this land and developed a Maasai frontier ethic. The official policy was to discourage the Maasai in the reserve from having contact with those in scheduled areas; permission to leave the reserve was issued under strict regulations and on legitimate business (Sandford, 1919). The restriction was enforced both by the Government and European settlers, often under the guise of stock control. For example, the government in 1917 CE imposed rigorous quarantine regulations to separate Maasai stock from European stock. These regulations denied the Maasai access to markets, especially in Northern Kenya, which had existed in the precolonial period. Outlets for selling surplus animals became non-existent as within Maasailand only a few markets operated and those that functioned offered the Maasai very low prices. This trade and exchange of stock, which also enabled the Maasai to practice selective breeding, was abolished in 1912 CE as efforts to improve their cattle by importing bulls from other regions were frustrated by the quarantine regulations (Masai Annual Report 1922) and opposed by the veterinary department (Leys, 1973; Sandford, 1919).

As at the end of the nineteenth century, recurrent droughts in the 1920s resulted in great losses of stock exacerbated by the lack of migratory adaptability. The district commissioners for Narok and Kajiado reported more than 30 and 15% loss of livestock, respectively, in the prolonged drought of 1927 CE. These growing problems led the Kenya Land Commission to create Maasailand in 1933 CE, although the commission failed to assess the intricacies of Maasai population pressure and instead blamed the Maasai for keeping too many cattle. The commission declared that the comparative low population densities in Maasailand existed because the Maasai were not cultivators! The colonial impact destabilised the Maasai traditional spatial and ecological order as the seasonal use of plains and uplands pastures was disrupted and, with grazing concentrated by the late 1930s, Maasailand was becoming seriously degraded (James, 1939). This shrinking of territory led to overstocking relative to resources and the pastoral populations exceeded the carrying capacity of the land (van Zwanenberg & King, 1975); primarily a result of the reduction in the availability of dry season grazing resources (Campbell & Migot-Adhola, 1979). The government pursued a policy of destocking Maasai herds and encouraging populations to abandon pastoralism for crop cultivation; again a trend that has continued through to the present day. The impact of the colonial era or pastoral populations and rangelands are stark: extensive alteration of traditional animal husbandry and cultural ecology led to severe land degradation. Compulsory and often half-hearted government programs to respond to this resource deterioration backfired. Silted dams, unrepaired pipelines, denuded and abandoned grazing improvement schemes, and resistance to sell surplus cattle (because prices were too low) all attest to failed government intervention in the pastoral system. Three processes, all of which began in the colonial period, have continued right through to the present: particularly around issues of wildlife reserves, encroachment of agriculture, and cultivator interactions.

5.4.2 Agricultural Lands

As seen in the section above, an increasingly mobile flow of people, goods, and services, combined with population growth and associated land use changes, have led to a general degradation of the landscape including deforestation and soil erosion (e.g., Johansson & Svensson, 2002; Snelder & Bryan, 1995). These are somewhat contradictory accounts on the depletion of forest resources before the colonial era. Some suggest that agricultural populations, like the Kikuyu, practised shifting cultivation by clearing and burning large patches of forest to grow crops that moved across the landscape (Ofcansky, 1984). As some of the accounts are based on information recounted by British forestry experts, these perspectives by the colonial forest departments may have been written to promote the interests of the settler community who needed productive land for settlement.

One of the most extensive newly imported crops grown in lowland Tanzania and Kenya were sisal plantations (Fig. 5.2) that became Tanzania's number one export. Sisal (*Agave sisilana*) was another Latin American import, this time introduced by the German agronomist Dr. Richard Hindorf, apparently young rootstock was imported in the belly of a crocodile! The growing global demand for fibre for ropes and sacking, combined with excellent growth, meant by the end of the colonial era Tanzania was the largest Sisal producer in the world and employed some 1,000,000 people. There were also quite extensive cotton plantations, particularly around the shores of Lake Victoria. Areas to the south, particularly along the coastal strip, support extensive Cashew trees that make Tanzania one of the biggest exporters of Cashew nuts in the world. The biggest direct ecological impact of the extensive sisal and cotton plantations would have been the clearing or large areas of savanna/miombo woodland.

In the upland areas much of the transformation was towards tea estates (Fig. 5.2) that also grow very efficiently. Tea was introduced by G.W.L. Caine in 1903 CE although it was not commercially expanded until 1924 CE. Today Kenya is the second biggest exporter of black tea and employs some 500,000 people in the mountain estates, particularly in the Mau highlands, in Kericho, across the Aberdare's, on the slopes of Mt Kenya, and in the Limuru, north of Nairobi. Despite the close proximity to Ethiopia, the origin of coffee cultivation and drinking, it was not until the late nineteenth century that coffee arrived in Kenya; attributed to the French Holy Ghost missions arriving from Reunion Island. Given the prevalence of smallholders and land tenure systems, the colonial administration introduced and encouraged a system of cooperatives to sell crops such as maize, coffee, tea, and tobacco. The cooperative system was introduced by the Governor David Gorden Hines in the 1950s in Uganda and has since expanded, particularly in the areas of coffee and tea production from a series of linked smallholders.

Tanzania has a similar production system with an increasing focus on Latin American imports of maize and potatoes with cassava and millets grown in drier areas. More niche crops like Tobacco were introduced in the 1920s and, although there is a significant amount impacted due to the Miombo and savanna land clearance and subsequent drying kilns (Jew et al., 2017), today it makes up around 1% of land cover In Tanzania. Potatoes are extensively, and ubiquitously, grown across highland East Africa. Ugandan highlands are quite different to those in Kenya and Tanzania that additionally grow extensive matoke (Plantain) that is more suited to the moist temperate climate of highland Uganda. Increasing intensive agriculture was introduced, for example, to

Loboi and Amboseli about 50 years ago, and cultivation has increased as more of the population has switched to subsistence or industrial farming (MEMR, 2012). This seems to be a common trend throughout East Africa as traditional smallholder and transhumance practices are curtailed by a growing and expanding population, increasing climate variability, and great access into markets and commercial transactions have made sedentary farming a more profitable livelihood. However, this has also come at a cost as the previously mobile impact of pastoral lifestyle, including the addition of nutrient on to the land, become concentrated in single locations rather than being spread across the wider transhumance landscape.

5.4.3 Forest Transformations

Land use transformation following colonial policy would have impacted directly on forest cover. Large-scale exclusion from previously communal forest and grazing lands, which were repurposed for private farming, conservation, hunting reserves, or forestry concessions, (Wynants et al., 2019) would have combined to significantly impact on forest composition and distribution. Historical data on forest cover is notoriously unreliable, and periodic estimates have estimated postcolonial forest cover at between 34 and 48% of mainland Tanzania; of this approximately 90% is miombo woodland (Mgaya, 2016). Early European explorers such as Joseph Thomson in 1883 spoke of trees in Southern Kenya that 'sprang up branchless eighty to one hundred feet before spreading out in a splendid umbrageous country'. Prior to the colonisation of Uganda, large parts of the country had well-developed monarchy system, notably the kingdoms of Buganda, Bunyoro, Ankole, and Toro (Were & Wilson, 1970). Within kingdoms, forests were either communally owned or used as an open access resource (Gombya-Ssembajjwe, 1995). People used forests for wood and non-wood forest products and had traditional ways of managing forest resources where clan members were free to collect forest products for domestic use such as firewood, grass for thatching, and clay for making pots (Turyahabwe & Banana, 2008). There were no written rules to describe forest management;

instead, the community grew up knowing how to use forest resources, which were passed on to future generations through oral and cultural traditions (Banana et al., 2008, 2014). Controls on forests considered sacred were presided over by a person whom the community appointed as a caretaker; if a person went to the forest without reporting the purpose of the visit to the clan head or the elders, or changed the purpose in the forest, he or she would have a spiritual punishment that was an effective deterrent to forest offenders (Beyaraza, 2004). They included getting lost in the forest (Gombya-Ssembajjwe, 1995), the offender's home being afflicted by insect invasions, crop failure, and infertility. The relatives of the offender had to cleanse the wrath of the spirits by making religious offerings in addition to fines in form of animals or food. Offenders who failed to comply with penalties were cursed by the elders and considered outcasts and despised members of their communities (Gombya-Ssembajjwe, 1995). Although the tide of history has long gone out, it would be interesting to see if the continuation of such a community-based management regime would have resulted in better conservation of forest resources had it not been the introduction of scientific forest management by the colonial administration (Ndemere, 1997)—clearly the job could not have been much worse as forests have generally been exploited, degraded and today are a shadow of their former glory, even within 'protected' areas (Fig. 5.6).

Similarly in Kenya, prior to the formal British occupation, there was substantially more closed forests than there are today (Ofcansky, 1984) with forest resources managed communally. For most communities, the rules were enforced by a council of elders who, through sanctions and fines, ensured the sustainable use of communal tree and forest resources. Characteristics of traditional systems of management were those pertaining to religious and cultural systems where sacred groves represented an excluded forest area in which elders conducted traditional religious ceremonies. Such ceremonies included sacrifices for harvests, for rain, thanksgiving, and rites of passage e.g. circumcision and burial sites. Land management in forest areas was closely regulated; for example around Mt. Kenya the Kikuyu and Embu, both agricultural communities, had evolved a system of land management in which forest land was owned by clans, but use was restricted up to two miles into the

Fig. 5.6 With the colonial administration came further forest clearance to open land for plantation and provide timber for the contribution and fuel for mechanisms transport. This growing demand for forest product let to extensive tree nurseries being established and (**b**) clear boundaries between forest reserves and agricultural land being formed (**c**)

forest (Castro, 1988); beyond this, the forest was off-limits and bringing new land into cultivation would only follow community consultation and consensus.

Similarly in Tanzania, there is little written information on the state of the environment (Ylhäisi, 2003) or on forest management and policy during the precolonial times (Zahabu et al., 2009), and this period was characterised by managing the forests through traditional institutions, low populations, and minimal forest resource exploitation (Zahabu et al., 2009). People lacked the mechanised forestry and markets to over-exploit natural resources and thus their impact on the environment in general was localised (Malimbwi & Munyanziza, 2004). The main anthropogenic impact on forest formations during precolonial times was probably modification of large savanna areas by fires (Harris, 1980) with increasing transformation of upland areas to cultivate maize and other crops for the growing market-based economies opened by the Caravan trade (Chapter 4). Due to low national population, the impact of shifting cultivation and grazing on the environment was quite limited and localised. In Tanzania, like in other East African countries, forests were the sources of food, medicines, clothes, water, and place for spiritual activities (CFA, 2007). Hunting was also a fundamental activity for many cultures as it contributed significantly to the community's food supply. Management and use of forest resources was controlled through customary institutions including beliefs, taboos, and customs (Grundy, 1990). Small scale commercial exploitation of forests and woodlands for timber in Tanzania started early in the sixteenth century when specific tree species were selectively harvested (Malimbwi & Munyanziza, 2004). One forest ecosystem where there would have been longer-term impacts was in the mangrove forests that were a focus of targeted clearance of the large trees that are ideal wood for the construction of boats and ships: being at the heart of the development of the Swahili coast (Chapter 4). As far back as 1898, von Bruch-Hausen, who had explored the Rufiji Delta, recommended a government ordinance to terminate unregulated exploitation of local mangrove forests (Mgaya, 2016), implying a demand for sustainable exploitation of forest recourses.

In addition to the use of forests for food, fuel, water, grazing, honey, saltlicks, or refuge, some also provided a home for forest-dwelling people such as the BaTwa pygmies populations of Southwest Uganda or the numerous groups collectively called 'Dorobo' by Maasai, an umbrella term for forest peoples living without cattle such as the Hazda. Forests utilised by forest dwellers may have prevented expansion of agriculture and provided informal forest protection. However, on the arrival of the colonial forest department, land was claimed without considering the rights of local inhabitants and strict regulations were imposed where native rights to the forests were not recognised, instead they were termed as either illegal squatters or tenants-at-will of the Crown. The displacement of indigenous peoples resulted in people being confined within reserves, under the Native Lands Trust Ordinance of 1930. An important consequence of confining indigenous populations and restricting access to large forest blocks was the depletion and over-exploitation of forests within reserves. One of the main factors undermining indigenous knowledge and management practices was colonial thinking that indigenous people were considered a threat to forests, and legislation was issued accordingly (Kajembe, 1994; Munyanziza & Wiersum, 1999). Fortress conservation secured economic output for the colonial regimes and protected game species (Mgaya, 2016). In the Usambara Mountains, central colonial forester officials blamed locals for 'encroaching' into the forest reserves, whereas district officers were more sympathetic towards the indigenous population's use of their natural resources (Conte, 2004). An era of intensive forest reservation began by the German administration in 1903, a time when East Africa was perceived to be in state of crisis (Sunseri, 2003). A forest conservation ordinance followed in 1904 and the reservation of forest areas began. There was an imposition of unpopular colonial legislation-imposed soil and watershed conservation schemes. Many reserves were gazetted in watershed areas from the point of view that they would be crucial to the fertility of the land and thus the livelihood of people in the future (Schabel, 1990; Sunseri, 2003). Across East Africa, traditional land use practices were largely regarded as detrimental to the environment by the colonial administration; land was also a way to establish control: as the State established protected areas, which restricted local people access to the natural resources upon which they depended for their livelihoods (Pendzich, 1994; Ylhäisi, 2003).

The Germans arrived in Tanzania with visions of scientific forestry derived from European templates of forest management that were premised on the creation of forest reserves (Fig. 5.6) without human settlement (Sunseri, 2005). In 1892, Eugen Kruger became the first professional forester who, along with Dr. Franz Stuhlmann, developed forestry in the territory (Schabel, 1990). The Crown Land Ordinance developed the first Tanzanian forestry act in 1895 that empowered the colonial state to create forest reserves (Mgaya, 2016). The ideal forest, according to the German model, was one of uniform tree species and size that could be quantified and harvested in set rotations to meet fiscal and industrial needs. The ability to create forest reserves not only provided the Germans with means to extract resources, but it also offered them a mechanism to control people. During the German colonial regime in Tanzania, there existed a preservationist approach to forest management

(Mgaya, 2016). The British administration operated under the League of Nations mandate: concerning natural resources management the mandate stated, 'in framing the laws relating to the holding or transfer of land and natural resources, the administrating authority should take into consideration native laws and customs and respect the rights of the native population' (Neumann, 1997). However, the British largely followed the German forestry policy in terms of adhering to the idea that the government ultimately held the ownership of the forests, and the primary goal of forest management was the generation of revenues (Fig. 5.7) (Neumann, 1997). Almost all forest lands, whether occupied or unoccupied, were declared public or Crown land. In addition, forest policies had direct and indirect impacts on resource control and access.

Fig. 5.7 Forest clearance was particularly acute in coastal areas where there was low lying ground suitable for large scale plantations and the timber resource was highly accessible. Some 95% of the coastal forests has been degraded or impacted on by forestry operations (**a**, **b**) and the ensuing conversion of wood to charcoal (**c**, **d**, **e**) for increasing energy demand (Photographs **a** and **b**: Antje Ahrends, Photographs **c**, **d** and **e**: Rob Marchant)

The exclusive discourse and practices of control that started under the German rule were intensified during the British rule (Conte, 1996). The British launched a policy allowing people to use forest resources to fulfil their household needs (Neumann, 1998), although there were simultaneously continued practices of enforcement and exclusion in forest control and the strict regulations were not accepted without resentment and forest reserve boundaries were challenged by herders and farmers (Mgaya, 2016). To try and appease these claims a new decentralised institution called Native Authority Forest Reserves was established in the early 1930s that demarcated forests for community use and for crown use. The institution not only served the policy of indirect rule, but also served to eliminate free issue of the indigenous people in a more acceptable way (Neumann, 1997). However, since the inception the amount of forest under state control rose by 1000s of square miles per year where the amount of forest under community use rose by an order of magnitude lower.

Settling of the land with plantations needed extensive clearing of the forested areas (Fig. 5.7) to allow plantations of Aloe, Cloves, and Coconuts to expand. Timber from forest reserves would generate income through concessions and royalties, and forest resources would contribute to growth in other sectors such as railways, agriculture, and mines (Forest Department, 1951; Mgaya, 2016). The colonial government needed more timber and fuel to cater for industrial development; particularly materials for the railway that was being constructed from Mombasa to Kisumu and wood fuel for steam ships and trains (Kamugisha, 1993; Turyahabwe & Banana, 2008). In early 1902 the colonial government appointed C. F. Elliott to the newly created post of conservator of forests—the introduction of exotic tree species for forest plantation such as Pine soon followed, as did Teak (*Tectona grandis*) that grew well along the coast. Black wattle, prized for its timber, was grown at several localities and other exotics were tested throughout the country including Mlanje cedar (*Widdringtonia whytei*), Bhutan cypress (*Cupressus torulosa*), Monterey cypress (*Cupressus macrocarpa*), California peppertree (*Schinus molle*), golden pine (*Grevillea robusta*), Jacaranda (*Jacaranda mimoscefolia*), and she-oak (*Casuarina*) (Ofcansky, 1984).

The fertile coastal strip was a particular focus of forest clearance (Fig. 5.7) as, under the British forest administration, the primary goal was the generation of revenues from the colony (Neumann, 1997); with some 95% of the coastal forest being converted to agriculture with a particular focus on commercial extraction of timber especially *Afzelia quanzensis, Combretum schumannii,* and *Milicia excelsa.* The exploitation of hardwood timber in East African coastal forest, such as the patch of forest remaining at Arabuko Sokoke forest, reaches back to the 1920s when *Brachylena huillensis, Afzelia quanzensis,* and *Manilkara sansibara* were systematically logged by European sawmills (Moomaw, 1960) (Fig. 5.7). Although there is no accurate data on the volume of wood extracted, the effects of past commercial selective logging and continued removal of large trees of several species have changed the species composition of the forest (Robertson & Luke, 1993). Remaining hard wood tree species like *Cynometra webberi* and *Brachystegia spiciformis* are still exploited for wood carvings for the local and regional (tourism) market (Habel et al., 2017; Wass, 1995). The impacts on the whole ecosystem would have been acute, particularly in coastal and montane areas where there are many endemic and range-restricted species (Virani et al., 2010). Montane areas were particularly impacted on where the target trees would have been the East African Camphor (*Ocotea usambarensis*) in Kenya and Tanzania, Mvule (*Chlorophora excelsa*) from Uganda and Tanzania, and the Mahoganies (*Khaya* spp. and *Entandrophragma* spp.) mainly from Uganda. (Rule, 1945). The African Pencil Cedar (*Juniperus procera*) was widely used by local housing ahead of becoming locally extirpolated and being replaced by expanding areas of exotic conifers, particularly Cypress (*Cupressus* spp.) in parts of Kenya (Rule, 1945). On first appearances supplies appeared to be considerable, particularly in Tanzania, but exploitation reveals that immature trees were not as common as first thought (Rule, 1945). Mvule supplies were getting scarce by 1945 and the exploitation areas in post WWII were mere remnants of once extensive areas cleared by fire and cultivation (Rule, 1945).

Formal management of forests in Uganda started in 1898 when the colonial government's Scientific and Forestry Department was established. The Forestry Department was established as a separate body in

1917 and renamed the Forest Department in 1927 (Forest Department, 1951). The first foresters were the British expatriates who arrived in the country in 1921. Working within the Forestry Department, the colonial government embarked on the process of reservation until the 1940s when the boundaries of Uganda's forest estate, like today, became established. In all crown land the control of resources was transferred from clan heads to chiefs and the rights of communities were lost or could only be exercised at the pleasure of the colonial Governor. The new regulations applied to both indigenous and foreigner populations, although indigenous people could procure some forest products free of charge provided the products were for domestic use. The first national forest policy was formulated in 1929 (Forest Department 1955) and stressed the need for more areas under Crown control, the reafforestation of more land, the management of forests for timber production, and the generation of adequate financial returns to the country. However, the needs of local people whose livelihoods depended on forests were not addressed. The process of acquiring forestry land by the colonial government was unsystematic and sometimes people, such as the indigenous Benet community in Mt Elgon and the Batwa in Mgahinga and Bwindi forests, were forcefully relocated. Large common lands including village grazing lands, community forests, and grasslands were gazetted as forest reserves through blanket notifications (Mugyenyi et al., 2005). In the process of gazetting forest reserves, the colonial authorities changed the public attitude towards forest management by undermining traditional rights to forest land ownership and subsequently undermining the claims of indigenous communities to forest resources (Turyahabwe & Banana, 2008). As rights to use forest resources were granted only to a few privileged individuals (Turyahabwe & Banana, 2008), this created a series of local elites who were educated, rich and, in return for land and power, supported colonial policies (Turyahabwe & Banana, 2008).

Reservation of forests clearly had negative impacts on communities who depended on forests for their livelihoods (Banana et al., 2014). As early as 1908, the protectorate government realised the growing conflict and established fuelwood plantations near large towns (Forest Department, 1951). Increasing population, resettlement patterns associated with urbanisation, introduction of modern economies, changes in local government, and a shift to western cultural practices had some impact on Kaya conservation. Castro (1991) indicates challenges to sacred groves as land appropriation by colonial administration and white settlers; the formation of a formal political hierarchy by the colonial government which eroded traditional authority of clan leadership, religious conversion to Christianity, mass education, and land privatisation, all of which diminished the status of sacred groves and traditional systems of management. The survival of some traditional strategies and their effectiveness in forest conservation serve to indicate the potential role they could play today (Mwangi et al., 2018). By 1925, there were a total of 212 forest reserves covering 3707 square miles (Neumann, 1998). In 1953, a forest policy was adopted emphasising the importance of protecting forest resources and managing them in the most productive way (Maddox et al., 1996). Many people ignored measures such as the prohibition of cultivation on steep slopes, control of bush fires, and timber exploitation.

The primary thrust of the 1938 forest policy that does not seem out of alignment with current thinking, stressed the important roles trees and forests play in the environment, in terms of economic benefits, climate amelioration, protecting water catchments, and minimising soil erosion (Turyahabwe & Banana, 2008). The underlying theme was that Uganda would benefit from a greater forest cover than it had, and hence the steady gazetting of more forest reserves during the period that followed. The policy laid guidelines for a two-tier system of forest management: whereby there were central forest reserves under the control of central government and local forest reserves under the proxy control of the local administration and the Buganda Government. It was argued then that the responsibility of meeting village-level wood requirements should rest on the local-level governments who were in a much better position to look after a large number of small forest reserves, costs would be less, and local administration involvement would help generate a vested interest in forestry (Forest Department 1955; Hamilton, 1984). Considerable areas of swamp were seen as wasted land, and subsequently drained and afforested for the dual purpose of reducing mosquito breeding sites and to provide fuel supplies. Large areas of swamp at Namanve near Kampala were afforested between 1930 and 1937 by means of loan funds to provide fuel for industries of that town (Turyahabwe & Banana, 2008).

Large quantities of timber (Fig. 5.7) were required to build schools and hospitals for the expanding population. Kenya's forests were also used by a small but growing lumber industry. The colony's forest products trade was originally centred around the Equator Sawmills of Ewart Grogan, who had access to 186,000 acres of Coastal Forest. In 1920 a network of twenty-four independent sawmills were scattered throughout Kenya, some for export and much for local consumption. Between 1905 and 1907 alone, the railroad's wood consumption rose from 27,947 tonnes to 47,309 tonnes—equivalent to around 400,000 trees! Fuel wood and timber shortages also spurred the introduction of *Eucalyptus* and the establishment of tree plantations in the railway's vicinity (HMSO, 1908). With the outbreak of the Second World War, forest conditions again deteriorated throughout Kenya. Camps for an expanding military were built at very short notice; some sawmills were compelled to double and treble their output, while others installed electric lighting to work around the clock to meet the demands. Altogether, the quantity of wood cut in the forest reserves increased from approximately 24,250 tonnes before the war to 46,700 tonnes in 1940 and 55,100 tonnes in 1941 (Ofcansky, 1984). These demands of forest resources continued with the result that many Crown forest reserves were extensively deforested through the colonial period with many of these production focussed mandates continuing through in the postcolonial administrations (Chapter 6).

5.5 Protected Area Foundations

Neumann (2005) calls the period after the Second World War 'the Conservation Boom in British Colonial Africa', as the period is characterised by a push for forest reserves without human settlements. Besides forest reserves, large areas of woodland and grassland were demarcated as game reserves and National Parks intended to protect the African fauna for purposes of tourism, hunting, and scientific study (Mgaya, 2016). The establishment of the National Parks Ordinance of 1959 was a final move in a process towards a rigid form of conservation that had been going on since the end of the caravan trade (Chapter 4). Early safari tourism in the form of hunting parties was promoted from the very beginning of the colonial period and provided a source of revenue, particularly for Kenya (Eliot, 1905; Sandford, 1919) and Tanzania (Miller & Yeager, 2018). There were numerous 'characters' involved in this trade, maybe most notable was Richard Cuninghame who established a hunting business from 1905, particularly supplying the mass trade in trophies for the worlds museums and naturalist societies such as the Smithsonian Institution, or Natural History Museum, London. Cuninghame led Carl Akeley from the Field Museum of Chicago, Theodore Roosevelt legendry hunting safari of 1909 that killed hundreds of specimens. Trophy hunting of big game soon reached proportions that endangered the survival of the large herds and the first game reserves in Kenya and Tanganyika were demarcated in 1896 (Fig. 5.8).

In Kenya, the huge Southern and Northern Game Reserves were established in the early 1900s, occupying most of the pastoral lands used by the Maasai and the Samburu pastoralists. Many other game reserves soon followed, and pastoralists were moved out (Enghoff, 1990). Following these game control ordinances at the beginning of the twentieth century, the first true National Parks in East Africa designated in the 1940s (Western 1997). This trend continued as more land was taken for the creating of hunting wildlife reserves and National Parks such as Serengeti, Amboseli, the Maasai Mara, Nairobi National Park, Samburu Reserve, Lake Nakuru National Park, and Tsavo. The forest laws became increasingly hostile to Tanzanians living inside the reserved lands, and it culminated in the British re-gazetting the Serengeti National Park in 1959 and resettling the Maasai outside the park (Nelson et al., 2007). There was strict regulation of access to protected forests and certain valuable tree species; again justified in terms of conservation and economic interests. Once the national park system was established, the number of parks and protected areas increased dramatically throughout the remainder of the colonial period resulting in large of East Africa being gazetted. Some parks were small, such as Nairobi National Park, while others were vast; the original proposal for Serengeti National Park encompassed 29,500 square kilometres (McCabe, 2003). By the end of the colonial period 43,673 km^2 or 7.5% of Kenya, was classified as protected in 36 National Parks or Game Reserves; 151,496 km^2 or 16%

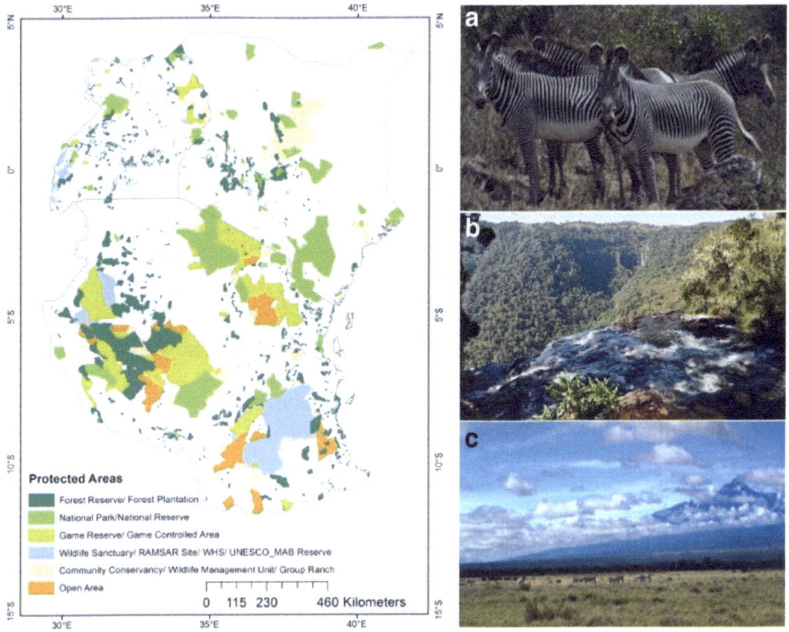

Fig. 5.8 The number of National Parks and Protected Areas increased dramatically throughout the colonial period and continue to do so through to the present day resulting in large of East Africa being gazetted. Some National Parks are small and shrinking, such as Nairobi National Park, while others were vast; resulting in large amounts of the area of East Africa being gazetted, particularly in Tanzania (All photographs: Rob Marchant)

of Tanzania is protected in 32 National Parks and Game Reserves; and 20,650 km^2 or 8.7% of Uganda protected in 26 National Parks and Game Reserves (Barrow et al., 2001). The needs and use of indigenous peoples were increasingly becoming recognised, and as we will explore in Chapter 7, the success of conservation policy is often linked to the incorporation of local communities in the conservation process (McCabe, 2003). There we also a series of negotiated schemes, for example, the Ngorongoro Conservation Area (Fig. 5.9) was established in Tanzania in 1959 where pastoralists, in exchange for vacating the Serengeti, were granted secure tenure to live alongside wildlife within a multiple land use protected area. In addition to the distribution of pastoral populations,

Fig. 5.9 The Ngorongoro Conservation Area consisting of the Ngorongoro Crater (**a**) and communal grazing land (**b**) extending to the Serengeti where the Maasai pastoralists have secure tenure to live alongside wildlife within a multiple land-use protected area (All photographs: Rob Marchant)

the altering of pastoral cultural ecology also had a significant impact on wider ecology and functioning of the savanna as the distribution of wildlife and stock ecosystem engineers were altered.

References

Anderson, D. (1984). Depression, Dust Bowl, Demography, and Drought: The Colonial State and Soil Conservation in East Africa during the 1930s. *African Affairs* [online], *83*, 321–343. https://www.jstor.org/stable/722351 [Accessed 11 June 2018].

Apuuli, K. (1994). *A Thousand Years of Bunyoro-Kitara Kingdom—The People and the Rulers* (p. 153). Fountain Publishers.

Banana, A., Bahati, J., Gombya-Ssembajjwe, W., & Vogt, N. (2008). Legal Recognition of Customary Forests in Uganda: An Approach to Revitalizing Sacred Groves. In M. J. Sheridan & C. Nyamweru (Ed.), *African Sacred Groves: Ecological Dynamics and Social Change* (pp. 195–206). James Currey and Ohio University Press.

Banana, A. Y., Byakagaba, P., Russell, A. J., Waiswa, D., & Bomuhangi, A. (2014). *A Review of Uganda's National Policies Relevant to Climate Change Adaptation and Mitigation: Insights from Mount Elgon.*

Barrow, E., Gichohi, H., & Infield, M. (2001). The Evolution of Community Conservation Policy and Practice in East Africa. In D. Hulme & M. Murphree (Ed.), *African Wildlife and Livelihoods: The Promise and Performance of Community Conservation*. James Currey.

Beyaraza, E. K. (2004). *Evolution of Property Rights in Uganda: A Legal and Philosophical Analysis of Past, Present and Future Trends*. Makerere University.

Bruner, E. M., & Kirshenblatt-Gimblett, B. (1994). Maasai on the Lawn: Tourist Realism in East Africa. *Cultural Anthropology, 9*(4), 435–470.

Campbell, David J., & Migot-Adhola, S. E. (1979). *Midterm Evaluation of Aspects of the Foot and Mouth Disease Control Programme in Kajiado and Narok Districts, Kenya*. Report to the Veterinary Department, Ministry of Agriculture by the Institute for Development Studies. University of Nairobi.

Castro, A. P. (1988). Southern Mount Kenya and Colonial Forest Conflicts. In J. F. Richards & R. P. Tucker (Ed.), *World Deforestation in the Twentieth Century*. Duke University Press.

Castro, A. P. (1991). Indigenous Kikuyu Agroforestry: A Case Study of Kirinyaga, Kenya. *Human Ecology, 19*(1), 1–18.

Conte, C. (1996). Nature Reorganized: Ecological History in the Plateau Forests of the West Usambara Mountains 1850–1935. In G. Maddox, J. L. Giblin, & I. N. Kimambo (Ed.), *Custodians of the Land: Ecology and Culture in the History of Tanzania*. James Currey, Ohio University Press.

Conte, C. (2004). *Highland Sanctuary: Environmental History in Tanzania's Usambara Mountains*. Ohio University Press.

Conte, C. A. (1999). The Forest Becomes Desert: Forest Use and Environmental Change in Tanzania's West Usambara Mountains. *Land Degrad Dev* [online], *10*, 291–309. https://doi.org/10.1002/(SICI)1099-145X(199907/08)10:4<291::AID-LDR363>3.0.CO;2-W

Eliot, C. (1905). *The East African Protectorate*.

Enghoff, M. (1990). Wildlife Conservation, Ecological Strategies and Pastoral Communities: A Contribution to the Understanding of Parks and People in East Africa. *Nomad. Peoples, 25–27*, 93–107.

Forest Department. (1951). *A History of the Uganda Forest Department 1898–1929* (Bulletin No. 3). D.L.Patel Press.

Forest Department. (1955). *A History of Uganda Forest Department 1930–1950* (Bulletin No .4). Government Printer.

Fungomeli, M., Githitho, A., Frascaroli, F., Chidzinga, S., Cianciaruso, M., & Chiarucci, A. (2020). A new vegetation-plot database for the Coastal forests of Kenya. *Vegetation Classification and Survey, 1*, 103.

Gillson, L. (2015). Evidence of a Tipping Point in a Southern African Savanna? *Ecological Complexity, 21*, 78–86.

Gombya-Ssembajjwe, W. S. (1995). Sacred Forests as a Traditional Arrangement in Modern Ganda Society. *Uganda Journal, 42*, 32.

Grundy, I. M. (1990). *The Potential for Management of Indigenous Woodlands in Communal Farming Areas of Zimbabwe with Reference to the Regeneration of Brachystegia Speciformis and Julbernardia Globiflora.* M. Sc. Thesis, University of Zimbabwe, Harare.

Gunther, J. (1955). *Inside Africa* (p. 411). Harper and Brothers. ISBN 0836981979.

Habel, J. C., Casanova, I. C. C., Zamora, C., Teucher, M., Hornetz, B., Shauri, H., Mulwa, R. K., & Lens, L. (2017). East African Coastal Forest under Pressure. *Biodiversity and Conservation, 26*(11), 2751–2758.

Hamilton, A. C. (1984). *Deforestation in Uganda.* Oxford University Press.

Harris, D. R. (1980). *Human Ecology in Savanna Environments* (p. 522). Academic Press.

Herskovits, M. (1926). The Cattle Complex in East Africa. *American Anthropologist, 28.*

HMSO. (1908). *Great Britain: Annual Report of the East Africa Protectorate, 1906–1907*. HMSO.

Homewood, K. (1995). Development, Demarcation and Ecological Outcomes in Maasailand. *Africa* [online], *65*, 331–350. https://doi.org/10.2307/1161050

Hydén, G. (1980). *Beyond Ujamaa in Tanzania: Underdevelopment and an Uncaptured Peasantry.* University of California Press.

Iliffe, J. (1967). The Organization of the Maji Maji Rebellion. *The Journal of African History, 8*(3), 495–512.

Iliffe, J. (1979). *A Modern History of Tanganyika.* Cambridge University Press.

Illife, J. (1987). *The African Poor*. Cambridge: Cambridge University Press.

James, L. (1939). The Kenya Masai: A Nomadic People under Modern Administration. *Africa, 12*(1), 49–73.

Jew, E. K. K., Dougill, A. J., & Sallu, S. M. (2017). Tobacco Cultivation as a Driver of Land Use Change and Degradation in the Miombo Woodlands of South-West Tanzania. *Land Degradation and Development, 28*, 2636–2645. https://doi.org/10.1002/ldr.2827

Johansson, J., & Svensson, J. (2002). *Land Degradation in the Semi-arid Catchment of Lake Baringo, Kenya: A Minor Field Study of Physical Causes with a Socio-economic Aspect.* Earth Sciences Centre.

Kajembe, G. C. (1994). *Indigenous Management Systems as a Basis for Community Forestry in Tanzania: A Case Study of Dodoma Urban and Lushoto Districts* (Tropical Resource Management Papers, No.6). Wageningen Agricultural University.

Kamugisha, J. R. (1993). *Management of Natural Resources in Uganda: Policy and Legislation Landmarks, 1890–1990*. SIDA Regional Soil Conservation Unit.

Kjekshus, H. (1996). *Ecology Control and Economic Development in East African History* (2nd ed.). James Currey.

Klopp, J. M. (2001). Ethnic Clashes and Winning Elections: The Case of Kenya's Electoral Despotism. *Canadian Journal of African Studies* [online], *35*, 473–517. https://doi.org/10.1080/00083968.2001.10751230

Knowles, D. (2016). The Lunatic Express. 1843. *The Economist*. Retrieved July 15, 2016.

Knowles, J. N., & Collett, D. P. (1989). Nature as Myth, Symbol and Action: Notes Towards a Historical Understanding of Development and Conservation in Kenyan Maasailand. *Africa, 59*, 4.

Lamprey, R., & Waller, R. (1990). The Loita-Mara Region in Historical Times: Patterns of Subsistence, Settlement and Ecological Change. In P. T. Robertshaw (Ed.), *Early Pastoralists of South-Western Kenya* (pp. 16–35). British Institute in Eastern Africa.

Lewis, E. A. (1934). Tsetse-flies in the Masai Reserve, Kenya Colony. *Bulletin of Entomological Research, 25*(4), 439–455.

Leys, C. (1973). Interpreting African underdevelopment: Reflections on the ILO report on employment, incomes and equality in Kenya. *African Affairs, 72*(289), 419–429.

Maddox, G., Giblin, J., & Kimambo, I. N. (1996). *Custodians of the Land, Ecology and Culture in the History of Tanzania*. James Currey Ltd.

Malimbwi, R. E., & Munyanziza, E. (2004). From Local People to Local People: Forest Management and Policy in Tanzania over 100 Years. *UNISWA Research Journal of Agriculture, Science and Technology, 7*(2), 101–108.

Mamdani, M. (2018). *Citizen and Subject: Contemporary Africa and the Legacy of Late Colonialism*. Princeton University Press.

Marchant, R. A. (2010). Understanding Complexity in Savannas: Climate, Biodiversity and People. *Current Opinion in Environmental Sustainability, 2*, 101–108.

McCabe, J. T. (2003). Disequilibrial Ecosystems and Livelihood Diversification among the Maasai of Northern Tanzania: Implications for Conservation Policy in Eastern Africa. *Nomadic Peoples, 7*(1), 74–91.

Mgaya, E. (2016). Forest and Forestry in Tanzania: Changes and Continuities in Policies and Practices from Colonial Times to the Present. *Journal of the Geographical Association of Tanzania, 36*(2), 45–58.

Migot-Adholla, S., Hazell, P., Blarel, B., & Place, F. (1991). Indigenous Land Rights Systems in Sub-Saharan Africa: A Constraint on Productivity? *World Bank Economic Review* [online], *5*, 155–175. https://doi.org/10.1093/wber/5.1.155

Miller, N., & Yeager, R. (2018). *Kenya: The Quest for Prosperity*. Routledge.

Moomaw, J. C. (1960). *A Study of the Plant Ecology of the Coast Region of Kenya Colony, British East Africa*. Goverment Printer.

Morgan, W. T. W. (1963). The "White Highlands" of Kenya. *The Geographical Journal, 129*(2), 140–155. https://doi.org/10.2307/1792632

Mugyenyi, O., Twesigye, B., & Muhereza, E. (2005). *Balancing Nature Conservation and Livelihoods: A Legal Analysis of the Forestry Evictions by the National Forestry Authority*. ACODE Policy Briefing Paper 13.

Munyanziza, E., & Wiersum, K. F. (1999). Indigenous Knowledge of Miombo Trees in Morogoro, Tanzania. *Indigenous Knowledge and Development Monitor, 7*, 10–13.

Mwangi, H. M., Lariu, P., Julich, S., Patil, S. D., McDonald, M. A. & Feger, K. H. (2018). Characterizing the Intensity and Dynamics of Land-use Change in the Mara River Basin, East Africa. *Forests, 9*(1), 8.

Ndemere, P. (1997). *Multiple Use Forest Values, Silvicultural Impacts of Timber Management, and Sustainability of Forest Management in Uganda's Budongo and Mabira Forest Reserves.* Ph.D. thesis, College of Environmental Science and Forestry, Syracuse, State University of New York, New York.

Nelson, F., Nshala, R., & Rodgers, W. A. (2007). The Evolution and Reform of Tanzanian Wildlife Management. *Conservation and Society, 5*(2), 232–261.

Neumann, K. (2005). The Romance of Farming: Plant Cultivation and Domestication in Africa. In A. B. Stahl (Ed.), *African Archaeology*. Blackwell.

Neumann, R. (1997). Forest Rights, Privileges and Prohibitions: Contextualizing State Forestry Policy in Colonial Tanganyika. *Environment and History, 3*(1), 45–68.

Neumann, R. P. (1998). *Imposing Wilderness: Struggles over Livelihood and Nature Preservation in Africa*. University of California Press.

Ochieng, W. R., & Maxon, R. M. (1992). *An Economic History of Kenya*. East African Publishers.

Ofcansky, T. (1984). Kenya Forestry under British Colonial Administration, 1895–1963. *Journal of Forest History, 28*(3), 136–143. https://doi.org/10.2307/4004697

Ofcansky, T. P. (1981). *A History of Game Preservation in British East Africa, 1895–1963*. Unpublished Ph.D. dissertation, West Virginia University.

Pendzich, C. (1994). *The Role of Alternative Conflict Management in Community Forestry* (Working Paper No. 1). Food and Agriculture Organisation of the United Nations.

Robertson, S. A., & Luke, W. R. Q. (1993). *The vegetation and conservation status of Kaya coastal forests in Kenya* (425 pp.). Nairobi: WWF. https://www.africabib.org/rec.php?RID=K00001940

Rule, A. (1945). East African Timber Production. *Empire Forestry Journal, 24*(1), 47–51. Retrieved March 25, 2021, from http://www.jstor.org/stable/42594332

Saitoti, T. O., & Beckwith, C. (1980). *Maasai*. Harper Collins.

Sandford, G. (1919). *An Administrative and Political History of the Masai Reserve* (p. 234). Waterlow and Sons.

Schabel, H. (1990). *Tanganyika Forestry under German Colonial Administration, 1891–1919*. A Special Issue on International Forest History.

Schneider, K. (1957). The Subsistence Role of Cattle among the Pokot of East Africa. *American Anthropologist, 59*, 278–300.

Smith, C .D. (1989). *Did Colonialism Capture the Peasantry? A Case Study of the Kagera District, Tanzania* (Vol 83). Uppsala: Scandinavian Institute of African Studies.

Snelder, D. J., & Bryan, R. B. (1995). The Use of Rainfall Simulation Tests to Assess the Influence of Vegetation Density on Soil Loss on Degraded Rangelands in the Baringo District, Kenya. *Catena, 25*(1–4), 105–116.

Stanley, H. M. (1889). *Through the Dark Continent: Or, the Sources of the Nile, Around the Great Lakes of Equatorial Africa, and Down the Livingstone River to the Atlantic Ocean*. Low, Marston, Searle and Rivington.

Sunseri, T. (2003). Reinterpreting a Colonial Rebellion: Forestry and Social Control in German East Africa, 1874–1915. *Environmental History, 8*(3).

Sunseri, T. (2005). 'Something Else to Burn': Forest Squatters, Conservationists, and the State in Modern Tanzania. *The Journal of Modern African Studies, 43*(4), 609–640.

Sutton, J. E. G. (1971). Temporal and Spatial Variability in African Iron Furnaces. In K. Haaland & P. L. Shinnie (Eds.), *African Iron Working, Ancient and Traditional* (pp. 164–196). Norwegian University Press.

Thompson, J. (1887). *Through Masai land: A Journey of Exploration among the Snowclad Volcanic Mountains and Strange Tribes of Eastern Equatorial Africa*. Low, Marston, Searle, and Rivington.

Turyahabwe, N., & Banana, A. Y. (2008). An Overview of History and Development of Forest Policy and Legislation in Uganda. *International Forestry Review, 10*(4), 641–656.

van Zwanenberg, R. M. A., & King, A. (1975). Nomadic Pastoralism: The Process of Impoverishment. In *An Economic History of Kenya and Uganda 1800–1970* (pp. 79–109). Palgrave Macmillan.

Virani, M. Z., Njoroge, P., & Gordon, I. (2010). Disconcerting Trends in Populations of the Endangered Sokoke Scops Owl Otus Ireneae in the Arabuko-Sokoke Forest, Kenya. *Ostrich, 81*, 155–158.

von Höhnel, L. R. (1894). *Discovery of Lakes Rudolf and Stefanie: A Narrative of Count Samuel Teleki's Exploring and Hunting Expedition in Eastern Equatorial Africa in 1887 and 1888* (Vol. 2). Longmans, Green and Co.

Waller, R. (1988). Emutai: Crisis and Response in Maasailand, 1883–1902. In D. Johnson & D. Anderson (Eds.), *The Ecology of Survival: Case Studies from Northeast African History* (pp. 73–114). Westview Press.

Wass, P. (Ed.). (1995). *Kenya's Indigenous Forests: Status, Management and Conservation.* IUCN.

Were, G. S., & Wilson, D. A. (1970). *East Africa Through a Thousand Years.* African Publishing Corporation.

Whitehouse, G. C. (1948, March). The Building of the Kenya and Uganda Railway. *The Uganda Journal. Kampala: The Uganda Society, 12*(1), 1–15.

Widgren, M., & Sutton, J. E. (1999). *'Islands' of Intensive Agriculture in the East African Rift and Highlands: A 500-year Perspective.* Departments of Human and Physical Geography, University of Stockholm.

Wynants, M., Kelly, C., Mtei, K., Munishi, L., Patrick, A., Rabinovich, A., Nasseri, M., Gilvear, D., Roberts, N., Boeckx, P., Wilson, G., Blake, W. H., & Ndakidemi, P. (2019). Drivers of Increased Soil Erosion in East Africa's Agro-Pastoral Systems: Changing Interactions Between the Social, Economic and Natural Domains. *Regional Environmental Change, 19*, 1909–1921.

Ylhäisi, J. (2003). Forest Privatisation and the Role of Community in Forests and Nature Protection in Tanzania. *Environmental Science and Policy, 6*, 279–290.

Zahabu, E., Eid, T., Kajembe, G. C., Mbwambo, L., Mongo, C., Sangeda, A. Z., Malimbwi, R. E., Katani, J. Z., Kashaigili, J. J., & Luoga, E. J. (2009). Forestland Tenure Systems in Tanzania: An Overview of Policy Changes in Relation to Forest Management [Eiendomsforholdene for skogarealer i Tanzania: en oversikt over policyendringer i skogforvaltningen]. *INA-Fagrapport, 14*, 24.

6

Postcolonial Transitions and Recent Political History

6.1 Postcolonial Pathways: Early Transitions

As would be expected, different nations, with different political structures and different histories, there have been very different postcolonial transitions that have taken place across East Africa. These range from African socialism (Tanzania) to capitalism with a strong anti-communist mandate (Kenya) to the instigation of the traditional rulers of power in Uganda, the Kabaka, albeit in a more ceremonial capacity than in precolonial Uganda (Bryceson, 2002). Behind these transitions were some charismatic leaders that took East Africa out of the colonial era (Jomo Kenyatta (Kenya), Julius Nyere (Tanzania), Apollo Milton Obote (Uganda), and the impact that these have had, good and bad, such as the mass evictions of Asians from Uganda during the 1970s, the rise of corruption, and massive national development agendas. The East Africa coast that links Kenya and Tanzania similarly underwent a transition. By the 1960s the unity of the Western Indian Ocean was collapsing as the newly independent states sought to limit trade with dhow and shipping movements were limited and controlled. Zanzibar, for example, outlawed port calls by vessels that had called previously in Arab ports

R. Marchant, *East Africa's Human Environment interactions*,
https://doi.org/10.1007/978-3-030-88987-6_6

(Gilbert, 2002). The new Kenyan and Tanzanian governments saw nothing positive in the wider Western Indian Ocean region and instead sought to identify themselves with a new continental Africa (Gilbert, 2002). The coast, with its history of regionalism and cosmopolitanism, became part of the Kenyan and Tanzanian mainland and centrally governed with the influence of Oman being largely restricted to the Zanzibar archipelago.

6.1.1 Kenya

Out of all the three countries, Kenya's passage from colonial rule was more fought for rather than being negotiated. Although the struggle for independence started in 1952, this followed years of growing unrest around lack of land, prosperity, and opportunity, culminating with the Mau Mau uprising. A precursor for the well-documented resistance in Central Kenya was the 'Dini ya Msambwa' or Cult of the Ancestors based across the foothills of Mt Elgon linking Kenya and Uganda. The 'Dini' movement was led by a charismatic prophet Elijah Masinde that rapidly spread into central Kenya and the Mau Mau uprising. At the heart of the Mau Mau resistance struggle, was the Kenya Land Freedom army that comprised mainly Kikuyu, Embu, and Meru ethnic groups but also with Maasai and Kamba. There were many roots to the uprising although at the heart of this were the issues of land and by 1948 the dominant ethnic group of some 1,250,000 Kikuyu owned 5200 km^2, while 30,000 British settlers owned 31,000 km^2! Through the lens of land control, the colonial administration also enacted control and power over livelihoods and effectively forced many Kenyans into forced labour and conscription, with treatment and conditions often amounting to slave labour. Many of the Kenyans at the heart of this mistreatment were ex British army soldiers that, when returning to Kenya, were not granted land, paid, or recognised for their service to their 'country'. These people found a home for protest in the Kenya Africa Union formed in 1946, which grew increasingly until it claimed its first victim in 1952, soon after a state of emergency was declared, and the legitimate Mau Mau struggle began. The struggle was long and drawn out, dividing communities and support, and resulting in at least 12,000 deaths and great expense to the colonial administration (Anderson, 2005). There are many interpretations of the term 'Mau Mau'—I use the interpretation by Wangari Maathai who states that when beginning a list in Kikuyu, one would say, '*maũndũ ni mau*', 'the main issues are…': the three main issues were land, freedom, and self-governance (Maathai, 2006). Rather than take a moderate line and seek advice from settlers who lived with the Kikuyu communities, the colonial government took a particularly combative approach by painting the Mau Mau as a savage and tribal cult and dedicated psychological and divisive warfare aimed to further marginalise the Mau Mau and Kikuyu (Mahone, 2006), and indeed the wider nation through a divide and rule campaign.

The British response was multifaceted in terms of direct conflict, fighting the Mau Mau in the forests of central Kenya, rounding up and putting on trial suspected leaders, and a process of villagisation. The rounding up of sympathisers/supporters of the cause started with a sweep of Nairobi where some 10,000 people were rounded up, included Jomo Kenyatta, the president in waiting, while the generals and combatants, such as Dedan Kimathi, fled to the forests around the Aberdare's, Mount Kenya, and the Mau Highlands where they carried out a series of reprisals against people loyal to the colonial administration. The active Kenya Land Freedom army operating out of these forest strongholds was supplied from supporters in Nairobi; to quash this supply and wider support base, the Kenyan population of Nairobi was segregated and questioned in 1954 with the help of a series of informers (Elkins, 2005; Mahone, 2006). In this massive undertaking called 'Operation Anvil' some 50,000 Mau Mau sympathisers/supporters were detained at Langata barracks or held in reserves (Elkins, 2005). In these reserves, that were essentially a form of gulag where forced labour, under-nourishment, interrogation, lack of sanitation, and ensuing disease were commonplace (Elkins, 2005), and the supporters were made to relinquish their allegiance to the Mau Mau.

One of the biggest impacts on land was the ensuing villagisation process that started after Operation Anvil in 1954: some eighteen months later, 1,050,899 Kikuyu had been resettled in 800 'villages' comprising a series of central buildings and accommodation in which entrances and exits were controlled and hence the supply of food to the Kenya Land Freedom army cut off. Indeed, the villagisation process resulted in more widespread animosity to the colonial administration due to dire conditions alongside commonplace food shortages and disease. There are massively varying accounts on the impact of the struggle: some suggesting the number of deaths was around 25,000 (Anderson & Anderson, 2005) others suggest 130,000 (Elkins, 2005; Maathai, 2006), but clearly there was a significant impact. In 1955 an offer of an amnesty was given to the Kenya Land Freedom army but not accepted. In 1956 land reform increased the holdings of the Kikuyu outside of the 'villages' and opened up the growing of cash crops like coffee that had previously been under control and the preserve of colonial farmers. These concessions resulted in a cessation of the Mau Mau uprising, although it was clear the wheels had been set in motion towards the end of the colonial period. The negotiated ending to colonial rule took place at the Lancaster House conferences of 1960, 1962, and 1963 where the constitutional framework and transition to self-rule were negotiated until independence in 1963. Following independence, it was the Kenya African National Union (KANU) party, led by Jomo Kenyatta who became elected and Jomo Kenyatta the first president. The Colony of Kenya and the Protectorate of Kenya each came to an end on 12 December 1963, with independence being conferred on all of Kenya. The Sultan of Zanzibar agreed that, simultaneously with independence, they would relinquish claim to the coastal strip.

6.1.1.1 Independence Challenges and the Greenbelt Movement

The Mau Mau struggle continued as their role in securing independence was not openly supported; ensuing land, position, and power afforded to those at the heart of the Kenya Land Freedom army were not forthcoming from the new administration. The conflict continued until 1965, and the battle for compensation until 2013, when the British Government compensated some 5000 claimants, and a statue was unveiled in their memory at Uhuru Park marking the role that the

Kenya Land Freedom army had in securing the path to independence. The central Kenya region was not the only struggle before Kenya got its independence. Somali ethnic people in the areas of Northern Frontier Districts petitioned for the UK Government not to be included in Kenya but to be aligned with Somalia—this is a continued area of contention manifested as recent conflicts in North-Eastern Kenya.

Under Jomo Kenyatta, corruption became widespread throughout the government, civil service, and business community. Kenyatta and his family were tied up with this corruption as they enriched themselves through the mass purchase of property and lands after 1963. Their acquisitions in the Central, Rift Valley, and Coastal provinces aroused anger among many landless Kenyans. The extended family used his presidential position to circumvent legal or administrative obstacles to acquiring property. The Kenyatta family also invested heavily in the coastal hotel business that was often supported with overseas soft funding. Following Kenyatta's death in 1978, Daniel Arap Moi became president, a change that initially came with significant hope but as is commonplace, electoral promises were again not met. Through both the Kenyatta and Moi regime there was growing distrust of politics and national policy that supported the patriarchal and political elite while smallholders saw declining yields and increasing land use conversion and environmental degradation.

In response to this decline in trust of politicians to lead, Wangarii Maathai established the Greenbelt Movement that sought greater equity, empowerment for women and, at the heart of the movement, was to see a growing reduction in environmental degradation. The non-violent environmental activism had a massive impact on getting the voice of women into a political structure that was completely absent at the transition to independence. Much of the work of the Greenbelt Movement focused on tree planting in communities where there were challenges to access the forest, water, and food resources. Following on from dedicated tree planting campaigns, the Greenbelt Movement became a mainstream voice advocating for environmentally sustainable practices and women's rights. A particular focus of the Greenbelt movement was to push back against harmful agricultural policies and practices leading to soil degradation and food shortage. Wider than the tree planting regimes and campaigns that had environmental impacts, the Greenbelt movement initiated 'Black feminism' which has become a movement of its own, resulting in feminist views, knowledge, and educational materials being increasingly spread throughout Kenya. In 1989 the Greenbelt movement took on the powerful business associates of President Daniel Arap Moi in a sustained protest against the construction of a business complex proposed for the heart of Uhuru Park in Nairobi. This clash with the political leaders resulted in the Kenyan government closing Greenbelt offices, twice jailing Wangarii Maathai who was subjected to a severe assault in 1992 by police while leading a peaceful protest against the imprisonment of several environmental activists. There were other high-profile successes; in 1991 a protest saved Jeevanjee Gardens from being turned into a multi-story parking facility. In 1998, the Movement led a crusade against the illegal allocation by the Minister of Environment, Jeremiah Nyagah, who had secretly allocated parcels of the forest to 64 different companies for housing projects in 8 km^2 Karura Forest, a vital water catchment on the outskirts of Nairobi. The projects resurfaced again in September 1998 when the Forest Department was issued a quit notice by private developers. There ensured a whole suit of protests, some of them violent, and Wangari Mathia was injured sparking international UN condemnation of the Kenyan Government and revealing the corruption behind the allocation and use of the forest water tower. The struggle was finally 'won' in 2003 when leaders of the newly elected NARC government under the presidential term of Mwoi Kibaki affirmed their commitment to the forest by planting trees in the area. However, this forest has continued to be a source of contention as parcels of land have been nibbled away for exclusive and expensive housing developments. The Green Belt Movement is now vibrant and has succeeded in achieving many of the goals it initially set out to meet. It has also provided a wide range of protection to natural resources and ecosystems around the world. This protection of the environmental and natural resources was achieved through tree planting, soil conservation measures, and sustainable management of the local environment and economy by boosting local livelihoods, particularly focused on women and children. The new Kenya Forests Act in 2005, and subsequent amendments in 2016, made it much more challenging for degazetting of forests, particularly catchment forests, in addition to stronger rights for local communities to have access to forest resources.

6.1.1.2 Uganda

Uganda also had a turbulent transition from the colonial administration, although this was very different turbulence from the militarised independence struggle in Kenya and stems from the historically ingrained forms of governance across the different kingdoms that the British had made alliances with (Chapter 5). Apollo Milton Obote, who led Uganda in 1962 from the British administration, became particularly invested in the independence cause while in Kenya where he became involved in the national movement. Upon returning to Uganda in 1956, he joined the Uganda National Congress, later becoming leader of the fractionally split Uganda National Congress that formed a coalition with the previous powerhouse of the country—the Buganda royalty. Similarly, to Kenya, the transition to independence was ultimately negotiated at the Ugandan Constitutional Conference held in Lancaster House in 1961. There were a series of meetings to craft the Uganda constitution and discuss the findings of the Uganda Relationships Commission report, that particularly was tasked with charting how the central administration could operate within the existing and traditional historical kingdoms of Uganda (Chapter 5). However, as previously, this was contested and the Kingdom of Bunyoro walked out of negotiations resulting in a relationship between the newly formed government and the Kingdom of Buganda that formed a coalition party where the Kabaka became the ceremonial President with Apollo Milton Obote as the Prime Minister. Even though the transition model was one of socialist leaning in the 'Common Man's Charter' or 'the move to the left campaign', this transition was turbulent and mired in corruption scandals, rapidly increasing payments to Government and army, coups, and eventual war with Tanzania. At the heart of the turbulence was corruption as the Government took 60% control in all private enterprises. Behind this financial regime was a reign of terror directed at certain classes and ethnic groups, particularly those with any influence and the Asian community, that was led by the secret police headed up by Obote's cousin, signally the nepotism that was at the heart of the new regime. In 1971, one of the coups led by Idi Amin was successful and he remained in power for the next eight years. The impact of Idi Amin's rule and systematic human rights abuses are legends, particularly against the Asian, Acholi, and Lango ethnic groups that were thought to have allegiances with the former ousted president. The impact was soon widespread, and it is thought that up to 500,000 people lost their lives during the eight years with millions more leaving Uganda (Keatley, 2003). Following deposition in 1979 Apollo Milton Obote came back into power for another term that was also marred by corruption and instability. The instability was a long-going campaign led by Yoweri Museveni who came to power in 1986. Initially hailed as a great departure and new breed of African leader, however, the last 35 years have been mired by increasing controversy, unfulfilled promises, and allegations of electoral corruption and ongoing abuse of power and lack of equitable development across the country.

6.1.1.3 Tanzania: Ujamaa and the Rise of African Socialism

Mwalimu Julius Nyerere, the first, and still very much-loved and revered postcolonial leader in Tanzania embarked on a very different policy from Uganda and Kenya—that of African socialism, or 'Ujamaa'. This policy, literally translated as 'extended family' or 'brotherhood', is a set of principles around cooperative economics, collective production, equalitarian values, and self-reliance. Country resources were to benefit the common good and secure the development of the country as a whole. President Nyerere published his development blueprint in 1967 as the five-part Arusha Declaration that was a road map and vision for nation state development outlining the principles of socialism and the role of government for national development. The Ujamaa ideology was deeply rooted in the concept of a self-reliant nation, which justified the massive governmental spending used to enhance production; central servicers were nationalised, and the Government became the largest employer in the country.

Ujamaa was first trialled in the village of Litowa that went on to serve as a model example of what the true concept of the initiative meant and how communal farming, community engagement in civil service, the spread of production practices, and modernisation of technical development skills (i.e. construction), could ultimately lead to effective and profitable farming systems that allowed resources to flow into the rapidly expanding Government for wider dissemination to the public good. Under the 'Ujamaa Vijiji' (villagisation) policy, people of different ethnic backgrounds were encouraged, in some cases forced, to live in a village and operate as a collective. Between 1973 and 1976 some eleven million people were moved to new settlements (Shao, 1986) that were constructed around a common blueprint—there was a nucleus with homes in rows, with central services such as a hospital, school, and a town hall (Fig. 6.1). These nucleated centres were surrounded by larger smallholdings where most families were assigned a 1-acre plot for

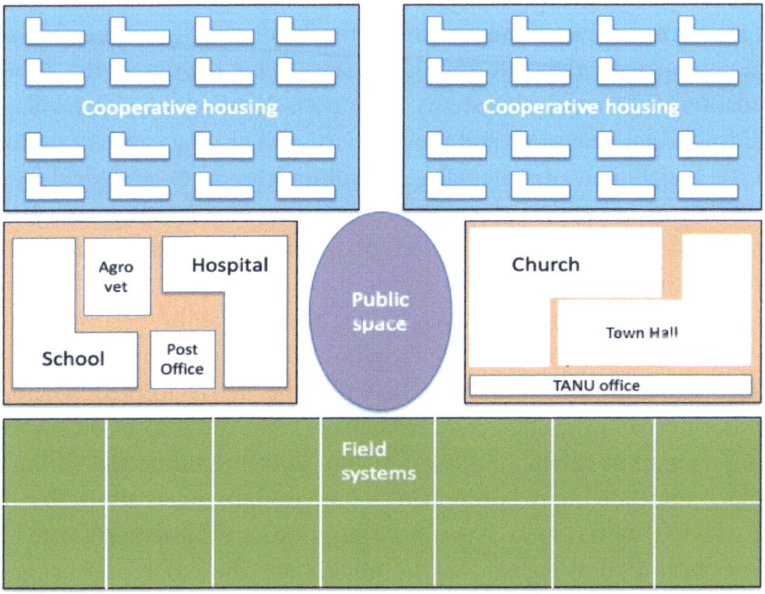

Fig. 6.1 Ujuma village were constructed around a common blueprint. There was a nucleus of public space and central services, political offices and a town hall. Homes tended to be a common size and design with set allotments of land for agricultural production

personal use; these were surrounded by the larger communal agricultural farms (Fig. 6.1). At the heart of the cooperative working system was the Tanganyika African National Union (TANU) party that strove to report success, such as greater gender equality; the TANU party created an entire government section that represented women's rights and equality within society. Most Ujamaa villages had a similar aim: production to build up the Tanzanian national identity and economy. Some schemes such as the Urambo scheme were successful in increasing farming yields but were shut down by the central authority because farmers were becoming increasingly wealthy and powerful, i.e. breaking the core socialist principles. Due to internal over-control and external shocks of oil crises in the 1970s, the collapse of export commodity prices, particularly sisal and coffee, and the onset of war with Uganda, in 1978 where Nyere went to the aid of his friend and former president Apollo Milton Obote and Yoweri Museveni the Ujamaa programme started to falter. Systemic contradictions began to foil the Ujamaa programme (Huizer, 1973); in places, the enforced Ujamaa policies disrupted the locally co-adapted agro-pastoral systems. The poor location of many villages regarding water provision, soil productivity, and grazing capacity prohibited the installment of sustainable agricultural practices, and yields were not sufficient to support the wider functioning of the settlement. Additionally, the forced sedentarisation of previously nomadic pastoralists increased the grazing pressures enormously around the village nucleus, leading to degradation. Moreover, lack of any land tenure security halted the production of perennial cash crops and capital investments in farms. Due to the complete imbalance between the newly formed 'Ujamaa' communities and the alien environment, systems often spiralled towards land degradation (Hydén, 1980; Sendalo, 2009). Despite the failing of Ujamaa, there are many lasting benefits today, Tanzania is a very flat and polite society with respect across positions: senior ministers will converse as one with a rural farmer. In the same vein, the enforced mixing of people from different ethnic backgrounds resulted in a weakening of the strong ethnic divisions along with tribal groups that still exist in Kenya and Uganda, and can overspill into internal conflict, such as the post-election violence in Kenya following the 2017 elections.

6.2 Colonial Legacies and New Forms of Land Management

Simplistic classification of weak or strong institutions fails to describe the intricacies of the contemporary East African political systems, which are plagued by apparent corruption and the lack of democratic responsibility to provide services that are deemed central to the modern state (Chabal, 2013). Ultimately this empowering of Government at all levels was an unintended consequence of allowing power, and hence corruption, to operate across the fledgling government structures. This corruption is not just about leaders or ministers squirrelling funds away to offshore bank accounts, but the favouring of certain sectors of society or regions of the country. This preferential treatment could deny access of certain ethnic groups or regions to basic state services, such as roads (markets), education, technology, and electricity, and prohibits communities from developing (Ananda & Herath, 2003). A good example is the case of Northern Kenya. Given the ethnic tensions that were apparent from the onset of colonial transition, with many in Northeast Kenya preferring allegiance to Somalia over the newly founded Kenya, pastoral societies remain largely marginalised. The diverse ethnic groups of Northern Kenya remain under-represented with many areas continually underdeveloped and isolated. The main North–South arterial road, that was an ambitious plan in the wave of optimism that accompanied postcolonial administration of the 1960s to construct a highway connecting Cape Town to Cairo. Unfortunately, the road stopped at Archers post and then continued again some 300 miles north at the Kenyan-Ethiopian border town of Moyale. The reasons given for not being able to construct the road by successive governments was a lack of water for construction; numerous administrations would 'try' and construct the road drilling deeper and deeper for water 'investing' millions going down 50, 100, 200, 400 m, to tap the aquifer. However, it was not until 2016 during the Chinese construction boom (Fig. 6.2) that ample subsurface water was found some 20–50 m below the surface and previous 'attempts' had done little but siphon off funds and maintain the marginalisation of North-Eastern Kenya! The road is now a reality and a journey from Nanuki to Moyale that could take a week can be done in a day.

Fig. 6.2 The Chinese led construction boom has included a large program of road construction with a new main road that now connects Nairobi to northern Kenya and Kenyan-Ethiopian border town of Moyale. There are also the development of new cities such as the Konzo technology city (c) and new road bridges in cities such as Dar es Salam (d) (All photographs: Rob Marchant)

Due to the lack of accountability of the political system towards certain communities, some policies have been downright exploitative, increasing poverty, inequality, and food insecurity (Chabal, 2013; Homewood et al., 2004). The differential political marginalisation of communities has been one of the legacies, particularly in Kenya where a system of ruling classes along tribal lines erupted in post-election violence in 2007. Following another contested election there were protests and targeted ethnic violence that quickly escalated from the disputed election with anger first directed mainly against Kikuyu people—the community of which President Mwai Kibaki is a member. The election result of 2007 was disputed between Mwai Kibaki and Railia Odinga, and this quickly divided the country with tensions arising from a deeper history of Kikuyu displacement and their growing influence across Kenya that was increasingly being pushed back by other ethnic groups. Around

1300 people died in the violence with some 600,000 people displaced in a matter of weeks. This ultimately led to a series of international agreements, a power-sharing deal, and the development of a new constitution that is in place at present. This new constitution essentially devolved power, budgets, and decision-making around local issues away from the central parliament to a series of country governments. Some of these have performed well and others have seen a new opportunity to waste funds and resources from the wider population to the privileged few. One of the main challenges of successive administrations that maintain division and undermine the wider and equitable benefit has been endemic corruption. Indeed, every election is full of promises to root out the causes and people at the heart of this but with little effect and most had failed. The late president Magufuli of Tanzania seemed to be making progress, although this was not widely appreciated given the vested interests of many individuals and sectors.

The other main development from the new Kenya constitution has been around the key issue of land and the issuing of title deeds to individuals and communities that have resulted in mass development and land speculation as land transitions to private ownership. As well as allowing developed power to operate at the local level this is also leading to unintended consequences (Chapter 7). The different post-independence nation-states in East Africa have pursued different directions in land policy. For example, in Kenya, private property rights were gradually introduced from 1956, this having been massively transformed under the new constitution and local title deeds. While in Tanzania, all land is state-owned, where individuals use land as tenants and the purchase, sale, and rental of land are limited within boundaries of the state (Pinckney & Kimuyu, 1994). As the regions' rural poor are heavily dependent on natural resources and have limited access to capital, fertilisers, and technology, options to invest in improved land management are restricted (Barron et al., 2003).

Another challenge linking the colonial to the independent governance is the blindness of the state to the complexities of the locally adapted agro-pastoral systems and appreciation of the deeper history of land management (Wynants et al., 2019). Mismatches between centralised agricultural policies and the diverse and dynamic East African environment (Chapter 2) often led directly to soil exhaustion, decrease in productivity, increased rates of erosion, and ultimately the depletion of soil resources, alongside the potential for conflict and corruption as alternate livelihoods are needed. Impacted communities will have difficulties adapting to changing demands and the pressures to sustainably manage their natural resources because of the underdevelopment of both social (social networks, political representation, agronomic and ecological knowledge, education, mobility) and economic (land rights and land access, capital, market access, infrastructure, fertilisers, agricultural technologies) domains. A lack of strong political, social, and economic structures impedes communities to adapt to the increasing demands of a rapidly growing population and changing environment. Exemplary to this is the absence of growth in agricultural productivity and livelihood options outside of agriculture, which forces communities to degrade and overexploit natural resources. Decreasing natural resources further increases social and economic pressures on ecosystems; a negative feedback that, in addition to having a social impact, has been one of the main drivers of increased rates of soil erosion and environmental degradation in East Africa (Blaikie, 2016). Land management will be presented in the three main spheres of the forest, agriculture, and pasture.

6.2.1 Forest Management

East African forests, both lowland and montane, are faced with continued impacts as this forms the frontier of land conversion between smallholder cultivators, multinational companies, and international or national conservation and development agencies (Pullan, 1988). The late colonial goal of creating a series of forest reserves had been achieved, albeit with very extensive forest clearance and selective deforestation for the high-value timber trees, but the forest estate did not remain static after independence (Mgaya, 2016). Though the decolonisation of British East Africa formally resulted in more widespread distribution of human and civil rights among East African citizens, many colonial forest management activities proceeded unaltered—and indeed some continue into the present. Many rural farmers saw this independence struggle as bringing access to land that was often located in areas that had recently been declared forest reserves. Similarly, deforestation following the disappearance of indigenous communal conservation regimes and the collapse or lack of strict state enforcement can be characterised as a tragedy of open access (Conte, 1999; Enfors & Gordon, 2007).

With the introduction of new land tenure laws combined with the partial application of indigenous tenure systems, East African communities have developed a complicated mix of both customary and formal land rights. As with the colonial administration, the restriction of access to the use of forest resources has resulted in conflicts between livelihood needs and nature and forest conservation. These conflicts are multifaceted and driven by global, local, and institutional factors (Habel et al., 2017). In the postcolonial context of modernisation and development, upland forests in East Africa took on significance as crucial water catchment areas for surrounding farms and other commercial enterprises (e.g. Kenya Forests Working Group [KFWG], 2004). Tanzanian foresters enthusiastically participated in the nationalist goal of increasing timber production for the nation state. In Uganda during the three-year development plan (1961–1964), the Forest Division had expected both 'productive' and 'protective' forests to be highly productive of timber. As a result, the Division put much effort in replacing slow-growing indigenous trees with fast-growing softwood and consequently, the forest division had, by 1968, expanded soft wood plantations to 22,000 ha (Mgaya, 2016), a considerable increase from 6000 ha at the end of the colonial rule (Sunseri, 2010). At this time the nationalist agenda suggested a high growth rate of development that would convert smallholders into modern farmers (Mgaya, 2016). However, the Ujamaa vision had, to a greater extent, degraded the extensive miombo woodland that covered almost half of Tanzania that was locally replaced by cultivated land, tree plantations, and exploitable forest reserves (Sunseri, 2005) that were supposed to preserve water catchments and guard against soil erosion.

The economic importance of forested watersheds was further underlined by a variety of hydroelectric and catchment irrigation initiatives, which relied upon forested ecosystems such as the Usambara or Mount Kenya forest complex for their commercial feasibility. Upland forest catchments and the regions' river systems are both important sources of water and energy production; hydroelectric generation accounts for some 16% of Tanzania's energy production and during periods of drought the whole economy is massively impacted as businesses must operate through the night when power is on or run on expensive diesel generators. Consequently, these economic incentives mean that the government has largely continued colonial policies of controlling indigenous populations' access to their customary territories. Although many of the post-independence issues regarding unsustainable land management can be attributed to the erosion of indigenous social structures during the colonial period (Chapter 5), combined with the sudden release from the strict colonial rules there has been significant impact on forest resources across the East Africa region. Good examples of this come from the Usumbara mountains in Tanzania, where the process of losing the 'indigenous' conservation ethic during the colonial administration, along with increasing population pressures and a sudden release from colonial forest enforcement, led to uncontrolled exploitation and encroachment following independence (Conte, 1999; Enfors & Gordon, 2007). As the farmers' demand for arable land was a major rallying point during independence struggles, the newly found Tanzanian state could not refuse claims for agricultural land. These resulted in rapid degradation of the new farmland, where the farmers often abandoned their newly gained plots after a couple of years (Conte, 1999; Lundgren, 1978; Lundgren & Lundgren, 1979). Large areas of tropical forest are degraded through human exploitation, fragmentation, pollution, exotic species invasion, and fire (Ahrends et al., 2021).

Through the long history of human forest interaction, there are no entirely natural forest systems remaining in East Africa with many of these being highly transformed and degraded (Ahrends et al., 2021; Marchant et al., 2018). While there is no globally agreed definition for forest degradation, it can be broadly defined as changes to a forest stand resulting in the long-term reduction of given attributes and functions, such as biodiversity, and the potential supply of goods and services (Ghazoul et al., 2015). Specific warnings on the need for forest conservation were given in 1966 by Hedberg and Hedberg (1968), and again by Mueller-Dombois (1971), but by this time the conservation of some humid forests was being attempted to protect water supplies and limit soil erosion. Although these conservation areas prevented agricultural incursions into forests with valuable timber and could control logging operations, they were never regarded as totally protected areas. The montane forests of Eastern Africa have been associated with tourism for many years. Even though forest tourism does not attract the same numbers of visitors and associated revenue, and hence on a fraction of the revenue, this is apart from the iconic touristic honey pots of Mount Kilimanjaro and the Bwindi Forest for the mountain trekking and the Mountain Gorilla populations respectively.

Approximately 37% of the forested land in Tanzania is classed as a forest reserve, controlled, and managed by the state Forestry and Beekeeping Division under the Tanzania Forest Services. The remaining forests, found outside the reserve network, lie on the village and general land (Mgaya, 2016). While most of these unreserved forests are poorly managed, traditional, and customary management practices have supported the conservation and maintenance of forest cover for sacred, religious, or social purposes in numerous localities across the country (Blomley & Iddi, 2009). The remaining humid montane forests in East Africa were relatively fragmented, having been impacted on by historical logging, and largely protected as forest reserves and National Parks, although there was push back around further demarking forest areas an internationally recognised protected areas due to local contention. Indeed, Kilimanjaro National Park (1972) is the only forest National Park created during the colonial period. The Eastern Arc Mountains (EAM) of Tanzania and Kenya are one of 35 global biodiversity hotspots and possibly the most threatened in the world (Mittermeier et al., 2011). Currently, non-protected forest comprises approximately 25% of all forest in the EAM (Platts et al., 2011) that have lost some 80–90% of their overall cover, and in some locations like the Taita Hills in Kenya, some 99% of the forest have been removed following exploitation for timber and conversion to agriculture (Chapter 4). Actively regenerating forests among the remaining fragments, as well as permitting open canopy forest to fill in, can play an important role in enlarging and linking the largest and closest forest fragments in the EAM (Newmark & McNeally, 2018). While there have been several initiatives over the last two decades to connect nature and forest reserve a much broader effort will be required to establish linkages and to provide protected area status to the remaining non-protected forest (Newmark & McNeally, 2018; Fig. 6.3).

Various successful activities have already been established during the past under the broader umbrella of 'Development through Conservation', ranging from Multiple-Use zones in Bwindi Forest (Wild & Mutebi, 1997) to the Kipepeo (Butterfly) project created by Birdlife to provide a platform for more than 400 butterfly farmers to sell pupae (Gordon & Ayiemba, 2003). Establishing tree nurseries, production of

Fig. 6.3 Around Bwindi Impenetrable Forest National Park in Uganda a series of Multiple Use zones were established around the National Park through a Development through Conservation project; people could enter the National Park and collect non-timber forest products such as basketry and medicinal plant resources and other income generating schemes such as honey production (All photographs: Rob Marchant)

extracts from Aloe plants, the cultivation of mushrooms, and establishing environmental education centres are further examples of how to create alternative income sources around the forest, and hence reduce demands and ensuing conflicts Fig. 6.4 (Gordon, 2003; Matiku et al., 2013; Sinclair et al., 2011). A recent strategy uses the 'domino effect' where revenues of successful community-based projects are used to start and invest into future community-based activities. This process is currently coordinated by the Arabuko Sokoke Forest Adjacent Dwellers Association (Fig. 6.4); an umbrella group, which also communicates problems of the surrounding villages, communities, and conservation groups (Habel et al., 2017) into the regional management.

Despite a steep rise in the number of these types of community initiatives, and participatory forest management plans in place across East Africa, there has been a small increase in the protected forest (Santoro et al., 2020) and the reality has been one of decreasing forest cover (Ahrends et al., 2021; Shaw et al., 2016). At the heart of the challenge, as with many, is corruption and the under-resourced nature of Forest Department and enforcement agencies and their subsequent openness to bribery at all levels, from district forest offices allowing logging of high-value timber species of 'protected' forests to the transport and subsequent export that ultimately led to a loss of resource at a national level, and the benefits of national natural resources ending up in the deep pockets of

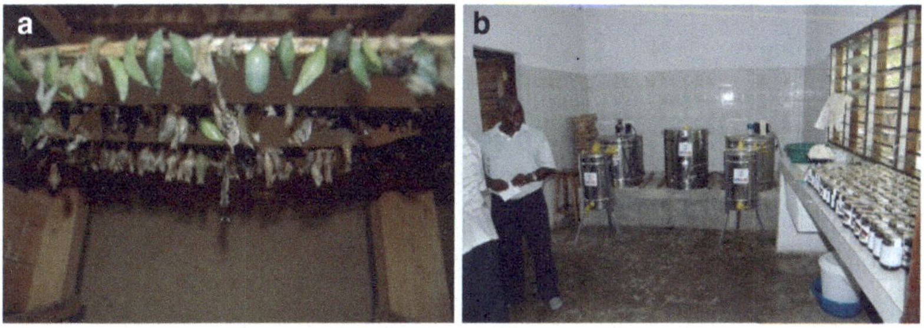

Fig. 6.4 To reduce the impact on forest resource there have been a wide series of income generating schemes such as developing butterfly farms to sell pupae (**a**) round the world. There is a long history of bee-keeping with honey a good alternative income source from forests (**b**) (All photographs: Rob Marchant)

the relative few (Milledge, 2009). The increasing militarisation of conservation heightened environmental law enforcement, and the wider appreciation of the wider economic valuation of forests and ecosystems has resulted in stronger management in recent years. This stronger management mandate comes at a social cost and ensuing legislation. Such as the 2005 Forests Act and the 2013 Wildlife Conservation and Management bill in Kenya, have criminalised many aspects of indigenous livelihoods, such as the subsistence use of wildlife and non-timber forest products. Attempts to access traditional lands and resources are likewise now interpreted as 'environmental crime'. Criminal attempts to undermine global efforts to conserve wildlife, biodiversity, and mitigate environmental change are threatening bilateral and multilateral investments in initiatives such as those coming out of the United National Framework Convention on Climate Change (UNFCCC), such as Reduce Emissions from Deforestation and Forest Degradation (REDD+), the Convention on International Trade in Endangered Species of Wild Fauna and Flora (CITES), and other efforts for strengthening Forest Governance, Law Enforcement, and Trade (FLEGT) that all of the East African Countries are signatories to.

From the 1980s onwards, Tanzania saw an increase in inclusive approaches to forest management (Mgaya, 2016). As a result, in 1998 the National Forest Policy of 1953 was reformulated, instigating a major reorientation in its approach to forest management; shifting from a centralised, state-led policy towards a greater emphasis on Participatory Forest Management (PFM) (Blomley et al., 2008). The new approach was enshrined in the Tanzanian Forest Act of 2002, which supports Community Based Forest Management (CBFM), on village land or land owned by communities (Mgaya, 2016). Under CBFM, villages (or groups within villages) may gazette village forest reserves, and thereby transfer management authority over village forest resources from the state to the community. This includes the right to collect fees on forest utilisation, and to impose and retain fines on illegal use (Mgaya, 2016). Until 1999, the Land Ordinance of 1923 was the principal governing statute regarding land tenure and management in Tanzania. This was replaced by two pieces of legislations, the Land Act No. 4, and Village Land Act No. 5 of 1999 (Zahabu et al., 2009). All of these transitions in Government policy, local implementation and historical events such as the gazetting of land will all impact the community decisions around how land will be used and have ramifications to the present day (Fig. 6.5) and future trajectories use (Chapter 7).

Increasing concern for wildlife, combined with the appreciation of the potential livelihood benefits, has led to the establishment of many conservation areas for large mammals. One of the significant challenges of natural areas is human–wildlife conflicts; the main tool against these conflicts are fences; either physical ones as in the case of the Aberdares National Park or thorn brush fences as in the case of a Masai boma. Although several studies indicate that fencing off areas prevents human–wildlife conflicts (Anthony et al., 2010), this comes with various ecological problems like the blockage of traditional migration routes of

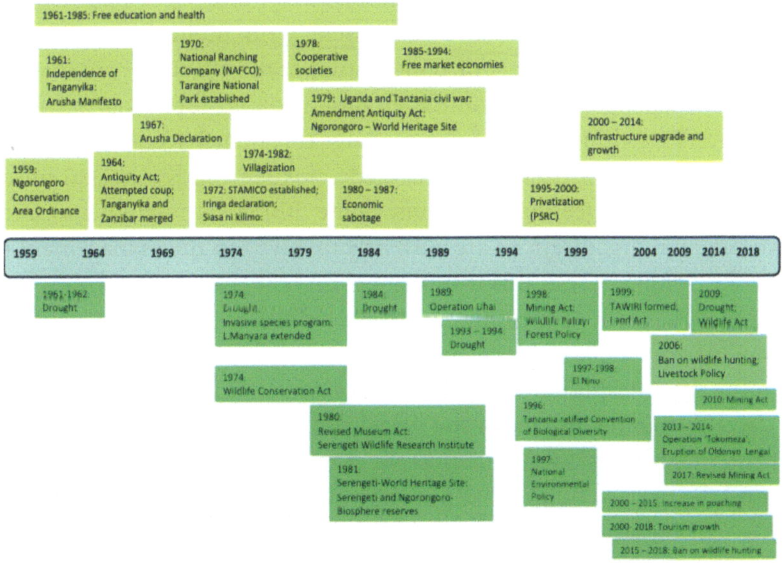

Fig. 6.5 Historical events, such as the gazetting of land or shifts in government policy, impact on community decisions around how land will be used and have ramifications to the present day. Timeline of key events that have shaped land use land cover change around the Ngorongoro Highlands (Kariuki et al., 2021)

large animal species. Arabuko Sokoke forest was fenced to avoid human–wildlife conflicts and to restrict elephant (*Loxodonta africana*) movement to the forest, however, the forest then became further degraded by the local and very dense elephant population producing high levels of ecosystem disturbance, intense tree debarking, and destruction of trees (Habel et al., 2017). Some National Parks, such as the Aberdares and Mount Marsabit and the Shimba Hills have physical fences while other areas have 'social fences' in place where the societies such as those living around the Kayas of Kenya, through a process of participatory forest management, determine the limits of forest use and the management of remaining sacred patches of forest. Although these are seen by some as a model for sustainable development of forest resources, the Kayas have been increasingly degraded through the rising demand for land, wood fuel, iron ore, and timber, both for construction and carving. Today some 52 Kayas remain that vary in size from around 10 to 300 hectares and suffer from complicated management with multiple institutions, organisations, and numbers of policies that make effective conservation strategies challenging. For example, Arabuko Sokoke forest management, the largest remaining area of the globally threatened Coastal Forest, is split into four forest regions (Jilore, Gede, Sokoke and Kararacha) under management of three forest stations. Some 300 governmental and non-governmental organisations are currently involved in the management of Arabuko Sokoke forest; this institutional diversity can lead to confusion, making it virtually impossible to assign a clear and consistent conservation management strategy and more crucially creates a climate for ongoing illegal exploitation of natural resources (Habel et al., 2017).

Recent attempts to curtail the illegal timber trade in East Africa have consistently been leaky and ineffective. A relatively new initiative is based around the understanding that forests contribute a global service by regulating carbon stores in their biomass and organic rich soils. This role has received growing traction on the global climate change agenda in an effort to 'Reduce Emissions from Deforestation and Forest Degradation' (REDD+); one of the policies coming out of the UNFCCC (Sandbrook et al., 2010). There are wider issue around the commodification of ecosystems' contributions to people through economic valuation

(Cavanagh et al., 2015), although arguably REDD+ presents a win–win opportunity by providing communities living close to forest with funds and technical support (Mgaya, 2016). Others suggest that REDD+ may further marginalise forest-dependent communities in East Africa, providing states with more incentives for exclusionary conservation and the recentralisation of control over natural resources (Cavanagh et al., 2015). Although REDD+—processes are in their early or 'readiness' stages in East Africa, the implementation of individual REDD+ projects and other voluntary carbon offset forestry schemes have reportedly resulted in the eviction of tens of thousands of people. Examples include Mount Elgon (Cavanagh & Benjaminsen, 2014; Himmelfarb, 2012), Namwasa (Grainger & Geary, 2011), and Bukaleba (Lyons & Westoby, 2014; Nel & Hill, 2013) in Uganda, the Cherangani Hills in Kenya (Otianga-Owiti et al., 2021), and the Rufiji Delta (Beymer-Ferris & Bassett, 2010), although the latter was hotly contested and was, as many of these forest issues are, much more complicated than a similar state-imposed eviction on the back of the global climate change agenda (Burgess et al., 2013). There are concerns that carbon offset forestry reinforces legacies of marginalisation and creates new incentives for the displacement of indigenous and forest-dependent communities. One must also consider the ulterior motives that East African states might potentially seek to pursue under the auspices of such concerns.

Kenya, Uganda, and Tanzania are all REDD 'Partner Countries', and while Tanzania has received substantial financial support from the programme in recent years, this will likely change. All three countries are part of the World Bank's Forest Carbon Partnership Facility (FCPF), and Tanzania has so far received US$36 million in REDD+ financing from the Norwegian International Climate and Forestry Initiative (NICFI). In addition to these REDD+ funds, each of the three states have also received large volumes of grants and loans to strengthen or expand the management of their forest estates, including the European Union, the Global Environment Facility, and various bilateral donors (Oiere, 2014). Arguably, REDD+ presents the institutional regulation of natural resources, which is increasingly growing traction as a necessary response to the global climate crisis (Mgaya, 2016). East Africa has already gone through a long history of deforestation and does not have the massive lowland forests found in the Congo Basin. By paying local communities to preserve forests to store carbon, carbon emerges as something tradable. If local communities can demonstrate an increase in carbon stocks, then they will correspondingly receive compensation and hence REDD+ could empower people to take control of the management and governance of their forests (Neumann, 2005). While REDD+ operates through technical and scientific rationalities and practices, the implications of the mechanism are political. This is also why any international mechanism that aims to govern the global forest carbon commons is simultaneously a mechanism that governs forest use on the local level. Although the general trend has so far shown how the approach to forest conservation in Tanzania has changed from fortress to participatory, the introduction of REDD+ in Tanzania is within the framework of participatory forest management in terms of their carbon-uptake value that will inevitably transform the relationship between people and forests. If handled inappropriately, the approach may promote local communities' reluctance to participate in forest management: communities have to be part of the entire conservation conversation so that objectives can be jointly developed, and appropriate management schemes applied.

6.2.2 Agricultural Transformations

Agriculture accounts for over 65% of full-time employment, 25–30% of gross domestic product (GDP), and over half of the total export earnings, and underpins the smallholder livelihoods as it remains the main source of household nutrition and income across East Africa (Khan et al., 2017). As we have seen, East Africa's agro-pastoral systems are shaped by millennia of co-adaptation, and feedback mechanisms between communities, environments, and ecosystems (Chapters 3 and 4). Agricultural communities have developed systems, which conserve or improve the soil properties (Berkes et al., 2000) and are inherently sustainable until shocked into a non-stable state by, for example, ill-thought through policy or climate changes. In some of the fertile areas such as the Mau complex, Loita Hills, Ngong Hills, Rukiga Highlands, and Mount Kilimanjaro the population levels in these relatively small upland areas are extremely high. Cultivation on hillsides, pasturing livestock on steep slopes with conservation practices such as intercropping of crops with different rooting depths or terraces, allow high productivity to be maintained and the areas to be intensively managed. Soil conservation measures such as terraces, mulching, and tree cover, which were enforced by the colonial government and rallied against by 'nationalist' movements in Kenya, Uganda, and Tanzania, were abandoned in some areas. Many people ignored measures such as the prohibition of cultivation on steep slopes, control of bush fires, and timber exploitation, an ignorance often supported by politicians who denounced colonial heritage. In the same way that traditional agricultural systems that practised soil and water conservation were not supported by the colonial administration, history repeated itself as soil conservation measures started to break down, leading to systematic degradation and loss of agricultural land (Anderson, 1984) as agriculture spread rapidly in the past decades (Fig. 6.6). Post-independence was increasingly ruled by centralised policies lacking the complexity and adaptability of local co-evolved agro-pastoral systems, often leading to national economic growth strategies which degraded the natural resources (Lane & Pretty, 1990) and hence were unlikely to deliver the anticipated benefits.

Sustainable agricultural intensification: producing more output from the same area of land while reducing the negative environmental impacts and at the same time increasing contributions to natural capital and the flow of environmental services (Conway & Waage, 2010), has been at the heart of all postcolonial government ambitions, though it remains challenging to achieve and it has largely expanded (Fig. 6.6). Sustainable agriculture maximises soil quality and crop productivity, adopting a systems approach (social, economic, and environmental) to agricultural development. There have been multiple initiatives such as the application of 'push–pull' technology, integrated pest management, integrated soil fertility management, agroforestry, aquaculture, water harvesting, and livestock integration or conservation agriculture with principles of minimum soil disturbance, continuous soil cover with a perennial cover crop and plant residue, as well as a diversified cereal–legume–fodder intercropping strategies (Khan et al., 2017; Midega et al., 2015).

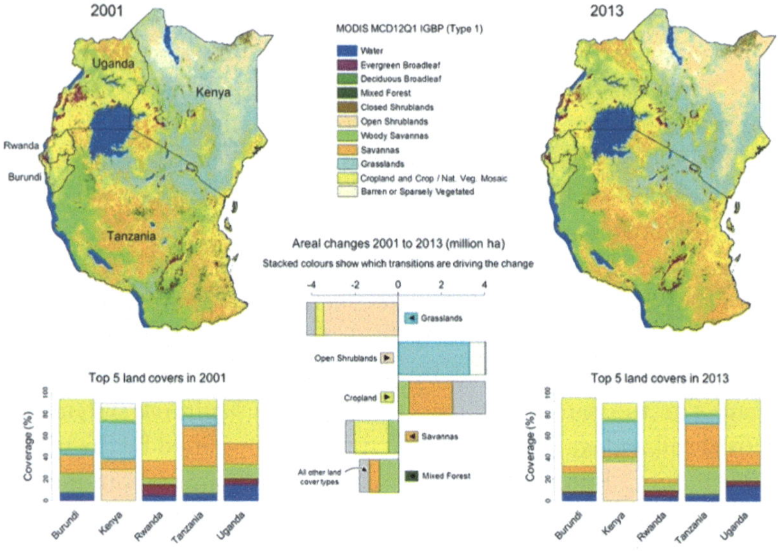

Fig. 6.6 Agriculture continues to spread rapidly in the past decades as depicted in a comparison of MODIS land cover from 2001 and 2013. Most notable is the increase in agriculture (yellow); particularly along the coastal strip and around Lake Victoria. Image analysis P. Platts

Currently, over 50,000 smallholder farmers in drier parts of Kenya, Tanzania, and Uganda have taken up the climate-smart push–pull agriculture and have effective control of one of the key pests of maize (stem borers) and striga weed infestation resulting in significant increases in grain yields of both maize and sorghum (Midega et al., 2015). Because farmer participation is built into the processes of push–pull agriculture research and dissemination, contextual changes encountered in the field can be communicated, discussed, and responded to. There is still a great need for adaptive agricultural practices that can cope with increasingly variable climatic conditions and still produce food for people and livestock (Khan et al., 2017) and this is likely to be one of the biggest future challenges (Chapter 7).

Another factor leading to pressure on land is the increase in the acreage under cultivation: the expansion and encroachment of cultivators have increased massively since independence in 1963, principally due to population growth. Unfortunately, this expansion has invariably been towards the more marginal land. For example, in Kajiado county, cultivators have occupied the areas around Loitokitok, Ngong Hills, and are beginning to buy or rent land in better-watered localities in the plains, for example at Kimama and Rombo leading to very rapid land use transformations; often too rapid to allow for proper assessment and planning. Much of this agriculture has been from an influx of the Kikuyu (Sindiga, 1984) and the subsequent expansion of the area under cultivation (Sindiga, 1984). As more marginal land is brought under cultivation, soil resources are being rapidly depleted by increased erosion, contributing to widespread land degradation, which further threatens food security, water, and livelihood security (Blaikie & Brookfield, 2015; Fleitmann et al., 2007). Soil degradation, erosion, and drying up of streams, cultivating on riverbanks leading to silting of streams and dams, unchecked gullying of cultivated slopes, and sheet erosion following bad grazing or agronomy practices is increasingly commonplace. Human societies are an integral part of soil erosion; even though this is a physical process, its underlying causes are also firmly rooted in the social, economic, and institutional environment in which land users make decisions (Ananda & Herath, 2003). As precipitated water has less time to infiltrate the soil and flowing water will move more rapidly on steeper slopes this subsequently higher energy will erode the land and lead to gully formation (Heckmann et al., 2014; Morgan, 2005).

Despite the aggressive growth of farms since 1973 (Fig. 6.6), most areas with arable potential had been settled by the late 1980s (Western & Manzolillo Nightingale, 2003), and new areas of cultivation tend to be increasingly marginal and thus prone to shocks, particularly those associated with drought and land degradation (Fig. 6.7). For example, marginal farms spread along the newly constructed Loitokitok-Sultan Hamud pipeline during the late 1980s and early 1990s and in areas between Tsavo and Amboseli National Parks (Fig. 6.7). Immigrant populations often dominate these newly opened landscapes during the early phase by buying or leasing farms. One of the recent challenges following the onset of the new constitution in Kenya and the devolution of power has been that under increasing land scarcity, conflicts over land are rising between individuals who obtained land rights through the

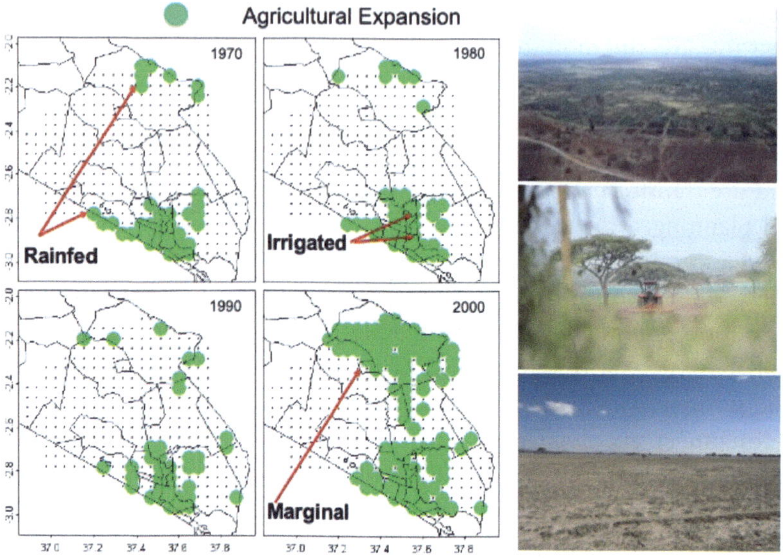

Fig. 6.7 Increased sedentarisation and agricultural expansion across southern Kenya around the Amboseli Basin. The concentration of animal impacts into single locations, rather than being spread across the wider transhumance landscape, can lead to degradation. New areas of cultivation tend to be increasingly marginal and thus prone to environmental shocks, particularly those associated with drought and land degradation (All photographs: Rob Marchant)

different mediums. Often the most powerful and educated people can best navigate the complex maze of bureaucracy and customary rights, perpetuating inequality and determining who owns land and who uses it for what; these asymmetries of power and access to information must be recognised and explicitly addressed (Kristjanson et al., 2009). Newly farmed plots, particularly those close to protected areas, have become common targets for elephants, buffaloes, and other wild herbivores with decreasing land security contributing to unsustainable management of soil resources in agricultural areas (Bluwstein et al., 2018).

One of the challenges of agricultural expansion has been the increase in invasive alien plant species (Fig. 6.8) that pose substantial threats to agriculture, biodiversity, and the delivery of a wider suite of ecosystem

Fig. 6.8 Invasive alien plant species. A particular challenging one has been the spread of *Opuntia* (**a**, **d**) prickly pear that was imported as fence species and now spread across large area. Ingested by animals, the thorns can kill and blind species that eat this, particularly during drought periods. *Pontederia crasipes*—water hyacinth—is an aquatic plant native to the Amazon basin, and is often a highly problematic invasive species in a number of large lakes such as Lake Naivasha (**b**). *Opuntia.* Although not an invasive *Acacia reficiens* (**c**)—false umbrella thorn—spreads extensively in savanna areas and shades out the grazing resource leading significant local hardship to pastoral populations (All photographs: Rob Marchant)

services. Unsuitable land uses and inappropriate land management practices such as slash and burn agriculture, timber and charcoal extraction, deforestation, overgrazing, cultivation on steep slopes, uncontrolled fires, and pollution of water resources (Dregne, 2002) all cause significant disturbances which facilitate plant invasions. With a few exceptions, comprehensive lists of alien plants that invade natural ecosystems are lacking even though they are known to impact negatively on the conservation of biodiversity as well as the livelihoods of rural people that depend heavily on natural resources in Eastern Africa (Maundu et al., 2009; Shackleton et al., 2017a, 2017b, 2017c). Several of the species that are now problematic were deliberately introduced by governments and aid agencies to augment existing natural resources, precipitating impacts that far outweigh any benefits they may have brought (Maundu et al., 2009; Mwangi & Swallow, 2008). For example, the Eastern African highlands, especially those in Kenya and Tanzania, have largely been invaded by black wattle (*Acacia mearnsii*) introduced from Australia, *Pinus patula* (Pinaceae), introduced from Central America, which has also escaped from cultivation in these areas, along with spiny shrubs such as various *Rubus* species. Other invasive taxa are 'escapees' such as *Maesopis eminii* and *Hura crepitans* in the East Usambaras that escaped from the Amani Botanical Gardens (Dawson et al., 2008; Sheil, 1994) whereas the coastal region is dominated by species such as the neem tree (*Azadirachta indica*).

Land degradation, pests, and weeds hamper efficient production of cereals, particularly maize and sorghum, the main staple, and cash crops for millions of smallholder farmers (Khan et al., 2017). Many fields are heavily infested with parasitic striga weeds, while insect pests, principally stem borers, devastate cereal crops, commonly causing a loss of over half the harvest. The main barriers to the effective management of invasive alien plants in Eastern Africa are the lack of appropriate policies and/or implementation. There is insufficient capacity especially about the identification and management of invasive plants, lack of awareness among government officials and other stakeholders as to the impacts of invasive plants on biodiversity, water resources, crop and pasture production, human and animal health, and economic development, and an absence of sufficient resources to tackle the issue at a national or regional level. The persistence of positive attitudes towards some invasive plants, some based on arguments perpetuated by international development agencies, continues to hamper the implementation of much-needed management interventions (Witt, 2017).

6.2.2.1 Pasture and Rangelands

Continuing the colonial trend there have been sustained policies, coupled with growing human populations, and competing demands for the use of land, that have resulted in the reduced capacity of rangelands to support livestock and livelihoods and continued marginalisation of many pastoralist societies (Maundu et al., 2009). The communal usage of land and resources, which is vital for pastoral communities, is heavily threatened by policies of privatisation and exclusion (Homewood et al., 2004; Western et al., 2009). In some cases, the policy of excluding smallholder farmers and pastoralists from their land has continued unabated from the colonial period, which is being repurposed for private agricultural enterprises under the guise of conservation and economic development (Homewood et al., 2004; Lane & Pretty, 1990). There had been an increase in restrictions imposed on pastoral populations that resulted in additional decreases in mobility (Fig. 6.7), further disrupting systems of rotational grazing and adaptability to rainfall fluctuations (Homewood, 1995). This systematic erosion of social and economic structures in agro-pastoral communities through the loss of access to land and natural resources, social organisation, knowledge, and mobility often led to increased land degradation.

Degradation of communal lands is often portrayed as a classic example of 'the tragedy of the commons', where an ever-increasing competition between the users of these lands drives its degradation. In these shared resource systems, individuals act to their self-interest and thus contrary to the common good (Ostrom, 2000). Therefore, the reality of overgrazing of communal lands can best be described as the 'tragedy of open access' rather than the 'tragedy of the commons' (Roth, 1996). However, the new governments that were very much focused on looking forward rather than learning from the past did not draw on historical evidence showing that pastoral communities developed effective systems of managing common resources in the long term. Pastoral livelihood strategies co-evolved with the unpredictable East African environment where mobility and a sufficient livestock herd acts as a buffer against droughts (Roth, 1996) (Chapter 5). Externally imposed limits to mobility threaten pastoral livelihood security and hence food security and this buffering (Rufino et al., 2013).

One of the biggest transitions has been the formalisation of tenure for many rangeland communities through either the legal assertion of state sovereignty in some countries (Tanzania) or through the imposition of private holdings in others (Kenya and Uganda) (Galaty, 2016). The Maasai have also rallied politically to secure title to their lands and discourage immigrants; pastoralist communities are increasingly reasserting informal rights over freehold and have reoccupied privately held or state lands, in some cases being repaid by their gaining legal title for their communities. Individual land titles put pastoralists on a par with other sectors of society in being guaranteed the sole returns of their efforts (Western & Manzolillo Nightingale, 2003) while also lengthening planning horizons and increasing investment in the land. Other benefits include the stabilisation of production, markets, and social services. Private land, in turn, becomes a commodity for sale or rent that often works to the unfortunate disadvantage of pastoralists who are soon persuaded to sell out under political coercion or economic hardship (Huggins, 2000; Southgate & Hulme, 2000). In response to the subdivision, privatisation and commercialisation of land, other communities and investors have immigrated into pastoral areas, establishing farms and irrigation schemes (Fig. 6.7), and using up scarce water resources (Southgate & Hulme, 2000). The emergence of Maasai NGOs and community-based organisations aimed at securing rights, raising financial assistance, building local capacity, and opening new opportunities, are changing the nature of the relationship between pastoral communities and their land. New sector-based institutions (farming, ranching, and wildlife) are also on the rise, promoting products and pursuing markets (Western & Manzolillo Nightingale, 2003). Given the lack of central government support, local initiatives have emerged to fill the resulting void and tackle the new environmental threats by promoting grazing associations and grass reserves among landowners on private and communal lands. Pastoralists continue to be marginalised by government policy that favours the dominant settled farming lifestyles (Horowitz & Little, 1987; Reid et al., 2016). This increasing impact on pastoral livelihoods is associated with one of the most significant ironies concerning the sustainability of pastoral livelihoods—by the latter part of the twentieth century many nomadic pastoralists in East Africa, who had depended almost exclusively on livestock, began to adopt cultivation, and decrease mobility (McCabe, 2003). Although pastoralism may be increasingly accepted as consistent with conservation goals, cultivation and a sedentary pastoral population is not.

6.3 Postcolonial Approach to Wildlife and Protected Areas

Both Mwalimu Julius Nyerere and Jomo Kenyatta put natural resource firmly at the heart of postcolonial development and policy; the former being particularly active in promoting the value of nature (Mgaya, 2016). This view is exemplified in Nyerere's often-cited Arusha Declaration of 1967, which laid the framework for wildlife and forest policies in the decades after independence: 'Tanzania declared that nature is important as a resource for Tanzania's development, and natural resource management should be based on a combination of tradition, experience, as well as external support'. Therefore, from the foundational Arusha Declaration, natural resources were at the heart of development in postcolonial Tanzania (Mgaya, 2016). Subsequently, wildlife is a central part of national policy and has contributed a significant amount of foreign currency and developmental income through tourism concessions, particularly in Tanzania and Kenya that are the safari hubs of the global tourism industry (Barrow et al., 2001; Western, 1994). Despite this wide acknowledgement of the importance of wildlife across East Africa, the enactment of laws to support wildlife have been adopted but enforcement is an ongoing challenge—there is no shortage of policy, but implementation is another issue (Ndangalasi et al., 2007).

Wildlife conservation promotes tourism, and tourism is one of the biggest earners of foreign exchange in Kenya, Tanzania, and to a certain extent Uganda (Enghoff, 1990). Justification for wildlife conservation is never held solely in terms of the economics of tourism but clearly this is a big motivation. East Africa is world famous for the National Park network that brings in vital tourist revenue across the region. The northern circuit of Tanzania consists of the following National Parks: Mt Kilimanjaro, Arusha, Serengeti, Lake Manyara, Mkomazi, Tarangire, and the Ngorongoro Crater Conservation Area (Fig. 6.9). Arguably, the most popular of these sites are Serengeti National Park and the Ngorongoro Crater. Established as a game reserve in 1929, a national park in 1951, and now a UNESCO World Heritage site, Serengeti National Park is the oldest and second-largest (14,763 km^2) national park in Tanzania. It is home to over 1 million wildebeest, 300,000 Thomson's gazelle, 200,000 zebra; all 'Big 5' species are present, as well as other charismatic megafauna such as hippo, giraffe, and cheetah. There are several mesofauna present such as hyenas, jackals, aardwolf, and servals, and 500 bird species. Serengeti National Park is also a site for the world-renowned wildebeest migration as they ungulate population, particularly the numerous wildebeest that move with the rains

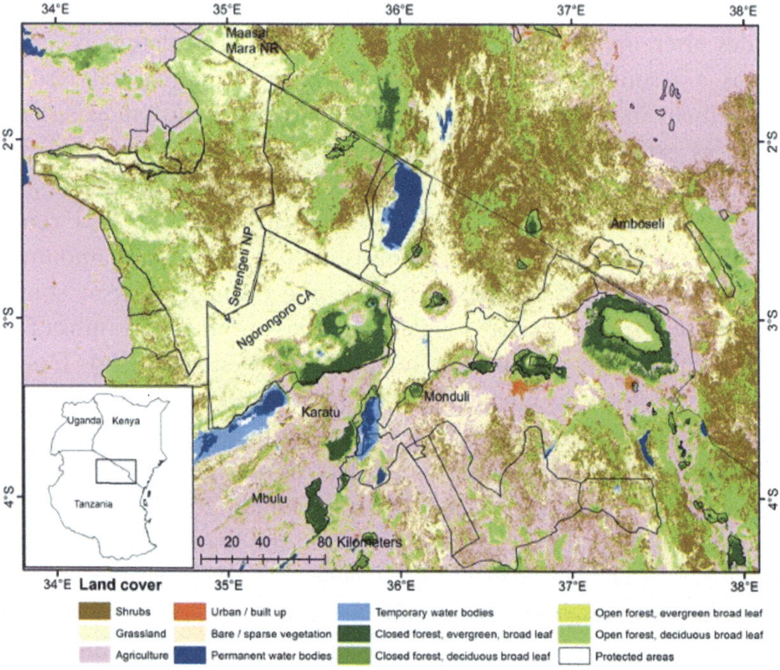

Fig. 6.9 The Kenya-Tanzania Borderland area is characterised by a wide range of land uses and a large proportion of the land dedication to conservation. The area comprises 14 protected areas including the Serengeti, Ngorongoro and Masai Mara

and the ensuing grazing areas that migrate across the Serengeti and the connected Maasai Mara (Fig. 6.9). The adjacent 8292 km^2 Ngorongoro Conservation Area (NCA), also a UNESCO World Heritage and Man and Biosphere Reserve, is administered by the Ngorongoro Conservation Area Authority that was established in 1959 and is a designated multiple-use area (Fig. 6.9). The wider Tanzania and Kenya Borderlands region (Fig. 6.9) is home to fourteen protected areas that attract millions of visitors per annum with billions of dollars in associated tourist revenue that is vital to the national economies.

Wildlife numbers have been in decline across most of the rangelands over the last thirty years (Blench, 2000; Grunblatt et al., 1996) due to poaching, predator poisoning, and the enclosure of water holes and pastures, although wildlife populations usually fare better in strictly protected areas such as National parks (Western et al., 2009). For example, 20 years of data from the Maasai Mara show a 49 and 72% decline in small and medium-sized ungulate species, respectively (Ottichilo et al., 2000). Some local extirpations have occurred as well: wild dogs (*Lycaon pictus*) have been particularly hard hit becoming extinct in 29 out of 38 reserves with spotted hyena (*Crocuta crocuta*) populations being lost from seven out of 35 reserves (Harcourt et al., 2001). Also, four local population extinctions have occurred in Tanzania's northern National Parks (Newmark, 1991): mountain reedbuck (*Redunca fulvorufula*) from Lake Manyara and Kilimanjaro National Parks, steenbok (*Raphicerus campestris*) from Arusha National Park, and Klipspringer (*Oreotragus oreotragus*) from Kilimanjaro National Park. These wildlife declines threaten wildlife in parks and in turn the lucrative tourist industries across East Africa (Gibson, 1999; Jones, 2001).

Long-term wildlife population monitoring is a crucial tool to track the changes in numbers and identify the drivers behind the change. One good example of long-term monitoring comes from Lake Manyara National Park, Tanzania where intermittent ground counts of the major herbivore species date back to 1959, the year before the area became a National Park in 1960 and indicated that the park supports one of the highest recorded large mammal biomass densities in the world (Prins & Douglas-Hamilton, 1987). However, these data have shown the impact of encroachment, particularly along its northern boundary, by a fast-growing human population, leading to further insularisation of the park (Kariuki et al., 2021; Msoffe et al., 2011). This is quite a common situation of population growth around protected areas (Pfeifer et al., 2012). And at the heart of this is that Tanzania has c. 40% of the nation protected; this is leading to a whole host of development and livelihood challenges. For example, as communities often extend their cultivation up to the park boundaries (Fig. 6.9), human–wildlife conflicts are exacerbated with ensuing calls to fence reserves and National Parks. To try and mitigate the human–wildlife conflicts, buffer zones are created around protected areas where income-generating schemes such as beekeeping or agroforestry aim to further reduce these human–wildlife conflicts. Given the challenge on the boundaries of parks, and in some cases encroachment into park boundaries, the creation of a National Parks and state ownership of wildlife has raised the temperature of human–wildlife conflicts (Western & Manzolillo Nightingale, 2003). Nature reserves and protected areas across East Africa have suffered from poaching and deforestation, as well as other forms of incursion and encroachment (Soulé et al., 1979).

The international conservation agenda developed momentum through the Rio Earth summit in 1992, and the Third World Parks Congress in Bali in the same year further emphasised the importance of conservation and consideration of local communities. This period also marked increasing the diversity of species behind the conservation mandate rather than just focusing on flagship species such as the charismatic 'big five'. A greater diversity of flagship taxa (e.g. insects, birds, amphibians) would, by default, encompass more niches, thereby extending the conservation benefits to more habitats (Skibins et al., 2016). The 'fortress' mentality of conservation within strict boundaries also began to be challenged at this time with initiatives such as those exemplified in the Amboseli National Park where the rights and needs of communities were recognised and in the future success of conservation policies will be reliant on the need to incorporate the local communities voice in the conservation process (McCabe, 2003). Although the 'fortress' mentality has not been abandoned, as witnessed by the expulsion of the Maasai from the Mkomazi Game Reserve in 1988 (Brockington,

1999), clearly, wildlife moves freely in and out of park boundaries onto pastoral lands with newly farmed plots becoming common targets for elephants, buffaloes, and other wild herbivores (Sindiga, 1999). As most wildlife resides in the pastoral rangelands outside park boundaries, there is growing conflict as pressures on the land build through title deed and spread of development and falling tolerance of wildlife (Eriksen et al., 1996). Wildlife Management Areas and Community conservancies (Fig. 6.10) are designed to deal with some of these conflicts by assigning

Fig. 6.10 Conservancies, such as the Siana, Nashulai are located around the Masai Mara where there are some 14 conservancies. Kalama, located close to Lewa in northern Kenya, clearly shows the link between pastoral communities and conservation with the milking of an elephant! Community conservancies around National Parks, such as the expanding number around Amboseli, are increasingly common as pastoral communities get land title deeds (All photographs: Rob Marchant)

lands delineated by one or more villages primarily for wildlife conservation and ecotourism businesses based on wildlife viewing. Local people are given the right to manage and use wildlife resources on this community land outside core protected areas (Kaaya & Chapman, 2017). This type of management approach, alongside benefits from wildlife across the wider landscape, is at the heart of the conversancy type model (Fig. 6.10) that has worked, to varying degrees of success, around the Maasai Mara. Amboseli National Park, gazetted in 1974, was one of the first National Parks that attempted to make wildlife an asset to indigenous occupants through revenue-sharing and wildlife utilisation (Western & Manzolillo Nightingale, 2003). Economic opportunity fits well with pastoral traditions, which saw wildlife as 'second cattle' to be used when livestock production faltered during droughts. Despite the significant advance in policy, conflicts remain due to limited rights and inequitable income flowing from wildlife (Western & Manzolillo Nightingale, 2003). Creation of National Parks and management of land for tourism and conservation often caused a rift between wildlife conservation and development, particularly pastoralism as will be explored in Chapter 7. There is currently a good deal more potential to diversify rangeland income through wildlife utilisation than there was traditionally, and this is clearly an evolving area. However, the massive impact on the travel and tourism industry imparted by the COVID-19 pandemic and the ensuing wiping out of conservancy fees, benefit-sharing, and the removal of the rationale to tolerate wildlife has resulted in increased conflicts between people and wildlife, questioning the resilience of the conservancy type approach.

A more participatory-based approach was also adopted as communities accessed information from diverse knowledge sources to tackle new challenges to provide healthy and profitable livelihoods for their families and conserve the rich heritage of wildlife they have lived with for several millennia (Reid et al., 2016). A good example of this is the Development through Conservation and Mountain Gorilla Project that was established in 1978 in the Volcanoes National Park (Rwanda) and Bwindi Impenetrable Forest National Park (BINP), Uganda (Fig. 6.3). Two of its main objectives were the development of tourism and conservation education. Prior to being gazetted as a park in 1991, BINP provided basketry and medicinal plant resources to the local community who had free access to forest resources, but local communities were denied access as a national edict bans people from National Parks. In 1994, supported by conservation and development organisations, Uganda National Parks (now Uganda Wildlife Authority-UWA) started a multiple-use programme for communities targeting the 21 parishes adjacent to BINP (Mutebi, 1994) to collect non-timber forest products (NTFPs) such as medical plants, poles, and berries. However, inadequate information on the ecology and physiology of NTFP species is the major drawback to such initiatives that aim to manage the resource base (Ticktin, 2004). Some species, especially those used as building poles, appear to be heavily harvested, particularly near villages (Ndangalasi et al., 2007). Harvesting of NTFPs negatively impacts plant species that are in high demand by local communities, thus having the potential to alter forest structure and integrity in the forests (Hall et al., 2003). Similarly, exploitation of trees for charcoal production in forest patches surrounding Kibale NP, Uganda, could have devastating effects on the overall ecosystem and local human communities (Naughton-Treves et al., 2007). Efforts such as planting of the most utilised species on adjacent farms should be encouraged as alternatives to NTFP extraction from the forest. Concentrating on NTFPs that are sufficiently valuable to local communities would likely enhance such practices. Forest enrichment planting and the setting of harvesting levels and cycles have been introduced in recent years in various areas (e.g. see Romero et al., 2014), but care should be taken to approach these practices against the background of potential ecological and socio-economic impacts (Ndangalasi et al., 2007), particularly surrounding Gorilla tourism that is a key source of income for Uganda National Parks. Gorilla groups had been habituated to tourist visitors and the demand for these specialised visits now exceeds the capacity of the park to meet it, even though fees have been greatly increased (Pullan, 1988) and were recently as high as US$500 per day.

Poaching for bushmeat remains a major threat to the conservation of wildlife around many protected areas (Lindsey et al., 2013). Unfortunately, the community wildlife management (CWM) initiatives set up have had only mixed results in limiting this illegal activity (Kaaya &

Chapman, 2017). One of the main reasons is a significant discrepancy that often exists between the very small material benefits ultimately distributed to households from most CWM programmes and the large direct costs from wildlife crop damages as well as the opportunity cost of lost arable and grazing land that many households near protected areas bear (Kideghesho, 2016). One of the first interventions was the Community Conservation Services programme, a community outreach programme established in 1988 by Tanzania National Parks (TANAPA) to promote conservation objectives by improving relations between National Parks and local people through the sharing of park benefits such as schools, water, and health services. In Kenya, conservation outside of government-protected areas has long been focused on incentives from tourism funded community-based conservancies (Western et al., 2015). The 'Parks Beyond Parks' campaign had a significant influence in promoting community-based conservation and ecotourism (Western, 2008). Firstly, it aimed to encourage local communities in prime wildlife areas to set up their own conservancies for their own benefit (Fig. 6.10). Second, through ecotourism, the tourist could stay on a community conservancy at a locally built lodge and watch animals free of tourist throngs. 'Parks Beyond Parks' spread the benefits, spread the load, and made local communities central in tourism rather than peripheral and simultaneously gave community conservation standing and recognition (Western, 2008). Parks Beyond Parks also focused on maintaining good grazing management and forage resource heterogeneity, both temporally and spatially, as a tool for supporting both livestock and wildlife (Tyrrell et al., 2017). Community-based conservation initiatives plan around the principles of livestock and grazing management, to maintain and exploit resource heterogeneity and facilitate wildlife–livestock coexistence by providing critical resource space for both livestock and wildlife (Fynn et al., 2016). Several tour operators and concessionaires have entered direct contractual arrangements with communities to establish partnership lodges with a few communities having exclusive ownership of ecotourism lodges. Several more progressive large lodges are also entering into business arrangements with communities and preferentially employing staff around the many protected areas (Fig. 6.11).

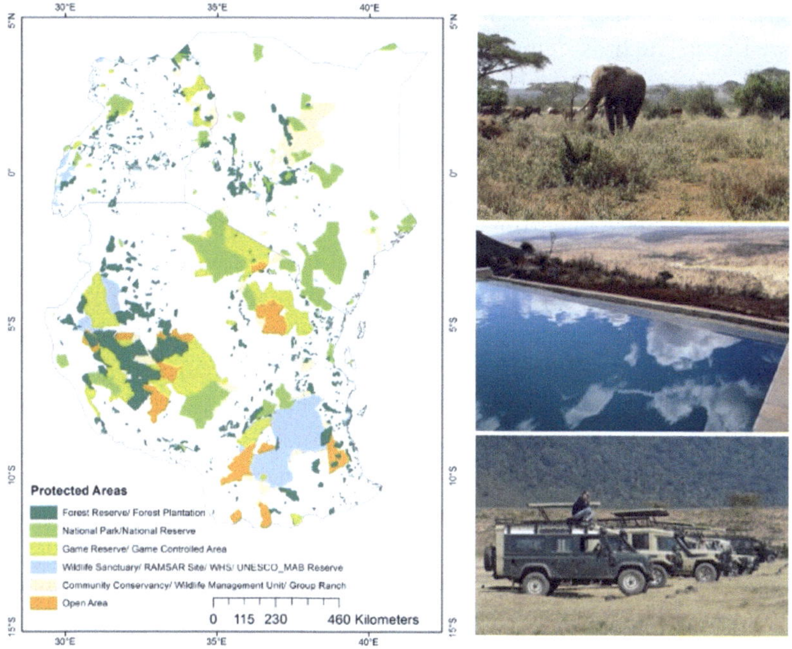

Fig. 6.11 The large number and area covered by National Parks and Protected Areas provide opportunities for employment and providing good and services to support the tourism industry (All photographs: Rob Marchant)

NGOs are actively exploring land trusts, conservation easements, and leasing fees with local communities to address environmental threats and turn conservation to economic and social advantage. Other organisations, including the Arid and Semi-Arid Lands projects and the North Rangeland Trust, are attempting to halt land fragmentation by improving on and diversifying pastoral strategies. Community Conservation Banks (COCOBA) were initiated by the Frankfurt Zoological Society (FZS) in Serengeti in 2009 to address the economic causes of poaching and improve livelihoods for people in the villages adjacent to the National Park (Fig. 6.11). These work by reducing poverty through the provision of low-interest loans to individuals for the establishment of small environmentally friendly business enterprises (Girabi & Mwakaje, 2013). These initiatives are aimed at spreading the benefits, so they feel less marginalised and view the WMA as less of a burden, even with its associated costs such as denial of access to grazing areas and crop damage problems caused by elephants (Sungusia, 2010). A study in Udzungwa National Park, Southern Tanzania, found that although communities surrounding the National Park had a clear understanding of community support provided, there was no evidence of a reduction of poaching and efforts are in place to ensure there is strong appreciation of the direct and indirect benefits provided by the National Park. However, such capacity building in Serengeti needs further development to ensure that local people can fully participate in the conservation programmes, including those designed to reduce poaching (Kaaya & Chapman, 2017).

However, in recent decades the impacts of changing and more unpredictable climate are being felt on the natural resources. For example, within the Amboseli National Park in Southern Kenya, the extensive drought from 2008 to 2009 resulted in a 90% mortality of the wildebeest populations, and corresponding reductions in the populations of other ungulates (Okello et al., 2015). Despite this decrease in ungulates, carnivore populations increased at this time, however, in the absence of wild ungulates the domesticated livestock of the pastoral communities such as cattle, sheep, and goats, came under pressure from predation instead (Manoa & Mwaura, 2016). To control this spiralling human–wildlife conflict, the Kenyan Wildlife Service spent approximately US$ 1,000,000 on importing 150 zebras from a private game reserve in Kenya to alleviate the pressure. Such declines in browsing species (e.g. due to changing grazing resource, disease outbreaks, or poaching) may release woody vegetation regeneration from strong herbivory and thus facilitate bush encroachment (Kiffner et al., 2017). Contemporary climate change, and how this will impact ecosystems, is highly uncertain. Likewise, the associated impacts on biodiversity, protected areas, and socio-economic benefits are largely unknown—futures that will be explored in Chapter 7.

6.3.1 Marine Protected Areas

Coastal marine management in Kenya and Tanzania has followed two different paths resulting from different philosophies of governance, socio-economic conditions, and associated policies. Globalisation, international influence, and urbanisation have had profound impacts on coastal landscapes throughout East Africa (Bavinck et al., 2017). For example, there is a strong Italian influence in coastal Kenya that can be dated back to early postcolonial periods where it was possible to channel finance from the grey economy to tourism investment, particularly around resorts like Mombasa and Watamu. Both countries began the process of declaring Marine National Parks in the 1970s. The Kisite-Mpunguti Marine National Park in Kenya was first legally gazetted in 1973, degazetted in 1978 to open a previously closed reef in Mpunguti, and managed by the government park service, Kenya Wildlife Service, (McClanahan et al., 2006). The Tanga Coral Gardens Marine Reserve was proposed in 1981, but not implemented as it was believed that dynamite fishing had already largely destroyed the proposed park area (Horrill et al., 2000). The greater economic success of nature and beach tourism in Kenya during the 1970s and 1980s, largely due to strong political interest and investment, created the conditions needed to support a closed-area management regime. In contrast, Tanzania had lower levels of coastal tourism and therefore primarily focused on community development and poverty alleviation (McClanahan, 1999). Since 1995, existing laws on illegal and destructive fishing practices have been enacted but rarely enforced. Illegal and destructive fishing practices including dynamite and cyanide fishing and beach seining were commonplace. The new enforcement units focused on monitoring the compliance with the government Fisheries Act, including boat and fishing licences, resulted in a reduction of dynamite fishing from about 180 blasts per month in 1995 to less than five blasts per month in 2003 (McClanahan et al., 2006). However, from a biodiversity conservation perspective, collaborative fisheries management was weak as the areas of closure are often small and can be periodically opened depending on the consensus of the resource users. This opening and closing would affect the degree of recovery that would need >20 years for coral reefs and fish populations to recover (McClanahan et al., 2006). Periodic harvesting results in a slow decline in biodiversity and ecosystem functions while maintaining resource and harvest levels (McClanahan et al., 2006). As with the situation on land, alternative livelihoods and management of change are crucial to influence local socio-economic and political philosophies, associated policies, and actions (McClanahan et al., 2006).

New coastal challenges have emerged around large infrastructure developments with new port facilities, either constructed or planned, at Lamu, Mombasa, Bagamoyo, and Dar es Salaam (Sect. 6.5). Like many African cities, the coastal hubs of Mombasa and Dar es Salaam have grown rapidly with the latter being the fastest-growing city on the African continent that brings its own challenges in terms of use of green space, urban road, communication, and service infrastructure. Similarly, the tourism hotspots of the Southern Kenya coast around Diani and the Zanzibar archipelago have experienced a massive transformation over the independent era as mass tourism and tourist resorts have been contrasted, airports upgraded, and infrastructure improved. Parallel with the expansion of the tourism industry there have been challenges to ensure communities benefit from these developments and livelihoods are maintained or enhanced. As with the safari industry, the massive impact on the travel and tourism industry imparted by the COVID-19 pandemic has altered the benefit-sharing and opportunity for employment, resulting in increased conflicts between people and coastal development, as societies have returned to traditional livelihoods of farming, fishing, and seaweed production.

6.4 The Rise and Fall of Fortress Conservation Through the Poaching Crises

Although elephant populations in East and Central Africa increased during the 1960s (Buechner & Dawkins, 1961; Caughley, 1976; Laws, 1970), these increases were within National Park boundaries and might have been accompanied by a decline over the wider area. In the early 1970s, up to 400 tonnes of ivory per year were officially (not accounting for illegal exports) exported from East Africa; this equates to c. 5000 elephants killed per year. The ivory trade was so well advanced in East Africa that few elephants were predicted to survive beyond 1995 outside high-security areas (Caughley et al., 1990). Although Ivory trade is nothing new to East Africa (Chapter 4) this new period of ivory demand saw the African elephant population reduced from 1.3 million to 600,000 individuals between 1979 and 1987 (Douglas-Hamilton, 1987). It was this loss of an estimated half of Africa's elephants in ten years that became an international crisis and prompted the Convention on International Trade in Endangered Species (CITES) to implement a ban on ivory trade in 1989 (Parker & Graham, 1989; Wasser et al., 2004). The elephant kills were particularly acute in Lake Manyara National Park where Elephant density decreased from about 6 per km^2 to 1 per km^2 between 1985 and 1991 (Prins & van der Jeugd, 1993).

There were multiple factors that precipitated the poaching crises, at the heart were corruption and poor working conditions of those in the National Parks services that were open to financial inducements, particularly as the records of ivory come principally from Customs records (Parker, 1989). Ultimately, the new wave of ivory trade was driven by 'need' in response to an interaction between environmental and social factors. There were a series of extreme droughts in the early 1970s, such as the drought of 1971, that divested the Karamojong pastoralists of more than five million cattle and left them destitute, prompting them to turn to the ivory trade as an alternate source of income (Parker & Graham, 1989). The number of elephants lost from the region are uncertain due to figures coming from official sources unlikely to be close to reality. The 'production' of ivory from East Africa was 99 tonnes in 1973, increasing to 433 tonnes in 1976, and then declining to 111 tonnes, although Luxmoore et al. (1989) advise treating this last figure with caution as all estimates of the volume of the ivory trade must be viewed as minima due to the unknown proportion of illicit trade. Because there are increased incentives to misreport ivory at customs now that most trade is illegal, a decline in ivory shipments reported by customs is likely to be a poor indicator of the effectiveness of a ban. The ivory leaving Africa since the imposition of stronger CITES controls in 1986 has, however, declined particularly steeply. This could be because of a genuine decrease in the hunting pressure on the elephant, or because of a decrease in the proportion of the trade reported in customs or CITES statistics.

Despite these challenges, reliable evidence of trends in the illegal ivory trade is important for informing decision-making, but it is difficult to obtain due to the covert nature of the trade (Underwood et al., 2013). There is evidence of increased poaching of elephants from the global monitoring programme MIKE (Monitoring the Illegal Killing of Elephants) that partners The Elephant Trade Information System (ETIS) mandated by CITES in 1997 to track the illegal ivory trade globally (Underwood et al., 2013). As more information becomes available these assessments provide information about the relationships between 'environmental crime'—such as the illegal ivory, timber, and charcoal trade and links to transnational organised crime, international terrorism, and various insurgencies. Groups like the Congolese militants and Al-Shabaab derive substantial resources from the trade in illegal ivory and charcoal, whereas the Lord's Resistance Army (LRA), other central African rebel groups, and Southeast Asian criminal syndicates thrive on brokering the trade in poached ivory. UNEP and INTERPOL (2014) estimated that the aggregate value of illegal markets for these commodities is enormous, worth between US$70 and 213 million yr^{-1}.

The illegal ivory trade remains a major threat to elephant populations and the illegal trade in elephant ivory has increased significantly in the past decade (Beale et al., 2018), with studies estimating the current rate of decline of regional African elephant populations to be as high as 8%, primarily due to poaching with savanna elephants experiencing a massive decline, particularly in Tanzania (Cerling et al., 2016). Indeed, some of the figures are quite alarming; the savanna elephant population in the Selous Wildlife Reserve in Tanzania experienced a 66% decline from 2009 to 2013 (Bennett, 2015). In Tanzania, this decline resulted in President Kikwete declaring a 'war on poaching' that sparked widespread allegations of human rights abuse and extra-judicial killings. For many, these and similar examples signal an ominous trend towards the militarisation of forest and biodiversity conservation in certain parts of East Africa with the so-called 'green militarisation' marking the growing use of military and paramilitary personnel, training, technologies, and partnerships in the pursuit of conservation efforts (Cavanagh et al., 2015). This ostensible militarisation of conservation is not just restricted to keystone conservation targets like gorillas or elephants but also forest areas; in

Kenya forested protected areas have also periodically served as 'sanctuaries' for fighters (Bennett, 2013). Clearly, the future conservation of iconic wildlife is going to be a challenge; creating space for wildlife while allowing for an expanding and developing population is one of the key challenges ahead that will be explored in Chapter 7.

6.5 Development of New Partnerships: The Belt and Road

As seen through this volume, there have been wide ranges of external factors that have left indelible footprints across the East African landscape. One of the most recent and quite broad influences has come from China's increasing role on the world stage. East Africa is at the heart of this given its previous connections with China through African socialism and indeed through more deep-rooted trade relationships (Chapter 4). China's Belt and Road Initiative (BRI), also known as *Yidaiyilu* or One Belt One Road, is an ambitious project launched by China's President Xi Jinping in late 2013 to build connectivity with over 60 countries and 4.4 billion people across the continents of Asia, Africa, and Europe through transnational infrastructural programmes (Chung, 2019). The first tri-annual ministerial conference between China and African states, which gave rise to the Forum on China-Africa Cooperation (FOCAC), took place in 2000 and now includes all African states except for Swaziland. The increasing influence of China is very much focused around trade and the speed of the rise has been exceedingly high: from US$1.3 billion in 2009, China's trade with Africa rose to US$170 billion in 2017, with investments focusing on infrastructure-building, manufacturing, finance, tourism, and aviation industries (Chen, 2018).

The BRI consists of two major initiatives: Silk Road Economic Belt (SREB), and the 21st Century Maritime Silk Road (MSR) that connects China's coast with Southeast Asia, South Asia, East Africa, and the Mediterranean. East Africa is at the heart of the MSR (Chen, 2018) with new port facilities either constructed or planned at Lamu, Mombasa, Bagamoyo, and Dar es Salaam from where arterial transport networks will connect the interior to the coast and beyond. In Kenya, the

Mombasa–Nairobi-Kisumu Standard Gauge Railway (SGR) linking the major cities and a series of dryland port facilities to the county's major port of Mombasa serves as a gateway to Central Africa and ultimately connects through the Congo Basin to West Africa (Fig. 6.12). Railway construction and operation contracts are a major business opportunity for Chinese companies and open up new markets for industries that have been affected by oversupply in China (Wissenbach & Wang, 2017). Essentially, the SGR is the most extensive railway project for Kenya and since the railway contrasted along a similar route by the British Colonial administration (Chapter 5) offers a triple win for China, it is a market for goods and production output from Chinese factories, it opens up new markets, and it increases efficient resource extraction. The first major SGR phase from Mombasa to Nairobi was constructed with a loan of US$3.2 billion from China to the Kenyan Government, the

Fig. 6.12 One of the flagship projects from the Chinese led construction boom through the Belt and Road Initiative has been the Standard Gauge Railway that is running from Mombasa through Kenya and towards Kisumu and onwards through Uganda and ultimately connecting East to West Africa. This bisects some of the National Parks (**a**), carrying people and freight (**b**) (All photographs: Rob Marchant)

SGR reduced transportation costs from \$0.20 to 0.08 a ton km^{-1} and significantly shortened travelling time between Nairobi and Mombasa from 15 to 4 h (Kacungira, 2017). Despite the stated benefits of MSR to the countries in East Africa, China's growing activities and influence in Africa were criticised by former US Secretary of State Rex Tillerson in 2018 for 'predatory loan practices' (Chen, 2018) by trapping African countries into incurring unserviceable debts with Chinese infrastructure investment loans. Indeed, there have already been a series of defaults and refinancing agreements and the long-term development of the new China–East Africa relationship will be an interesting phenomenon to chart over the coming years.

In Tanzania, one of the showcases MSR projects is the 680-m seven-laned Nyerere Bridge in the capital city of Dar es Salaam (Fig. 6.2), the largest cable-stayed bridge in East Africa that will open up the city and hopefully increase traffic flow. There's also new railway and road infrastructure and new port terminal, particularly focused on fertiliser that would feed agricultural intensification/development, such as through the Kilombero Agricultural Growth Corridor. In addition to these hard infrastructure projects, an Internet Data Centre has been built by the China International Telecommunication Construction Corporation in Dar es Salaam that was part of Tanzania's national fibre-optic broadband cable network with a total length of approximately 7500 km and built with a \$250 million loan from China (Boyle, 2012). The Internet Data Centre has provided high-speed broadband connectivity and is behind the Government claim that the Internet Data Centre will provide specialised Information and Communication Technology services as well as being an engine for economic development.

China Merchants Holdings is about to start developing Bagamoyo, a small fishing port, into Africa's biggest container port in the next 10 years, with piers and docks extending along 10 miles of coastline and the capacity to handle 20 million containers a year. The area south of Bagamoyo, seen by China as a new Shenzhen-type Special Economic Zone, will have factories in a fenced-off industrial area, possibly an international airport, and apartment blocks to accommodate the estimated future population of 75,000 (Van Mead, 2018). If successfully pursued, Bagamoyo will be the largest and most significant Chinese project in

Tanzania. However, there are several aspiring projects like this that have very grand ambitions, such as the Konzo Technological city in central Kenya (Fig. 6.2) or the Turkana resort city in Northern Kenya, that keep on stalling and are mired in controversy over the financing, the environmental impacts, and ultimately equity and justice issues of the beneficiaries.

The SGR project is one of many Chinese-financed infrastructure initiatives in East Africa, which have been subject to broader debates over their modalities, motivations, finances, and environmental impact. One of the major but unstated reasons for MSR's interest in Kenya is crude oil, not from Kenya itself, but the oilfields of South Sudan and Northern Uganda. Access will need the construction of a further transport corridor with railway, highway, and oil pipeline plans to connect South Sudan's capital of Juba through Northern Kenya to emerge at a new port developed at Lamu, the first berth of which was completed in June 2019 (Kazungu, 2018). These infrastructure developments are not without controversy. One of the most public has been the SGR railway route that cuts through the middle of Nairobi National Park, with the ensuing complaint around noise and dust raised by its construction adversely affecting the wildlife and vegetation of the park, which is Nairobi's main tourist attraction (Ambani, 2017). Kenyan conservation advocates and NGOs have been largely overruled by their own government. Similarly, the SGR cuts through the Tsavo National Park, and to reduce the railway's disturbance of the migration route of elephants, giraffes, and other wildlife between the two sides of the National Park, CRBC designed wildlife corridors under the railway dam. Animals can also pass under the Voi River Super Bridge (Wissenbach & Wang, 2017), although there are challenges with the construction and use of these and there is increased wildlife conflict with these developments (Fig. 6.13) (Nyumba et al., 2020). The bridge cuts through the Tsavo National Park's elephant migration corridor. Conservationists are concerned that the frequent passage of trains may have a negative impact on some species and eventually affect the National Park's tourism. However, politicians such as the governor of Taita Taveta argue that 'animals will have to adapt just like people' (Wissenbach & Wang, 2017). A similar attitude prevails in the discussions about the extension (phase 2A) from Nairobi

Fig. 6.13 New linear infrastructure, such as roads and railways, are not without its challenges. Although there are mitigations measures in the design of bridges, noise screens and underpasses to reduce negative impact on wildlife, how effective these mitigation measures is yet to be determined both directly (**a**) and through the growth of cities (**d**) and facilitating the spread of invasive plants (**c**) (All photographs: Rob Marchant)

to Naivasha, which may raise tourism issues for the Nairobi National Park. Consultations with Kenya Wildlife Service, conservationists, Kenya Rail, the government, and the CRBC seem to have produced a new design of bridges with noise screens to reduce negative impact on wildlife and tourism, although how effective these mitigation measures will be are yet to be determined (Wissenbach & Wang, 2017). The CRBC tried to address community concerns by hiring liaison officers and creating a vocational training facility, but the company has demonstrated less flexibility on the main contract provisions, explained in part by the Kenyan government's pressure to finish construction on time and within budget.

These challenges are particularly acute due to the speed of the development, and the normal process of social and environmental impact assessments are not carried out and digested into the planning process before the project begins to evaluate and mitigate the negative impacts on the local population (Tilt et al., 2009). The wider environmental impact of Chinese companies' projects has also been a topic of debate and research, particularly around wildlife conservation, which has been a key concern for the SGR project (Nyumba et al., 2020). Nevertheless, the SGR was finished on time during the first term (2013–2017) of President Kenyatta and has been presented as a flagship project of a government committed to economic development and a symbolic closing of reliance on colonial infrastructure. To mark the significant move from colonial influence to one of Chinese influence, the SGR began service on Madaraka Day, Kenya's national holiday commemorating self-rule on 1 June 1963 (Wissenbach & Wang, 2017).

In all countries, the level of construction and associated transport has led to a surge of Chinese firms and consumer goods on the market that are competitive, although with numerous complaints on the quality. Alongside the goods on sale, there has been an influx of Chinese traders across East Africa that are starting to impact local businesses. In addition to its significant commercial investments and infrastructure of ICT, China operates more than 40 Confucius Institutes at the Universities of Nairobi, Dar es Salaam, and Makerere (Kampala) that teach courses on Chinese language and culture. Additionally, there are growing numbers of exchange student scholarships, compared to those of traditional education providers such as the United States and the United Kingdom that have seen declines (Xinhuanet.com, 2017).

These recent large infrastructure, extractives, and agriculture-focused projects have raised concerns surrounding issues of displacement, environmental damage, and impact on the livelihoods of African host communities. Kenyan politicians and planners clearly need to improve their communication with local communities to better convey the rationale behind these major projects, especially in the early planning phases where adjustments are possible. Such consultation needs to address social and environmental concerns and requires mitigation measures to be put in place. Local consultation and participation in discussions of land issues and labour recruitment would also ensure that the planned extension phases of the SGR could benefit from lessons learned from the current project, such as managing expectations of different stakeholders, localisation of supplies, and skills training for workers, as well as day-to-day project implementation issues. The jury is still very much out on how these new and developing East Africa–China relationships will evolve and who the winners and losers of the partnerships will be. As with all previous relationships, the wider social, ecological, and environmental realities will take time to emerge, and the result is something that will play out in the future—issues to be explored in Chapter 7.

References

Ahrends, A., Bulling, M. T., Platts, P. J., Swetnam, R., Ryan, C., Doggart, N., Hollingsworth, P. M., Marchant, R., Balmford, A., Harris, D. J., & Gross-Camp, N. (2021). Detecting and Predicting Forest Degradation: A Comparison of Ground Surveys and Remote Sensing in Tanzanian Forests. *Plants, People, Planet,* 3, 268–281. https://doi.org/10.1002/ppp3.10189

Ambani, M. M. (2017). *GIS Assessment of Environmental Footprints of the Standard Gauge Railway (SGR) on Nairobi National Park, Kenya*. Msc., University of Nairobi. http://erepository.uonbi.ac.ke/bitstream/han-dle/11295/101271/MildredAmbani_Final

Ananda, J., & Herath, G. (2003). Soil Erosion in Developing Countries: A Socio-Economic Appraisal. *Journal of Environmental Management, 68*, 343–353. https://doi.org/10.1016/S0301-4797(03)00082-3

Anderson, D. (1984). Depression, Dust Bowl, Demography, and Drought: The Colonial State and Soil Conservation in East Africa During the 1930s. *African Affairs* [online], *83*, 321–343. https://www.jstor.org/stable/722351. Accessed 11 June 2018.

Anderson, D. (2005). *Histories of the Hanged* (p. 252). W. W. Norton and Company.

Anderson, D., & Anderson, D. M. (2005). *Histories of the hanged: The dirty war in Kenya and the end of empire*. WW Norton & Company.

Anthony, B. P., Scott, P., & Antypas, A. (2010). Sitting on the Fence? Policies and Practices in Managing Human-Wildlife Conflict in Limpopo Province, South Africa. *Conservation and Society, 8*, 225–240.

Barron, J., Rockström, J., Gichuki, F., & Hatibu, N. (2003). Dry Spell Analysis and Maize Yields for Two Semi-Arid Locations in East Africa. *Agricultural and Forest Meteorology* [online], *117*, 23–37. https://doi.org/10.1016/S0168-1923(03)00037-6

Barrow, E., Gichohi, H., & Infield, M. (2001). The Evolution of Community Conservation Policy and Practice in East Africa. In D. Hulme & M. Murphree (Eds.), *African Wildlife and Livelihoods: The Promise and Performance of Community Conservation.* James Currey.

Bavinck, M., Berkes, F., Charles, A., Dias, A. C. E., Doubleday, N., Nayak, P., & Sowman, M. (2017). The Impact of Coastal Grabbing on Community Conservation—A Global Reconnaissance. *Maritime Studies, 16*(1), 8.

Beale, C. M., Hauenstein, S., Mduma, S., Frederick, H., Jones, T., Bracebridge, C., Maliti, H., Kija, H., & Kohi, E. M. (2018). Spatial Analysis of Aerial Survey Data Reveals Correlates of Elephant Carcasses Within a Heavily Poached Ecosystem. *Biological Conservation, 218*, 258–267.

Bennett, E. L. (2015). Legal Ivory Trade in a Corrupt World and Its Impact on African Elephant Populations. *Conservation Biology, 29*(1), 54–60.

Bennett, H. (2013). *Fighting the Mau Mau: The British Army and Counter-Insurgency in the Kenya Emergency.* Cambridge University Press.

Berkes, F., Colding, J., & Folke, C. (2000). Rediscovery of Traditional Ecological Knowledge as Adaptive Management. *Ecological Applications* [online], *10*, 1251–1262. https://doi.org/10.1890/1051-0761(2000)010(1251:ROTEKA)2.0.CO;2

Beymer-Farris, B. A., & Bassett, T. J. (2012). The REDD menace: Resurgent protectionism in Tanzania's mangrove forests. *Global Environmental Change, 22*(2), 332–341.

Blaikie, P. (2016). *The Political Economy of Soil Erosion in Developing Countries.* Routledge.

Blaikie, P., & Brookfield, H. (2015). *Land Degradation and Society*. Routledge.

Blench, R. (2000). *Extensive Pastoral Livestock Systems: Issues and Options for the Future.* FAO-Japan Cooperative Project "Collection of Information on Animal Production and Health". www.fao-kyokai.or.jp/edocuments/document2.pdf

Blomley, T., & Iddi, S. (2009). *Participatory Forest Management in Tanzania: Lessons Learned and Experiences to Date, URT*. Ministry of Resource and Tourism, Forest and Beekeeping Division.

Blomley, T., Pfliegner, K., Isango, J., Zahabu, E., Ahrends, A., & Burgess, N. (2008). Seeing the Wood for the Trees: An Assessment of the Impact of

Participatory Forest Management on Forest Condition in Tanzania. *Oryxen, 42*(3), 380–391.
Bluwstein, J., Lund, J. F., Askew, K., Stein, H., Noe, C., Odgaard, R., Maganga, F., & Engström, L. (2018). Between Dependence and Deprivation: The Interlocking Nature of Land Alienation in Tanzania. *Journal of Agrarian Change* [online], *18*, 806–830. https://doi.org/10.1111/joac.12271
Boyle, J. (2012). Tanzania's Invisible Web Revolution. *BBC World News*, October 2. https://www.bbc.com/news/world-africa-19451044
Brockington, D. (1999). Conservation, Displacement and Livelihoods: The Consequences of the Eviction for Pastoralists Moved from the Mkomazi Game Reserve, Tanzania. *Nomadic Peoples, 3*, 74–96.
Bryceson, D. F. (2002). The Scramble in Africa: Reorienting Rural Livelihoods. *World Development* [online], *30*, 725–739. https://doi.org/10.1016/S0305-750X(02)00006-2
Buechner, H. K., & Dawkins, H. C. (1961). Vegetation Change Induced by Elephants and Fire in Murchison Falls National Park, Uganda. *Ecology, 42*, 752–766.
Burgess, N. D., Mwakalila, S., Munishi, P., Pfeifer, M., Willcock, S., Shirima, D., ... Marchant, R. (2013). REDD herrings or REDD menace: Response to Beymer-Farris and Bassett. *Global Environmental Change, 23*(5), 1349–1354.
Caughley, G. (1976). The Elephant Problem: An Alternative Hypothesis. *East African Wildlife Journal, 14*, 265–283.
Caughley, G., Dublin, H., & Parker, I. (1990). Projected Decline of the African Elephant. *Biological Conservation, 54*(2), 157–164.
Cavanagh, C., & Benjaminsen, T. A. (2014). Virtual Nature, Violent Accumulation: The 'Spectacular Failure' of Carbon Offsetting at a Ugandan National Park. *Geoforum, 56*, 55–65.
Cavanagh, C. J., Vedeld, P. O., & Trædal, L. T. (2015). Securitizing REDD+? Problematizing the Emerging Illegal Timber Trade and Forest Carbon Interface in East Africa. *Geoforum, 60*, 72–82.
Cerling, T. E., Barnette, J. E., Chesson, L. A., Douglas-Hamilton, I., Gobush, K. S., Uno, K. T., Wasser, S. K., & Xu, X. (2016). Radiocarbon Dating of Seized Ivory Confirms Rapid Decline in African Elephant Populations and Provides Insight into Illegal Trade. *Proceedings of the National Academy of Sciences, 113*(47), 13330–13335.
Chabal, P. (2013). *Africa: The Politics of Suffering and Smiling*. Zed Books.

Chen, X. (2018). Why Fears of China's Neocolonialism in Africa Ring False in the Face of Numbers That Tell a Different Tale. *South China Morning Post*, September 25.

Chung, C. P. (2019). China's Maritime Silk Road Initiative: Strategic and Economic Implications for East Africa. *China and the World, 2*(02), 1950011.

Conte, C. A. (1999). The Forest Becomes Desert: Forest Use and Environmental Change in Tanzania's West Usambara Mountains. *Land Degradation and Development* [online], *10*, 291–309. https://doi.org/10.1002/(SICI)1099-145X(199907/08)10:4<291::AID-LDR363>3.0.CO;2-W

Conway, G. R., & Waage, J. (2010). *Science and Innovation for Development*. UK Collaborative on Development Sciences (UKCDS).

Dawson, W., Mnolwa, A. S., Burslem, D. F. R. P., & Hulme, P. E. (2008). Assessing the Risks of Plant Invasions Arising from Collections in Tropical Botanical Gardens. *Biodiversity and Conservation, 17*(8), 1979–1995.

Douglas-Hamilton, I. (1987). African Elephants: Population Trends and Their Causes. *Oryx, 21*(1), 11–24.

Dregne, H. E. (2002). Land Degradation in the Drylands. *Arid Land Research and Management, 16*, 99–132.

Elkins, C. (2005). *Imperial Reckoning: The Untold Story of Britain's Gulag in Kenya*. Henry Holt and Company. ISBN 978-0-805-07653-0.

Enfors, E., & Gordon, L. (2007). Analysing Resilience in Dryland Agro-Acosystems: A Case Study of the Makanya Catchment in Tanzania Over the Past 50 Years. *Land Degradation and Development* [online], *18*, 680–696. https://doi.org/10.1002/ldr.807

Enghoff, M. (1990). Wildlife Conservation, Ecological Strategies and Pastoral Communities: A Contribution to the Understanding of Parks and People in East Africa. *Nomad Peoples, 25–27*, 93–107.

Eriksen, S., Ouko, E., & Marieka, N. (1996). Land Tenure and Wildlife Management. In C. Juma & J. B. Ojwang (Eds.), *In Land We Trust: Environment, Private Property and Constitutional Change* (Environmental Policy Series, No. 7). Nairobi: Initiatives Publishers and London: Zed Books.

Fleitmann, D., Dunbar, R. B., McCulloch, M., Mudelsee, M., Vuille, M., McClanahan, T. R., Cole, J. E., & Eggins, S. (2007). East African Soil Erosion Recorded in a 300 Year Old Coral Colony from Kenya. *Geophysical Research Letters, 34*, L04401.

Fynn, R. W. S., Augustine, D. J., Peel, M. J. S., & de Garine-Wichatitsky, M. (2016). REVIEW: Strategic Management of Livestock to Improve Biodiversity Conservation in African Savannahs: A Conceptual Basis for

Wildlife-Livestock coexistence. *Journal of Applied Ecology* [online]. https://doi.org/10.1111/1365e2664.12591
Galaty, J. (2016). Reasserting the Commons: Pastoral Contestations of Private and State Lands in East Africa. *International Journal of the Commons, 10*(2), 709–727.
Ghazoul, J., Burivalova, Z., Garcia-Ulloa, J., & King, L. A. (2015). Conceptualizing Forest Degradation. *Trends in Ecology and Evolution, 30*(10), 622–632.
Gibson, C. (1999). *Politicians and Poachers: The Political Economy of Wildlife Policy in Africa.* Cambridge University Press.
Gilbert, E. (2002). Coastal East Africa and the Western Indian Ocean: Long-Distance Trade, Empire, Migration, and Regional Unity, 1750–1970. *The History Teacher, 36*(1), 7–34.
Girabi, F., & Mwakaje, A. E. G. (2013). Impact of Microfinance on Small-Holder Farm Productivity in Tanzania: The Case of Iramba District. *Asian Economic and Financial Review, 3*(2), 227–242.
Gordon, I. (2003). Harnessing Butterfly Biodiversity for Improving Livelihoods and Forest Conservation: The Kipepeo Project. *The Journal of Environment and Development, 12*, 182–198.
Gordon, I., & Ayiemba, W. (2003). Harnessing Butterfly Biodiversity for Improving Livelihoods and Forest Conservation: The Kipepeo Project. *The Journal of Environment and Development, 12*, 82–98.
Grainger, M., & Geary, K. (2011). *The New Forests Company and Its Uganda Plantations.* Oxfam International.
Grunblatt, J., Said, M. Y., & Wargute, P. (1996). *National Range Report: Summary of Population Trends and Estimates of Wildlife and Livestock.* Ministry of Planning and National Development. DRSRS, Nairobi.
Habel, J. C., Casanova, I. C. C., Zamora, C., Teucher, M., Hornetz, B., Shauri, H., Mulwa, R. K., & Lens, L. (2017). East African Coastal Forest Under Pressure. *Biodiversity and Conservation, 26*(11), 2751–2758.
Hall, J. S., Harris, D. J., Medjibe, V., & Ashton, P. M. S. (2003). The Effects of Selective Logging on Forest Structure and Tree Species Composition in a Central African Forest: Implications for Management of Conservation Areas. *Forest Ecology and Management, 183*(1–3), 249–264.
Harcourt, A. H., Parks, S. A., & Woodroffe, R. (2001). Human Density as an Influence on Species/Area Relationships: Double Jeopardy for Small African Reserves? *Biodiversity and Conservation, 10*, 1011–1026.

Heckmann, M., Muiruri, V., Boom, A., & Marchant, R. (2014). Human-Environment Interactions in an Agricultural Landscape: A 1400-yr Sediment and Pollen Record from North Pare, NE Tanzania. *Palaeogeography, Palaeoclimatology, Palaeoecology, 406*, 49–61.

Hedberg, I., & Hedberg, O. (1968). Conservation of Vegetation in Africa South of the Sahara. *Acta Phytogeographica Suecica, 54*, 98–195.

Himmelfarb, D. (2012). *In the Aftermath of Displacement: A Political Ecology of Dispossession, Transformation, and Conflict on Mt. Elgon, Uganda.* PhD Dissertation, University of Georgia, Athens.

Homewood, K. (1995). Development, Demarcation and Ecological Outcomes in Maasailand. *Africa* [online], *65*, 331–350. https://doi.org/10.2307/1161050

Homewood, K., Coast, E., & Thompson, M. (2004). In-Migrants and Exclusion in East African Rangelands: Access, Tenure and Conflict. *Africa* [online], *74*, 567–610. https://doi.org/10.3366/afr.2004.74.4.567

Horowitz, M. M., & Little, P. D. (1987). *In Drought and Famine in Africa: Denying Drought a Future* (pp. 59–82). Cambridge University Press.

Horrill, C., Kamukuru, A. T., Mgaya, Y. D., & Risk, M. (2000). Northern Tanzania, Zanzibar and Pemba. In T. R. McClanahan, C. R. C. Sheppard, & D. O. Obura (Eds.), *Coral Reefs of the Indian Ocean: Their Ecology and Conservation* (pp. 167–198). Oxford University Press.

Huggins, C. (2000). Local Strategies for Coping with Drought in Kenya: Constraints and Opportunities. Summary of a presentation by Dr David Western. *ACTS Policy Insights, 2*(1), 1–4.

Huizer, G. (1973). Theujamaa Village Program in Tanzania: New Forms of Rural Development. *Studies in Comparative International Development, 8*(2), 192. https://doi.org/10.1007/bf02810000

Hydén, G. (1980). *Beyond Ujamaa in Tanzania: Underdevelopment and an Uncaptured Peasantry*. University of California Press.

Jones, B. (2001). The Evolution of a Community-Based Approach to Wildlife Management at Kunene, Namibia. In D. Hulme & M. Murphee (Eds.), *African Wildlife and Livelihoods: The Promise and Performance of Community Conservation.* James Currey.

Kaaya, E., & Chapman, M. (2017). Micro-Credit and Community Wildlife Management: Complementary Strategies to Improve Conservation Outcomes in Serengeti National Park, Tanzania. *Environmental Management, 60*(3), 464–475.

Kacungira, N. (2017). *Will Kenya Get Value for Money from Its New Railway?* BBC Africa, June 8. https://www.bbc.com/news/world-africa-40171095

Kariuki, R. W., Western, D., Willcock, S., & Marchant, R. (2021). Assessing interactions between agriculture, livestock grazing and wildlife conservation land uses: A historical example from East Africa. *Land, 10*(1), 46.

Kazungu, K. (2018). Kenyans to Wait Longer for First Berth at Lamu Port. *Daily Nation* (Nairobi), October 1.

Keatley, P. (2003). Idi Amin. *The Guardian.*

Kenya Forests Working Group (KFWG). (2004). *Changes in Forest Cover in Kenya's Five "Water Towers" 2000–2003.* KFWG, Nairobi.

Khan, Z. R., Midega, C. A., Pittachar, J. O., Murage, A. L. I. C. E., & Pickett, J. O. H. N. (2017). Climate-Smart Push-Pull: A Conservation Agriculture Technology for Food Security and Environmental Sustainability in Africa. In *Conservation Agriculture for Africa: Building Resilient Farming Systems in a Changing Climate* (pp. 151–166). BABI, Wallingford.

Kideghesho, J. (2016). Reversing the Trend of Wildlife Crime in Tanzania: Challenges and Opportunities. *Biodiversity and Conservation, 25*, 427–449.

Kiffner, C., Rheault, H., Miller, E., Scheetz, T., Enriquez, V., Swafford, R., Kioko, J., & Prins, H. H. (2017). Long-Term Population Dynamics in a Multi-Species Assemblage of Large Herbivores in East Africa. *Ecosphere, 8*(12), e02027.

Kristjanson, P., Reid, R. S., Dickson, N., Clark, W. C., Romney, D., Puskur, R., Macmillan, S., & Grace, D. (2009). Linking International Agricultural Research Knowledge with Action for Sustainable Development. *Proceedings of the National Academy of Sciences of the United States of America, 106*(13), 5047–5052.

Lane, C. R., & Pretty, J. N. (1990). *Displaced Pastoralists and Transferred Wheat Technology in Tanzania* (Vol. 20). International Institute for Environment and Development.

Laws, R. M. (1970). Elephants as Agents of Habitat and Landscape Change in East Africa. *Oikos, 21*, 1–15.

Lindsey, P. A., Balme, G., Becker, M., Begg, C., Bento, C., Bocchino, C., Dickman, A., Diggle, R. W., Eves, H., Henschel, P., Lewis, D. (2013). The Bushmeat Trade in African Savannas: Impacts, Drivers, and Possible Solutions. *Biological Conservation, 160*(4), 80–96.

Lundgren, B. (1978). *Soil Conditions and Nutrient Cycling Under Natural and Plantation Forests in Tanzanian Highlands.* Swedish University of Agricultural Studies.

Lundgren, L., & Lundgren, B. (1979). Rainfall, Interception and Evaporation in the Mazumbai Forest Reserve, West Usambara Mts., Tanzania and Their Importance in the Assessment of Land Potential. *Geografiska Annaler: Series*

A, Physical Geography [online], *61*, 157–178. https://doi.org/10.1080/04353676.1979.11879988

Luxmoore, R., Caldwell, J., & Hithersay, L. (1989). The Volume of Raw Ivory Entering International Trade from African Producing Countries from 1979 to 1988. In *The Ivory Trade and the Future of the African Elephant.*

Lyons, K., & Westoby, P. (2014). Carbon Colonialism and the New Land Grab: Plantation Forestry in Uganda and its Livelihood Impacts. *Journal of Rural Studies, 36*, 13–21.

Maathai, W. (2006). *Unbowed: A Memoir* (p. 63). Alfred A. Knopf. ISBN 0307263487.

Mahone, S. (2006). The Psychology of Rebellion: Colonial Medical Responses to Dissent in British East Africa. *The Journal of African History, 47*(2), 241–258. https://doi.org/10.1017/s0021853706001769

Manoa, D. O., & Mwaura, F. (2016). Predator-Proof Bomas as a Tool in Mitigating Human-Predator Conflict in Loitokitok Sub-County, Amboseli Region of Kenya. *Natural Resources, 7*(01), 28–39.

Marchant, R., Richer, S., Boles, O., Capitani, C., Courtney-Mustaphi, C. J., Lane, P., Prendergast, M. E., Stump, D., de Cort, G., Kaplan, J. O., Phelps, L., Kay, A., Olago, D., Petek, N., Platts, P., Punwong, P., Widgren, M., Wynne-Jones, S., Ferro-Vázquez, C., … Wright, D. (2018). Drivers and Trajectories of Land Cover Change in East Africa: Human and Environmental Interactions from 6000 Years Ago to Present. *Earth-Science Reviews, 178*, 322–378.

Matiku, P., Caleb, M., & Callistus, O. (2013). The Impact of Participatory Forest Management on Local Community Livelihoods in the Arabuko-Sokoke Forest, Kenya. *Conservation and Society, 11*, 112–129.

Maundu, P., Kibet, S., Morimoto, Y., Imbumi, M., & Adekar, R. (2009). Impact of *Prosopis juliflora* on Kenya's Semi-Arid and Arid Ecosystems and Local Livelihoods. *Biodiversity, 10*(23), 33–50.

McCabe, J. T. (2003). Disequilibrial Ecosystems and Livelihood Diversification Among the Maasai of Northern Tanzania: Implications for Conservation Policy in Eastern Africa. *Nomadic Peoples, 7*, 74–91.

McClanahan, T. R. (1999). Is There a Future for Coral Reef Parks in Poor Tropical Countries? *Coral Reefs, 18*, 321–325.

McClanahan, T. R., Marnane, M. J., Cinner, J. E., & Kiene, W. E. (2006). A Comparison of Marine Protected Areas and Alternative Approaches to Coral-Reef Management. *Current Biology, 16*(14), 1408–1413.

Mgaya, E. (2016). Forest and Forestry in Tanzania: Changes and Continuities in Policies and Practices from Colonial Times to the Present. *Journal of the Geographical Association of Tanzania, 36*(2), 45–58.

Midega, C. A. O., Bruce, T. J., Pickett, J. A., & Khan, Z. R. (2015). Ecological Management of Cereal Stemborers in African Smallholder Agriculture through Behavioural Manipulation. *Ecological Entomology, 40* (Suppl. S.1), 70–81.

Milledge, S. (2009). Forestry Governance and Trade Transformations: Experiences from Tanzania and Implications for Sustainable Development. In *Governing Africa's Forests in a Globalized World* (pp. 303–319). Routledge.

Mittermeier, R. A., Turner, W. R., Larsen, F. W., Brooks, T. M., & Gascon, C. (2011). Global Biodiversity Conservation: The Critical Role of Hotspots. In *Biodiversity Hotspots* (pp. 3–22). Springer, Berlin, Heidelberg.

Morgan, R. P. C. (2005). *Soil Erosion and Conservation* (Vol. 3). Blackwell Science.

Msoffe, F. U., Said, M. Y., Ogutu, J. O., Kifugo, S. C., De, J., Van Gardingen, P., & Reid, R. S. (2011). Spatial Correlates of Land-Use Changes in the Maasai-Steppe of Tanzania: Implications for Conservation and Environmental Planning. *International Journal of Biodiversity and Conservation, 3*, 280–290.

Mueller-Dombois, D. (1971). Planned Utilization of the Lowland Tropical Forests. *Nature and Resources, 7*, 18–22.

Mutebi, J. (1994). Assessing the patterns of use of forest resources by local communities: Field experience in and around Bwindi (Impenetrable) National Park in Southwestern Uganda. *Proceedings on non-timber forest products*.

Mwangi, E., & Swallow, B. (2008). *Prosopis juliflora* Invasion and Rural Livelihoods in the Lake Baringo Area of Kenya. *Conservation and Society, 6*, 130–140.

Naughton-Treves, L., Kammen, D. M., & Chapman, C. (2007). Burning Biodiversity: Woody Biomass Use by Commercial and Subsistence Groups in Western Uganda's Forests. *Biological Conservation, 134*(2), 232–241.

Ndangalasi, H. J., Bitariho, R., & Dovie, D. B. (2007). Harvesting of Non-Timber Forest Products and Implications for Conservation in Two Montane Forests of East Africa. *Biological Conservation, 134*(2), 242–250.

Nel, A., & Hill, D. (2013). Constructing Walls of Carbon—The Complexities of Community, Carbon Sequestration and Protected Areas in Uganda. *Journal of Contemporary African Studies, 31*(3), 421–440.

Neumann, K. (2005). The Romance of Farming: Plant Cultivation and Domestication in Africa. In A. B. Stahl (Ed.), *African Archaeology: A Critical Introduction*. Blackwell.

Newmark, W. D. (1991). Tropical Forest Fragmentation and the Local Extinction of Understory Birds in the Eastern Usambara Mountains, Tanzania. *Conservation Biology, 5*(1), 67–78.

Newmark, W. D., & McNeally, P. B. (2018). Impact of Habitat Fragmentation on the Spatial Structure of the Eastern Arc Forests in East Africa: Implications for Biodiversity Conservation. *Biodiversity and Conservation, 27*(6), 1387–1402.

Nyumba, T. O., Emenye, O. E., & Leader-Williams, N. (2020). Assessing Impacts of Human-Elephant Conflict on Human Wellbeing: An Empirical Analysis of Communities Living with Elephants Around Maasai Mara National Reserve in Kenya. *PloS one, 15*(9), e0239545.

Oiere, S. (2014). Kenya: EU Signs $39 Million Agreement for Water Tower Conservation. *Water* [online], *21.* http://www.iwapublishing.com/template.cfm?name=news2281. Accessed 9 Nov 2014.

Okello, M. M., Kenana, L., Maliti, H., Kiringe, J. W., Kanga, E., Warinwa, F., Bakari, S., Gichohi, N., Ndambuki, S., Kija, H., & Sitati, N. (2015). Population Status and Trend of Water Dependent Grazers (Buffalo and Waterbuck) in the Kenya-Tanzania Borderland. *Natural Resources, 6*(2), 91–114.

Ostrom, E. (2000). Collective Action and the Evolution of Social Norms. *Journal of Economic Perspectives* [online], *14*, 137–158. https://doi.org/10.1257/jep.14.3.137

Otianga-Owiti, G. E., Okori, J. J. L., Nyamasyo, S., & Amwata, D. A. (2021). *Governance and challenges of wildlife conservation and management in kenya. wildlife biodiversity conservation* (pp. 67–99).

Ottichilo, W. K., de Leeuw, J., Skidmore, A. K., Prins, H. H. T., & Said, M. Y. (2000). Population Trends of Large Non-Migratory Wild Herbivores and Livestock in the Masai Mara Ecosystem, Kenya, Between 1977 and 1997. *African Journal of Ecology, 38*, 202–216.

Parker, I. S. C. (1989). *The Raw Ivory Trade 1979–1989* (Report to CITES). Lausanne.

Parker, I. S., & Graham, A. D. (1989). Elephant Decline: Downward Trends in African Elephant Distribution and Numbers (Part II). *International Journal of Environmental Studies, 35*(1–2), 13–26.

Pfeifer, M., Disney, M., Quaife, T., & Marchant, R. (2012). Terrestrial ecosystems from space: a review of earth observation products for macroecology applications. *Global Ecology and Biogeography, 21*(6), 603–624.

Pinckney, T. C., & Kimuyu, P. K. (1994). Land Tenure Reform in East Africa: Good, Bad or Unimportant. *Journal of African Economies* [online], *3*(1), 28. https://doi.org/10.1093/oxfordjournals.jae.a036794

Platts, P. J., Burgess, N. D., Gereau, R. E., Lovett, J. C., Marshall, A. R., McClean, C. J., Pellikka, P. K. E., Swetnam, R. D., & Marchant, R. (2011). Delimiting Tropical Mountain Ecoregions for Conservation. *Environmental Conservation, 38*(3), 312–324.

Prins, H. H., & van der Jeugd, H. P. (1993). Herbivore Population Crashes and Woodland Structure in East Africa. *Journal of Ecology, 81*, 305–314.

Pullan, R. A. (1988). Conservation and the Development of National Parks in the Humid Tropics of Africa. *Journal of Biogeography, 15*, 171–183.

Reid, R. S., Nkedianye, D., Said, M. Y., Kaelo, D., Neselle, M., Makui, O., Onetu, L., Kiruswa, S., Kamuaro, N. O., Kristjanson, P., & Ogutu, J. (2016). Evolution of Models to Support Community and Policy Action with Science: Balancing Pastoral Livelihoods and Wildlife Conservation in Savannas of East Africa. *Proceedings of the National Academy of Sciences, 113*(17), 4579–4584.

Romero, C., Dovie, D. B., Gambiza, J., Luoga, E., Schmitt, S., & Grundy, I. (2014). Effects of Commercial Bark Harvesting on Adansonia Digitata (Baobab) in the Save-Odzi Valley, Zimbabwe, with Considerations for Its Management. *Advances in Economic Botany, 17*, 95–114.

Roth, E. A. (1996). Traditional Pastoral Strategies in a Modern World: An Example from Northern Kenya. *Human Organization* [online], *55*, 219–224. https://www.jstor.org/stable/44127165

Rufino, M. C., Thornton, P. K., Mutie, I., Jones, P., Van Wijk, M., & Herrero, M. (2013). Transitions in Agro-Pastoralist Systems of East Africa: Impacts on Food Security and Poverty. *Agriculture, Ecosystems & Environment* [online], *179*, 215–230. https://doi.org/10.1016/j.agee.2013.08.019

Sandbrook, C., Nelson, F., Adams, W. M., & Agrawal, A. (2010). Carbon, Forests, and the REDD Paradox. *Oryx, 44*(3), 330–334.

Santoro, A., Venturi, M., Bertani, R., & Agnoletti, M. (2020). A review of the role of forests and agroforestry systems in the FAO Globally Important Agricultural Heritage Systems (GIAHS) programme. *Forests, 11*(8), 860.

Sendalo, D. (2009). *A Review of Land Tenure Policy Implications on Pastoralism in Tanzania*. Ministry of Livestock Development and Fisheries, Dar es Salaam

Shackleton, R. T., Witt, A. B, Nunda, W., & Richardson, D. M. (2017a). *Chromolaena odorata* (Siam weed) in Eastern Africa: Distribution and Socio-Ecological Impacts. *Biological Invasions, 19*(4), 1285–1298.

Shackleton, R. T., Witt, A. B., Piroris, F. M., & van Wilgen, B. W. (2017b). Distribution and socio-ecological impacts of the invasive alien cactus Opuntia stricta in eastern Africa. *Biological Invasions, 19*(8), 2427–2441.

Shackleton, R. T., Witt, A. B., Aool, W., & Pratt, C. F. (2017c). Distribution of the invasive alien weed, Lantana camara, and its ecological and livelihood impacts in eastern Africa. *African Journal of Range & Forage Science, 34*(1), 1–11.

Shao, J. (1986). The Villagization Program and the Disruption of the Ecological Balance in Tanzania. *Canadian Journal of African Studies, 20*(2), 219–239. https://doi.org/10.2307/484871

Shaw, K., Ruyonga, G., & Nicholson, M. (2016). Enhancing Tree Conservation and Forest Restoration in East Africa. *BGjournal, 13*(2), 28–31.

Sheil, D. (1994). Invasive Plants in Tropical Forests: Warnings from the Amani Botanic Gardens, Tanzania. *Botanical Gardens Conservation News, 2*(3), 23–24.

Sinclair, A. J., Collins, S. A., & Spaling, H. (2011). The Role of Participant Learning in Community Conservation in the Arabuko-Sokoke Forest, Kenya. *Conservation and Society, 9*, 42–53.

Sindiga, I. (1984). Land and Population Problems in Kajiado and Narok, Kenya. *African Studies Review, 27*(1), 23–39.

Sindiga, I. (1999). Alternative Tourism and Sustainable Development in Kenya. *Journal of Sustainable Tourism, 7*(2), 108–127.

Skibins, J., Powell, R., & Hallo, J. (2016). Lucky 13: Conservation Implications of Broadening "Big 5" Flagship Species Recognition in East Africa. *Journal of Sustainable Tourism, 24*(7), 1024–1040.

Soulé, M. E., Wilcox, B. A., & Clarke, H. (1979). Benign Neglect: A Model of Faunal Collapse in the Game Reerves of East Africa. *Biological Conservation, 15*, 259–272.

Southgate, C., & Hulme, D. (2000). Uncommon Property: The Scramble for Wetland in Southern Kenya. In P. Woodhouse, H. Bernstein, & D. Hulme (Eds.), *African Enclosures? The Social Dynamics of Wetlands in Drylands.* James Currey.

Sungusia, E. (2010). Community Based Conservation in Tanzania, Getting the Incentives Right. *Tropical Resources: Bulletin of the Yale Tropical Resources Institute, 29*, 54–58.

Sunseri, T. (2005). 'Something Else to Burn': Forest Squatters, Conservationists, and the State in Modern Tanzania. *The Journal of Modern African Studies, 43*(4), 609–640.

Sunseri, T. (2010). *Wielding the Ax: State Forestry and Social Conflict in Tanzania, 1820–2000 Athens*. Ohio University Press.

Ticktin, T. (2004). The Ecological Implications of Harvesting Non-Timber Forest Products. *Journal of Applied Ecology, 41*(1), 11–21.

Tilt, B., Braun, Y., & He, D. (2009). Social Impacts of Large Dam Projects: A Comparison of International Case. Studies and Implications for Best Practice. *Journal of Environmental Management, 90*, 249–257.

Tyrrell, P., Russell, S., & Western, D. (2017). Seasonal Movements of Wildlife and Livestock in a Heterogenous Pastoral Landscape: Implications for Coexistence and Community Based Conservation. *Global Ecology and Conservation, 12*, 59–72.

Underwood, F. M., Burn, R. W., & Milliken, T. (2013). Dissecting the Illegal Ivory Trade: An Analysis of Ivory Seizures Data. *PloS one, 8*(10).

UNEP & INTERPOL. (2014). *The Environmental Crime Crisis: Threats to Sustainable Development from Illegal Exploitation and Trade in Wildlife and Forest Resources*. UNEP/GRID, Arendal, Norway.

Van Mead, N. (2018). China in Africa: Win-Win Development, or a New Colonialism? *The Guardian*, July 31.

Wasser, S. K., Shedlock, A. M., Comstock, K., Ostrander, E. A., Mutayoba, B., & Stephens, M. (2004). Assigning African Elephant DNA to Geographic Region of Origin: Applications to the Ivory Trade. *Proceedings of the National Academy of Sciences, 101*(41), 14847–14852.

Western, D. (1994). Ecosystem Conservation and Rural Development: The Case of Amboseli. In D. Western & R. M. Wright (Eds.), *Natural Connections: Perspectives in Community Based Conservation* (pp. 15–52). Island Press.

Western, D. (2008). *Ecotourism, Conservation and Development in East Africa: How the Philanthropic Traveller Can Make a Difference*. Paper commissioned for The Centre for Responsible Tourism, Stanford University.

Western, D., & Manzolillo Nightingale, D. L. (2003). *Environmental Change and the Vulnerability of Pastoralists to Drought: A Case Study of the Maasai in Amboseli, Kenya.*

Western, D., Groom, R., & Worden, J. (2009). The Impact of Subdivision and Sedentarization of Pastoral Lands on Wildlife in an African Savanna Ecosystem. Biological Conservation [online], *142*, 2538e2546. https://doi.org/10.1016/j.biocon.2009.05.025

Western, D., Mose, V., Worden, J., & Maitumo, D. (2015). Predicting Extreme Droughts in Savannah Africa: A Comparison of Proxy and Direct Measures in Detecting Biomass Fluctuations, Trends and Their Causes. *PLoS One* [online], *10*. https://doi.org/10.1371/journal.pone.0136516

Wild, R., & Mutebi, J. (1997). *Bwindi impenetrable forest*. Uganda: Conservation.

Wissenbach, U., & Wang, Y. (2017). *African Politics Meets Chinese Engineers: The Chinese-Built Standard Gauge Railway Project in Kenya and East Africa* (Working Paper No. 2017/13). China Africa Research Initiative, School of Advanced International Studies, Johns Hopkins University, Washington, DC. http://www.sais-cari.org/publications

Witt, A. B. R. (2017). Use of Non-Native Species for Poverty Alleviation in Developing Economies. In M. Vilá & P. E. Hulme (Eds.), *Impact of Biological Invasions on Ecosystem Services: Invading Nature-Springer Series in Invasion Ecology 12.* Springer.

Wynants, M., Kelly, C., Mtei, K., Munishi, L., Patrick, A., Rabinovich, A., Nasseri, M., Gilvear, D., Roberts, N., Boeckx, P., Wilson, G., Blake, W. H., & Ndakidemi, P. (2019). Drivers of Increased Soil Erosion in East Africa's Agro-Pastoral Systems: Changing Interactions Between the Social, Economic and Natural Domains. *Regional Environmental Change, 19*, 1909–1921.

Xinhuanet.com. 2017. *Chinese-Built Broadband Gives Tanzania Premium Speed*, December 3, 2016.

Zahabu, E., Eid, T., Kajembe, G. C., Mbwambo, L., Mongo, C., Sangeda, A. Z., Malimbwi, R. E., Katani, J. Z., Kashaigili, J. J., & Luoga, E. J. (2009). Forestland Tenure Systems in Tanzania: An Overview of Policy Changes in Relation to Forest Management [Eiendomsforholdene for skogarealer i Tanzania: en oversikt over policyendringer i skogforvaltningen]. *INA-Fagrapport, 14*, 24.

7

Using the Past to Chart Future Pathways?

7.1 Introduction

Fundamental to managing our future is understanding that humans have shaped all aspects of East Africa through a series of nested human-environment interactions. These interactions have involved migrations from within the African continent, such as Bantu agricultural communities and technologies spreading from their proto homeland in West Africa and pastoralists migrating from the north. International interactions have extended back to Ancient Greece and Roman periods with strong connections through maritime trade to and from Arabia. Farther afield interactions with India, China, and more recently North America European countries including the United Kingdom, France, Germany, and Portugal have dominated, particularly over the past 300 years or so. One of the key rationales in exploring the past linkages between land cover change and human interaction is to examine the potential applications of past insights into contemporary issues and how these can be used to guide future pathways (Fig. 7.1) that could be more informed and hence more inclusive, sustainable, just, and equitable. For example, understanding the differences, direction, and intensity of environmental

R. Marchant, *East Africa's Human Environment interactions*,
https://doi.org/10.1007/978-3-030-88987-6_7

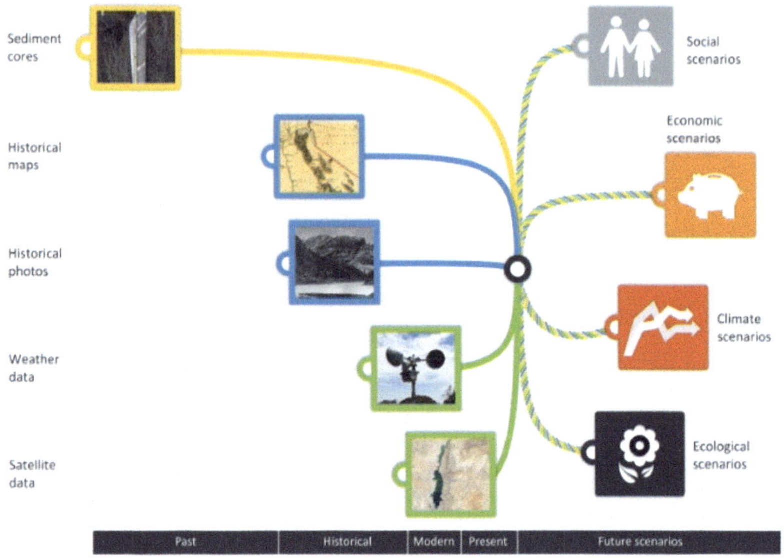

Fig. 7.1 One of the key rationales in exploring the past linkages between land cover change and human interaction is to examine the potential applications of past insights into contemporary issues, and how these past insights can be used to guide future pathways. By combining data from a wide range of evidence it is possible to explore social–environmental interactions past, present and future

change in different parts of the region can help to enhance our future predictions. Such potential application falls within a wide and increasing number of areas, but this chapter will focus on four diverse examples: understanding human/environmental interactions, ecosystem and climate modelling, ecosystem services and biogeochemical cycling, and finally conservation and sustainable development. The range of topics has been chosen to highlight the potential breadth for the utility of the past for the provision of insights into a variety of different disciplines and is not exhaustive.

7.2 Developing Data and Methodologies to Better Understand Human Environmental Interactions

As seen by accessing the existing archaeological and palaeoecological insights, East Africa has been forged by past interactions between environments, ecology, and people (Ekblom et al., 2017; Marchant et al., 2018). As new synthesises, data, and methods emerge we must rethink the relationship between nature and humans in the past so that this can frame our ability to future looking into (Fig. 7.1). The *mélange* of information preserved in sediment cores, or at an archaeological site, is complicated and messy. Not least as climate change and human activity are operating in tandem; identifying cause and effect is highly challenging. For example, a transition to a phase of dry climate might result in a similar ecosystem response as the removal of large numbers of elephants from the landscape during the ivory trade. Careful selection of sites and an open full consideration of all factors that are likely to have affected the proxies, and also forced change within the sedimentary basin catchment, is key so that cause and effect can be attributed. As methods applied on sediments such as plant macrofossils (Birks & Birks, 2000), phytoliths (Alexandre et al., 1997), non-pollen palynomorphs (Ejarque et al., 2011; Ekblom & Gillson, 2010), and a suite of geo- and biochemical techniques (Crowther et al., 2017) remain the exception rather than the norm in East Africa, there are clearly many new insights and new understandings waiting to be uncovered.

7.2.1 Data Gaps and How to Fill Them

Although the number of sites across East Africa has grown significantly, as the archaeology and palaeoecology disciplines move rapidly from a descriptive to a much more interpretative framework, the invaluable insights that can be gleaned from the past remain hampered by the relatively coarse spatial resolution of the current database. Although the data and the spectrum of techniques applied have also greatly increased over the last couple of decades, many spatial and/or temporal gaps remain (Fig. 7.2). Furthermore, the evidence is biased towards certain key sites often due to the historic legacy of researchers working in areas where their supervisors worked. These choices may reflect themes that are currently of interest to the research team and the practicalities of conducting fieldwork, with many reported sites being close to present-day road networks (Fig. 7.2).

Particularly interesting sites may have had multiple studies conducted on them; for example, Engaruka was first visited by L.S.B. Leakey on his way to Olduvai in 1935 and then worked on by a succession of scholars (e.g. Kabora et al., 2020; Leakey, 1936; Robertshaw, 1986; Sassoon, 1967; Stump, 2006; Sutton, 1978, 2004; Westerberg et al., 2010). Large areas that are also more inaccessible, such as South-Western Tanzania,

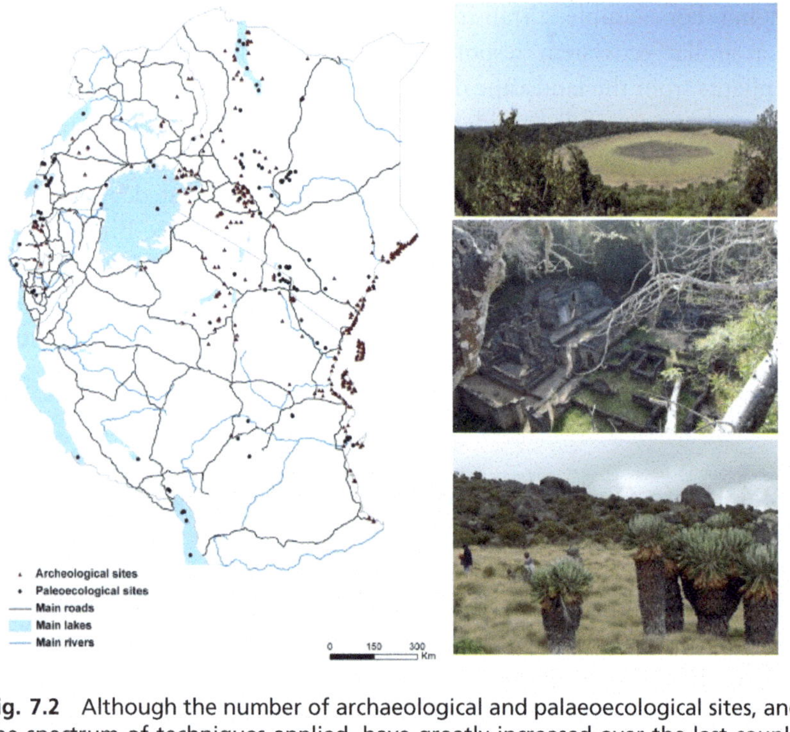

Fig. 7.2 Although the number of archaeological and palaeoecological sites, and the spectrum of techniques applied, have greatly increased over the last couple of decades, many spatial and/or temporal gaps remain and biases in past human-environmental interaction (All photographs: Rob Marchant)

North-Eastern Kenya, and Northern Uganda, remain relatively under-researched. Some of these transitions are due to the rise of cultures; for example, archaeological evidence along the coast reflects the expansion as Swahili trade developed (Wynne-Jones, 2016). While this spread of data and insight across the region may represent the choices of people in the past, we should keep in mind that it is potentially reflecting the choices of people in the present! There is also a difference in the visibility of the research: almost all the palaeoecological and climatic research that has been undertaken in East Africa is dated and published in widely circulated scientific journals. The same cannot be said of contemporary or historical archaeological research, where a lot of undated and unpublished work has been undertaken.

While the geographic gaps in the datasets (Fig. 7.2) need to be addressed, additional challenges come with combining insights from archaeological and paleoenvironmental research, the key is that we are to understand how people have interacted with the environment and transitioned and adjusted to climate shocks and transitions of the past; vital insights if we are to use this information to underpin future adaption to a rapidly changing climate. Targeted future research that employs higher-temporal-resolution and multiproxy methodologies while focusing on areas and/or time periods where our understanding of the interactions between people, the environment, and land cover change are most contentious and/or poorly resolved is certainly needed. But this is not a call for doing more of the same, however useful that would be! In addition to combining methods and insights from different disciplines (Fig. 7.3), the scientific community should aim to identify innovative trans-disciplinary projects that integrate archaeology, paleoethnobotany, archaeozoology, physical geography, and ecology from the planning phase (Eggert, 2005). A greater number of sites, together with a trans-disciplinary approach integrated with GIS and modelling frameworks, will quickly nudge the disciplines into new domains with greater potential utility and insight.

Indeed, one key issue is the differing chronological scales that underpin our understanding; these vary from the seasonal to annual response of farming communities to the decadal to centennial scale resolution of paleoenvironmental studies, with most pollen records

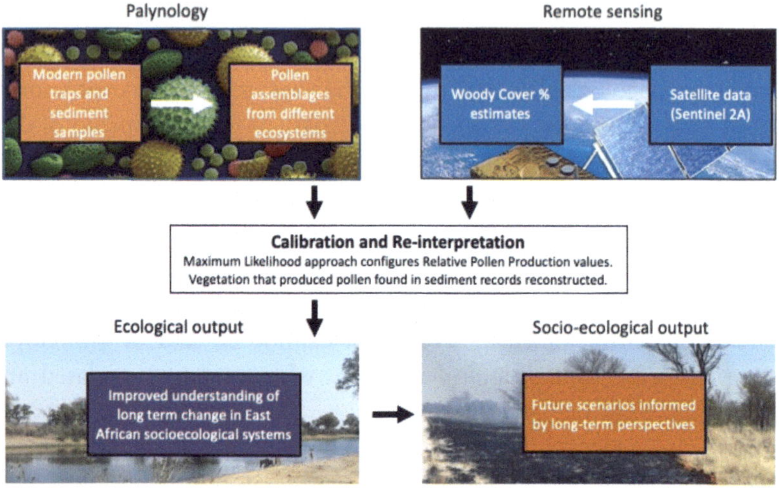

Fig. 7.3 There is a need to combine different disciplines such as palynology, ecology and remote sensing to truly understand the challenges facing ecosystems and how these have evolved so that future scenario tools can be based on enhanced understanding of ecosystem–human interactions

having a temporal resolution >50 years, and some of the longer-term large lake records having a temporal resolution >1000 years between samples. This is often reflected in the different implicit meanings of commonplace phrases such as 'long-term', 'sustainable', and 'resilient' which may be understood by scholars from different disciplines in quite different ways (Bollig, 2014; Lane, 2015; Stump, 2010). A better understanding of the processes that have led to sediment formation (e.g. taphonomic processes) is required to optimally address issues of sampling resolution and chronological control and how these can be represented. By employing high-resolution analysis in conjunction with robust chronology construction it should be possible to make interpretations at decadal to annual scale and begin to connect to human decision-making processes and perspectives (Finch et al., 2017). These common timeframes allow us to link records and engage with contemporary issues that are specifically relevant to ecologists and anthropologists. This is only achieved by the careful initial selection of samples, gaining a sufficiently large number of radiocarbon determinations, utilising fine-scale radiometric techniques (e.g. ^{210}Pb and ^{137}Cs) on the upper sections of cores (e.g. Finch et al., 2017; Ssemmanda et al., 2005), and employing predictive (as well as retrospective) Bayesian age-depth modelling techniques to provide more robust and precise chronologies (e.g. Blaauw et al., 2011; Öberg et al., 2013).

7.2.2 Harmonising Datasets and Interlinking Databases and Disciplines

The body of archaeological and paleoenvironmental data is rapidly growing, with new techniques allowing previous interpretations and contexts to be revisited. The highly dispersed nature of the datasets for East Africa means that previous syntheses (e.g. Ambrose, 1984; Courtney-Mustaphi & Marchant, 2016; Mgomezulu, 1981) and paleoenvironmental databases (Marlon et al., 2016; Sánchez Goñi et al., 2017) remain incomplete and some are relatively static with few updates. The increased number of sites, combined with a growing need for longer-term perspectives, underlines the necessity and requirements for data archiving, accessibility, and on-demand visualisation of complex datasets for multiple end-user audiences for the region. Data from these different disciplines are currently stored in multiple real and virtual locations with independent structure such as the African Pollen Database or Museum catalogues. Although these individual data collections can improve visibility and accessibility (locally and internationally), there is a need for these to 'talk' to each other so the potential use and application to other disciplines, such as ecologists, National Environmental management agencies, and other non-specialists can be realised. The access and use of these data should not exclusively be to the scientific community but should be made available to a broader audience in the private and public sectors, non-governmental organisations, and private citizens for research, context, policy and management guidance, entertainment, education, and training (Courtney-Mustaphi et al., 2014; Keeso, 2014). Part of this process requires the need to move away from showing change in highly specialised forms such as stratigraphic pollen diagrams and developing cartoons of landscape change (Fig. 7.4).

Attempts to categorise land use patterns in prehistoric sub-Saharan Africa relied on existing socio-cultural categorisations and showed how the footprint of a society can be quantified at various scales based on their diet, technology, and economy (Kay & Kaplan, 2015; Phelps et al., 2020). Subsequent work has focused on the specific effects and terminology used to describe pastoral societies (Phelps & Kaplan, 2017), and mapping agricultural livelihoods (Kay et al., 2016; Morrison et al., 2021). This work must acknowledge the multi-scalar nature of the data and how these impacts the ensuing reconstruction. For example, the

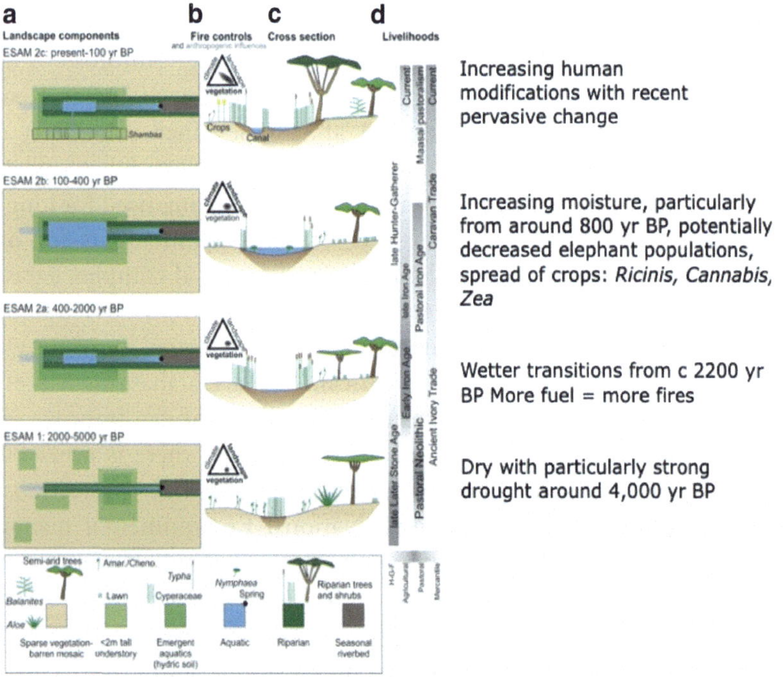

Fig. 7.4 Understanding of past environmental change and how people have been enmeshed within this to shape our contemporary ecosystem can be shown in a number of ways. This stylised carton of landscape and ecosystem evolution for the Amboseli basin can help make palaeoecological insights accessible to a wider user community

presence of a particular domestic crop in a region does not mean that the only land use was farming. Many livelihoods incorporate farming or stock keeping to varying degrees, and as today, multiple land uses would be present in the landscape at any one time. Categorisation schemes have been developed in order to map and quantify land use across space and time, whether on global (e.g. Morrison, Gaillard et al., 2016; Morrison et al., 2021; Phelps & Kaplan, 2017) or regional scales (Githumbi et al., 2021; Kay & Kaplan, 2015; Phelps et al., 2020) that are helping to use these data in new and worthwhile ways to better understand how humans have not just responded to, but also influenced, the dynamic environments of East Africa over the last thousands of years (Fig. 7.4).

In tandem with these new databases and visualisation tools, researchers from a variety of disciplines have begun to develop methodologies and conceptual frameworks that breakdown rigid distinctions between humans and the environment, and which go beyond the ecological concepts of 'carrying capacity', the 'balance of nature', and 'human impacts', that have long underpinned models of ecosystem functioning and indeed conservation. Integrating information from different disciplines, with independent ontologies, methodologies, vocabularies, and interpretative frameworks, is not a straightforward task but will enhance our understanding of past environmental change and how people have been enmeshed within this process (Fig. 7.5). We will never know exactly how people in the past perceived their environments, or the changes that occurred within them, but documenting the evidence of these complex interactions between people and the environment, understanding the mechanisms by which the signals of activity are recorded in the sedimentary records, and working to theoretically combine different types of evidence (Richer & Gearey, 2017a, 2017b; Richer et al., 2019) have to be undertaken before we can fully start to inform future responses to conservation and land use strategies (Fig. 7.5).

This growing number of disciplines that combine to understand landscape evolution can be placed under the banner of 'Biocultural Heritage' (Fig. 7.5). This is a combination of research disciplines concerned with the interactions through time between societies and environments, and the consequences of these interactions for understanding the formation

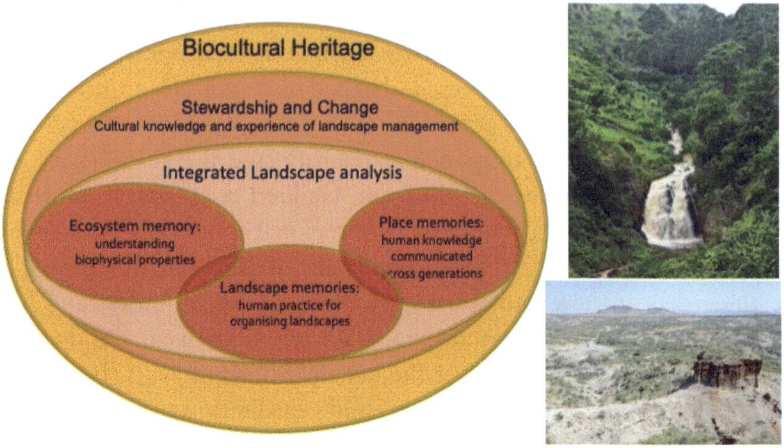

Fig. 7.5 A number of disciplines and insights combine to understand the role of societies in shaping landscape evolution. 'Biocultural Heritage' can be used to inform contemporary thinking and decision-making and rooted in the understanding of the past this will be very different in different landscapes such as Olduvai Gorge, vs. newly inhabited landscapes such as the Pare Mountains (All photographs: Rob Marchant)

of contemporary and past cultures, habitats, and landscapes (Crumley, 1994). Historical ecologists are often concerned with how contemporary landscapes came into being and how these landscapes were shaped by constantly changing human-environmental interactions. Historical ecology is increasing in adoption by many researchers across the spectrum of earth sciences, the humanities, and social sciences largely because it offers both conceptual and practical tools for joining very different kinds of information into an assessment of human-environmental interaction. Historical ecology is fundamentally interdisciplinary and typically involves the use and integration of multiple data sets and sources, spanning everything from pollen records and soil geochemistry, to archaeological remains, historical archives, cartographic evidence, oral histories, and the use of archive photography. The diverse range of methods employed across these varied sources typically also necessitates the creation of multi-disciplinary research teams. Historical ecology requires integrative tools such as GIS and remote sensing approaches to provide a common platform to bring data together from hitherto disparate disciplines. Given the inherently complex nature of trying to assess the future, a harmonising theme is needed with 'land' being able to provide the lens within which to view challenges past (Kariuki et al., 2018), present (Kariuki, Western et al., 2021b), and future (Capitani et al., 2016, 2019; Kariuki, Munishi, et al., 2021; Kariuki, Western, et al., 2021a). Adopting a more holistic conceptualisation of human-environment interaction which acknowledges that few, if any, habitats have been unaffected by human-driven events and processes, has also encouraged the critique of many long-established narratives concerning human agency in shaping the landscape; especially those construed in terms of harmonious adaptation or environmental mismanagement (Lane, 2010). Understanding the historical ecology of a landscape entails the gathering of contemporary and antecedent environmental and cultural evidence to identify the key variables, and the relationships between these, that have shaped the landscapes (Lane, 2010).

7.3 Linking the Past to the Future Through Modelling Frameworks

Although it is interesting to describe and understand the past in its own right and, hopefully, learn from our history, due to our inability to change what has happened there is quite rightly a focus on the future, as is admirably demonstrated through the current global climate change emergency. Climatic conditions in East Africa, as the rest of the world, have varied through time with the only constant being that of change! Climates can change relatively suddenly and would have been highly noticeable to human populations and, similarly to today, would have simultaneously offered both difficulties and opportunities. We cannot know how people perceived these climatic changes in the past but by the same measure, researchers worldwide have yet to fully consider the potential implications of the sudden environmental changes that would have occurred as the result of both climatic and anthropogenic drivers. In addition to improving our ability to reconstruct and understand the processes of past environmental change, and to explain mechanisms that might be responsible for such variability, there are many potential benefits of combining data-based with model-based interpretations of change (Jolly et al., 1998; Marchant et al., 2018). These will be explored in terms of climate, ecosystem, and land use models.

7.3.1 Climate Models

Climate models are *the* tools used to understand future potential direction and magnitude of change in the Earth's climates. There are some 23 General Climate Models (GCMs) currently used in the IPCC assessments (IPCC, 2021) that are based on physical laws of thermodynamics with energy moving from one climate model grid cell to the next. How these GCMs are implemented and how they consider land-surface feedbacks, ocean–atmosphere-land interactions, and indeed, human climate interaction varies considerably between models. Clearly models that investigate the impact of future climate change on our ecosystems, crop systems (Thornton et al., 2007), and the ensuing potential of nature-based solutions for development, must be testable. The only data to provide such a test of model performance are those data sets generated from the past such as explored here. To understand the impacts of these earth system feedbacks requires coordinated inter-comparison projects alongside independent data—with that data being presented in the form of palaeo archives. Hence, an understanding of past environmental history is crucial for validating global and regional climate model simulations when these are hindcast (Braconnot et al., 2012). Although there remain numerous discrepancies between modelled and data-based views of past climate variability in East Africa (Rowell et al., 2015) a better understanding of the mechanisms that drive abrupt changes in climate is resulting from combining the climate modelling and paleoenvironmental communities (Braconnot et al., 2012; Clark et al., 2002). An important aspect of the recent United National Framework Convention on Climate Change (UNFCCC) has been plans to integrate climate change perspectives into new policies and programmes that aim to enhance adaptive capacity, strengthen resilience, and reduce vulnerability to climate change. The currently anticipated changes to the planet's climate pose threats to biodiversity and natural resources, human health, food and water security, public health, natural resources, and biodiversity (Swallow et al., 2009).

Current GCM projections for East Africa suggest a climate that will be warmer and wetter (Fig. 7.6), with enhanced net primary productivity, associated carbon storage, and increased run-off. Such a broad-scale prediction across the region that is characterised by high environmental and ecosystem complexity (Sect. 2.3) is clearly open to error. Indeed,

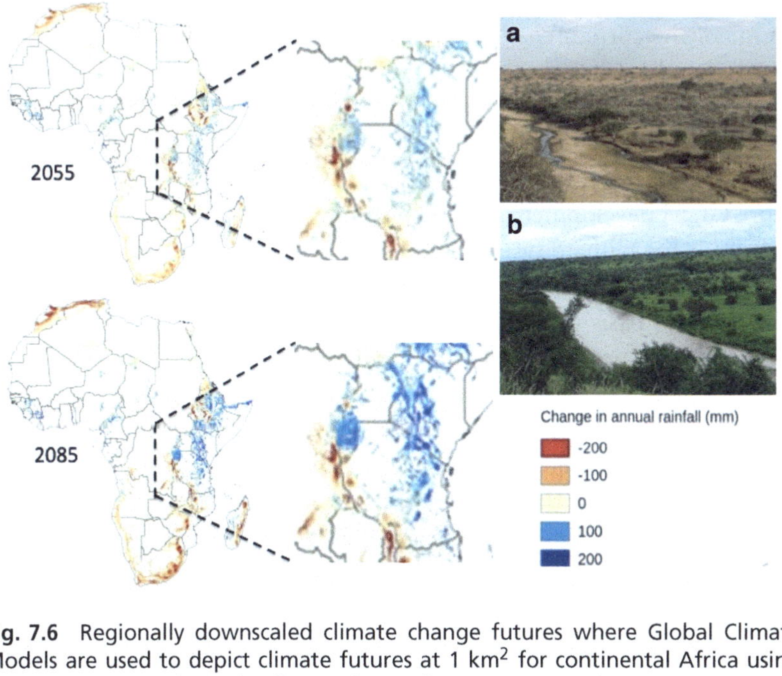

Fig. 7.6 Regionally downscaled climate change futures where Global Climate Models are used to depict climate futures at 1 km^2 for continental Africa using the AFRICLIM product. The figure shows changes in precipitation at 2055 and 2085. While these are produced for a wide range of climate parameters, as in the past, future timing and quantity of rainfall are going to key and have the biggest impact on ecosystems—as shown by pictures of Tarangire in the dry (**a**) and wet (**b**) season. Climate predictions suggest some areas getting drier and others getting wetter; trends are likely to be enhanced in the future (Platts et al., 2015) (All photographs: Rob Marchant)

climate predictions are generally at odds with the instrumental data and those gathered from participatory research that suggest a drying climate, or at least one becoming more variable (IPCC, 2021). It is clear from the paleoenvironmental data (Chapter 3) that spatial variability and complexity existed in the past, with some areas getting drier and others getting wetter and that this is likely to continue in the future (Fig. 7.6). Many more regionally focused modelling approaches are needed to capture the ranges in spatial variability (Platts et al., 2015) and combine these with local perspectives on the nature of changing climates and how ecosystems interact with a changing climate, raised levels of atmospheric CO_2, and changing densities and character of important ecosystem engineers, such as elephants, sheep, goats, and cows (Fig. 7.7).

One of the current challenges that drives the discrepancy between the observed and predicted is how dynamic land sources and rapidly varying land cover are incorporated into climate models; partly because vegetation and climate often change in a nonlinear fashion that will alter albedo feedbacks (Friedlingstein et al., 2006). One key focus area for future GCM development is to incorporate vegetation changes and feedbacks into the climate system (Chevalier et al., 2017; Richardson et al., 2013). Within East Africa, rapid switching between wet and dry phases on a seasonal basis (Fig. 5.8) will also drive changes in vegetation, soil, and albedo (Claussen et al., 1999; Friedlingstein et al., 2006; Knorr et al., 2001). An example of the albedo effect can be seen in the annual conversion of savannah grassland mosaic from brown to green with the ensuing rains (Fig. 7.6). Longer changes to surface albedo, surface roughness, leaf area index, and fractional vegetation coverage are apparent as the number of trees in the savanna landscapes are driven by a factor complex comprising reduced herbivory, increased levels of carbon dioxide, greater water-use efficiency, decreased burning, and increased sedentarisation (Fig. 7.7). Collectively, these changes can alter water and energy exchange between the ground surface and the atmosphere, which in turn affects the timing and intensity of the monsoons (Claussen et al., 1999; Demenocal et al., 2000; Friedlingstein et al., 2006).

Given this growing recognition that human activities have been one of the main drivers of land cover change, especially over the last

Fig. 7.7 Ecosystems are shaped by the interaction of a changing climate, raised levels of atmospheric CO_2, burning regime and changing densities of ecosystem engineers such as elephants, sheep, goats, and cows. Reduced levels of herbivory and fires combined with increased levels of carbon dioxide lead to greater water-use efficiency of drought-adapted trees that is resulting in the spread of woody biomass in savannas as shown by this pair of images from Laikipia taken in 1935 and 2005

3000–4000 years, there is an increasing realisation that more comprehensive and resolved information on the nature of coupled land-human systems is required to drive the next generation of Earth System models. Perspectives brought together from past environmental and archaeological archives can provide spatially constrained, long-term land cover and land use reconstructions that can feed into fully coupled, next-generation climate and Earth system models. More and better data about how people have influenced landscapes will lead to an enhanced predictive capability and an improved match between model-based and data-based reconstructions of past climate and environments. At present, those Earth System Models that take anthropogenic land cover change into account generally rely on simple model-based scenarios of a poorly defined single variable called 'land use' from e.g. the HYDE (Klein Goldewijk et al., 2010, 2011, 2017) or KK10 scenarios (Kaplan & Krumhardt, 2011). These scenarios are generally additive views of land cover and population change that tend to do a poor job of reflecting the nature and magnitude of land use and anthropogenic land cover change, particularly at regional scales. To meet this need the PAGES LandCover6k initiative is developing data-based global surfaces of land cover at 6000 yr BP, CE 1500, and CE 1850 relating, respectively, to the mid-Holocene 'optimum', the precolonial period, and the pre-industrial time (Morrison et al., 2021; Gaillard et al., 2015).

7.3.2 Ecosystem Models

The terrestrial ecosystem is directly related to a variety of ecosystem services, or natures contributions to people, such as water purification, soil retention, and carbon sequestration (Cuni-Sanchez et al., 2021; Wei, Wang, Fu, Pan, et al., 2018), and understanding how these change can have a direct impact on livelihoods. A variety of models are available to investigate individual species or ecosystem responses to climate change (Platts et al., 2013, 2015) and land use change interactions (Jung et al., 2017). To date, one of the most popular methods for investigating plant response to climate change has been species niche modelling (Fig. 7.8) (Araújo & Peterson, 2012; Platts et al., 2011). Correlating the known distribution of the species in question creates distribution envelopes with various ecological and climatic parameters: generating a niche representation of the boundaries within which that species can survive. As the ecological distribution and climatic conditions are known for the present, these can subsequently be predicted for the future to assess how the species distribution may change in relation to climate change (e.g. Maiorano et al., 2013; Platts et al., 2015). Combining palaeoecological and metacommunity approaches to understand impacts of climate change on species distribution, one of the key challenges of our time, has the potential to be a very fruitful combination to comprehend

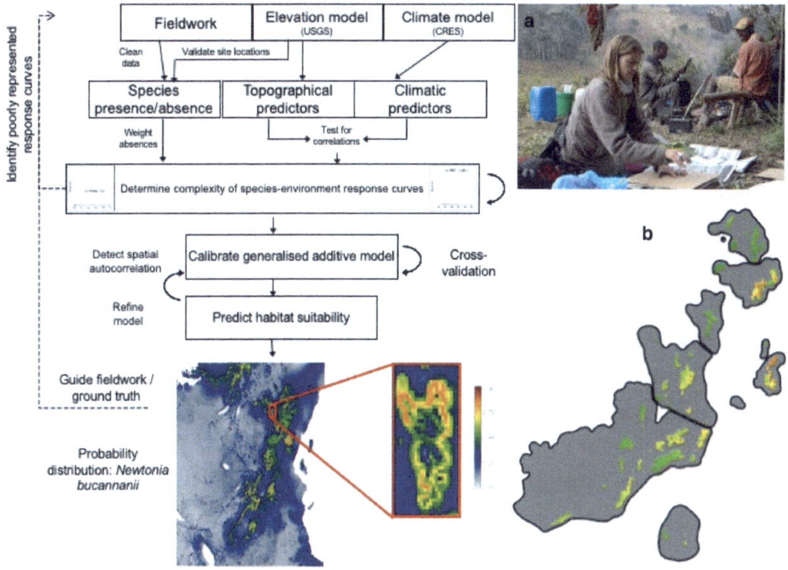

Fig. 7.8 Niche modelling schematic from collating of data to running the models and describing species distribution across the landscape—run here for *Newtonia bucannanii*. In addition to being a good tool for assessing impacts of climate change on species distribution, niche models can be used to guide botanical fieldwork (**a**) and combined over multiple distributions to show concentrations of diversity. **b** shows the globally important Eastern Arc Biodiversity Hotspot where 570 niche models are combined to assess diversity patterns (Platts, 2012)

ecological pasts and futures (Svenning & Sandel, 2013). For example, such a combination will allow us to understand transition mechanisms in savannah ecosystems; as to how they change from woodland to apparently stable grassland 'phases', how they can be resilient to disturbance, and how they can persist over the long-term and still withstand disturbance (Li, 2002; Prins, 1971; Prins & van der Jeurg, 1993). Fire and/or herbivory, primarily by elephants, are usually the causes of disturbance (Fig. 7.6) that are responsible for the transition of woodland to grassland savannah and need to be incorporated into niche models as predictor variables (Bond & Midgley, 2000) to capture and understand how factors such as disease or hunting affect herbivore numbers, as evidenced during the ivory trade (Chapter 3), and how these control the nature of transition from grassland to woodland (Dublin, 1995).

In addition to impacts from climate systems and changes in disturbance regimes, and the numbers and distribution of ecosystem engineers like elephants, changes in atmospheric composition are shifting ecosystem-environment relationships. Under the present situation of increased CO_2 concentrations, there is an increasingly 'closed' savannah with woody growth becoming apparent in many dryland ecosystems around the world. This expansion of woody growth is not a result of CO_2 'fertilisation' caused by higher CO_2 concentrations, but as a result of increased water-use efficiency and the ensuing growth rates (Midgley & Bond, 2015); where increased water-use efficiency means the plants have to have their stomata open less and therefore lose less water during the process of photosynthesis (Fig. 7.6). In water-limited environments, such as savannahs, increases in atmospheric CO_2 concentrations can promote additional increases in net primary productivity, and enhanced carbon storage (Midgley & Bond, 2015). These transitions driven by ecosystem processes and how these respond to changing atmospheric composition can be investigated by biogeochemical models, such as LPJ-Guess, to assess impacts of changing CO_2 concentrations on ecosystem dynamics (Jolly & Haxeltine, 1997).

7.3.3 Land Use Scenarios

Land use systems are shaped by many factors operating at multiple scales. Understanding how people interact with their environments at the present, and in the recent past, is possible through a participatory scenario approach as people retain social memory that can be used to reveal potential land use futures (Capitani et al., 2016). Participatory scenario methods have emerged as a key tool developed in recent years and can be used to investigate land cover change from national (Capitani et al., 2016) to local levels (Capitani et al., 2019a; Kariuki, Munishi, et al., 2021). The participatory scenario process starts by bringing together participants from a diverse array of potential stakeholders (e.g. private industry, conservation organisations, researchers, civil society, and municipal officials) with knowledge on land use characteristics and the drivers behind creating this (Fig. 7.9).

Participants are tasked with imagining future land covers and the drivers that are behind land use transformation, be it natural, policy-induced, or through human agency (Fig. 7.10). The futures can both envisage positive outcomes that can inspire change and undesirable futures that can ultimately serve as warnings of what to avoid (Oteros-Rozas et al., 2014). As they explore multiple alternatives, participatory scenarios can also account for uncertainty around future events and the impact of different development pathways. Indeed, participatory scenario approaches can be useful in planning adaptation strategies as they offer the opportunity to consider multiple dimensions, such as how projected climate change impacts, socio-economic trends, and local

Fig. 7.9 Communication is key. Due to the diversity and often highly context-specific nature of challenges we need to start the research process with a conversation that brings the different communities and perspectives together. Conversations are at the heart of understanding the scale of the apparent challenges, the key questions that need addressing and what are the potential solutions (All photographs: Rob Marchant)

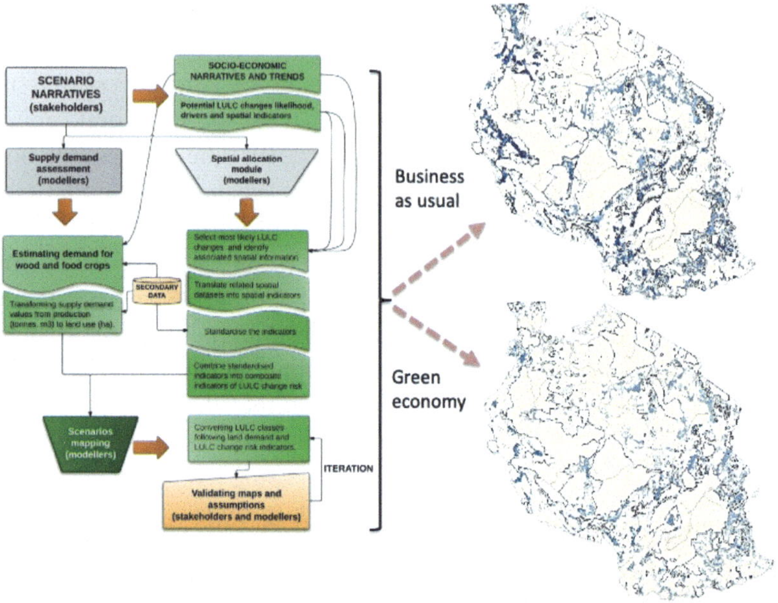

Fig. 7.10 KESHO participatory scenarios has emerged as a key tool to investigate potential futures through the lens of potential land cover change. KESHO has been applied from national to local levels, looking at coastal development, pastoral-agricultural-conservation interactions, coffee futures, and impacts of new railways. Participants are tasked with imagining future land cover and the drivers behind land use transformation, be it natural, policy-induced, or through human agency. In the application shown future forest/carbon cover is shown at 2030 under a Green Economy and a Business as Usual scenario—the difference are massive and could equate to US$10 billion in carbon payments

production and governance systems interact (Capitani et al., 2019a). One participatory scenarios tool developed ad applied at a range of different scales and a range of different foci has been KESHO (meaning tomorrow in Swahili). For example, KESHO has been used to look at how national-level carbon budgets could change under 'Business and Usual', vs 'Green Economy' future (Fig. 7.10).

Participatory scenarios can provide clarity on the interlinkages between societal and ecological processes, such as habitat fragmentation and human–animal conflict, and are coming under increasing pressure to ensure the livelihoods of people currently living in these landscapes

(Marchant, 2010). The development of generalised land cover change summaries generated from diverse sources can be useful for supporting dialogues to frame issues around sustainable and inclusive development, particularly when communicated to several audiences (Kariuki, Munishi, et al., 2021). Trying to divine the future and understand what challenges are coming down the road is difficult and hampered by the relatively coarse spatial resolution of our current data, the large gaps in understanding, and the rapidly changing nature of the world. As we have seen throughout this volume, land and land use is highly dynamic, complex, and ever-changing due to the varying drivers of environment, climates, and human agency in changing things like elephant populations. These rapid fluctuations and known unknowns have been clearly played out with the current COVID-19 pandemic that are having an impact on how people use land and the benefits that can be derived from it. But it is precisely these complex relationships that have shaped the productive and varied land cover that we see today and why a diverse range of need to people come together to produce the participatory scenarios. As such, policy/decision-makers can use participatory scenario planning outputs to develop short- and long-term adaptation strategies, keeping in mind that other global processes (population growth, globalisation, climate change) are occurring at different speeds (Fig. 7.11).

Exploring the impact of changing management decisions and how these would impact on land use systems are of high social and ecological interest but are challenging to capture. Challenges include heterogeneous societal actors, a diversity of knowledge types, data paucity, and communication gaps, which all benefit from incorporating stakeholder insights (Kariuki, Munishi, et al., 2021). Participatory planning approaches can consider and target the multiple stressors that contribute to vulnerability of communities to change impacts such as climate changes, population fluxes, land use changes, poverty, management practices, access to technology and markets, and the provision of services, including ecosystem service delivery. These can then be used to promote sustainable land management, appropriate water management systems, support the development of agroforestry systems and mosaic agricultural landscapes, investigate how high-quality agricultural products, such as coffee, can preserve landscape multi-functionality and enhance community

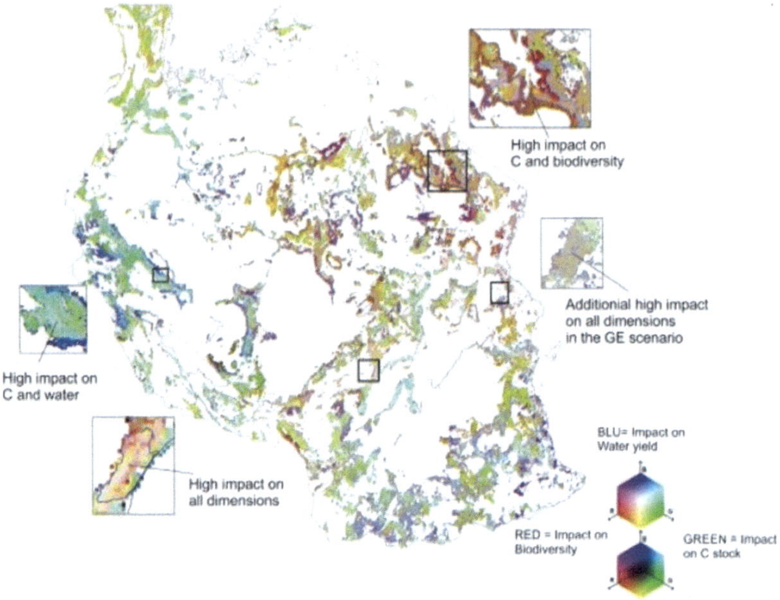

Fig. 7.11 Land use ultimately underpins key ecosystem services and hosts biodiversity. Future land use transitions can be used to explore the Interactions between biodiversity, water and carbon. Red–Green–Blue plot showing the combined impacts on carbon stocks (black to bright green), biodiversity (black to bright red) and water balance (black to bright blue)

resilience to the changing climate and ensure development aims are achieved, or at least avoid and safeguard the continuation of impactful futures. Results can further be used to monitor progress, for example, towards the Sustainable Development Goals or the African Union 2063 Development Agenda, and ultimately, sustain interventions for equitable, resilient, social-ecological futures.

7.3.4 Remotely Sensed Land Use Transformations

More recent changes in land surface and contemporary analysis of land cover change are possible since the launch of Landsat in 1972 from remotely sensed images. Over the past 50 years or so the timing, direction, and rate of land cover change can be assessed with multiple and a growing number of remotely sensed products (Pfeifer et al., 2012). Important in the context of this book, where there is a focus on longer-term perspectives, remote sensing data allows the deep time to connect to the present day and the encompassing raft of contemporary issues. For example, two paired MODIS land cover images taken in 2001 and 2013 (Fig. 6.6) clearly show the continued expansion of agriculture, particularly around the eastern shores of Lake Victoria and the coastal strip. Another useful application has been to investigate the impacts of given land designations such as how populations have evolved and changed surrounding the protected areas (Pfeifer et al., 2012) or recent large infrastructure developments such as roads and railways (Thorn et al., 2021). However useful these remotely sensed views of land cover are it is vital to remember that land cover change did not start in 1972 and earlier perspectives are essential to provide context and interpretation of the drivers and nested patterns of land cover change.

7.4 Future Challenges for East Africa

There are multiple challenges and opportunities ahead for the East African region. Along with population growth (Canning et al., 2015) and anthropogenically induced land use and land cover changes (Capitani et al., 2019b), climate change adaptation is among the greatest future challenges for many African countries (Niang et al., 2014). Understanding past interactions between ecosystem dynamics, movements of people, and land use strategies, and how these can provide salutary lessons that must be learnt is paramount if we want to understand the relationship between human societies and their landscape. We will explore some of these challenges around the interlinked themes of agricultural expansion, climate change, and population growth (Fig. 7.12).

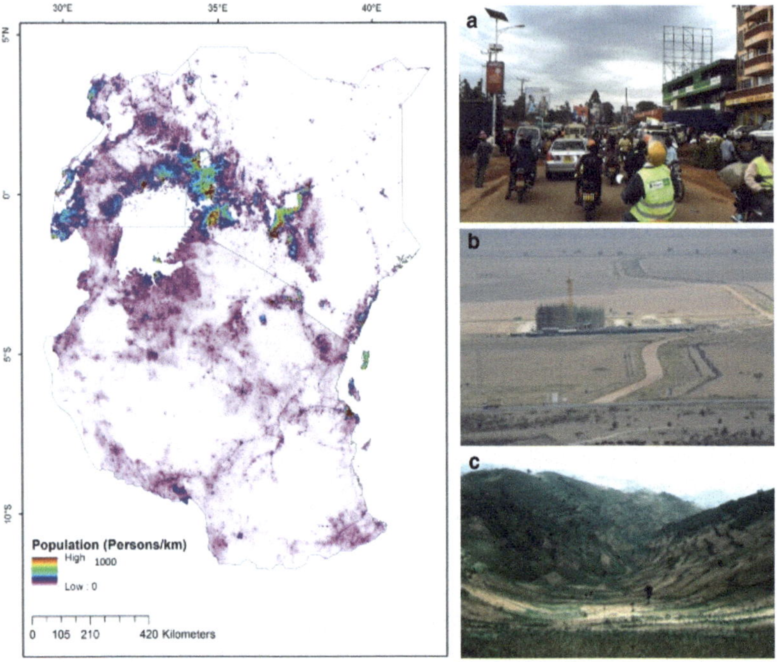

Fig. 7.12 Population distribution in East Africa derived from WorldPop. It is clear the highest population densities are centred along the coastal strip, around fertile mountains (**c**), the low-lying areas around the large lakes and expanding urban centres (**a**, **b**) (All photographs: Rob Marchant)

7.4.1 Agricultural Expansion and Consequences of Land Conversion

Agriculture dominates the economies of all East African countries; it is not just underpinning the national economy; agriculture is the base for rural livelihoods across the region. The living conditions for agricultural dependent households in the highlands of East Africa have been deteriorating owing to the declining quality and quantity of agricultural land resources, which has led to low agricultural productivity, frequent drought, poorly distributed rainfall, pests and diseases, insecurity associated with land tenure, high population growth, soil erosion, land degradation, deforestation, weak or non-functioning institutional support, overgrazing, and a lack of strong rural organisations (Ng'ang'a et al., 2020). Characterising agricultural transitions and the rapid spread of crops across East Africa is one of the most challenging research topics for the future (Neumann, 2005), particularly as land conversion to agriculture has outpaced human population growth in recent decades. Food production in Kenya, for example, has increased steadily from around the turn of the century, but because of population increase, the food supply in calories per head fell (Maitima et al., 2009).

Expansion of agricultural land is one of the main causes of land use change (Fig. 6.6). Cropland has increased as forests and grassland continue to be converted (Mwangi et al., 2018). There are many drivers behind this expansion including reduced soil fertility, soil organic carbon, degradation of soil structure, reduction in the availability of major nutrients (N, P, K), micro elements, and an increase in toxicity due to acidification and salinisation, especially in irrigated farming systems; all compounded as farm sizes become smaller because of subdivision where smallholdings commonly are 1 ha or less (Maitima et al., 2009). Indeed, land is a scarce resource especially in the highlands where high populations and ever declining soil fertilities and plot size are resulting in declining yields. Soils are severely degraded and have low organic matter because of continuous monocropping (Khan et al., 2017). Nitrogen levels in the wider environment are particularly elevated because of the use of fertilisers and crop farming strategies, and this is seen clearly as pollution in many East African lakes (Sutton, 2004), with an increase in nutrients run-off and organic matter leading to algal blooms and expansion of aquatic plants like the water hyacinth (Fig. 6.8); in turn, reducing the ability of the lakes to provide viable fisheries and deliver wider ecosystem services. Again, the sedimentary archives have been useful for tracking pollution, both spatially and temporally. Using stable isotope analysis of nitrogen ($\delta^{15}N$) from the sedimentary record provides the wider community with a baseline for pollution control (Jeffers et al., 2015).

Developing adaptable and productive agricultural systems that are resilient in the face of the risks and shocks associated with long-term climate variability is essential to maintaining food production into the future. In response to this challenge, there have been multiple efforts to promote sustainable land use and increase agricultural production and ensure sustainable use of natural resources through implementing land management practices and rehabilitating degraded land. Such 'climate-smart agricultural systems', as well as being resilient, need to protect and enhance natural resources and wider ecosystem services (Khan et al., 2017). Agronomic initiatives can increase yields and micronutrient contents of crops and ultimately require integrated soil fertility management (Warinda et al., 2020). For example, the adoption of soil carbon enhancing practices is one low-cost sustainable land management technology with numerous benefits of enhancing soil fertility and increasing farm productivity while offering opportunities to increase farmers income and food security (de Jong Cleyndert et al., 2020). The establishment of woodlands and nutrient management through the application of compost, green manure, sludge, and mulching can all enhance soil carbon and quality (Lal, 2004).

The solution to multiple challenges lies in sustainable agricultural intensification in terms that go beyond just meeting basic human needs to embrace a broader vision of sustainability as fairness: enhancing human well-being to meet the needs of both current and future generations more equitably. Communities or cooperatives should set their own sustainability goals and choose how to pursue them (Clark & Harley, 2020). Solutions are not just focused on international agreements and developing supportive regulations and markets. Local adaption has grown as divergent groups of farmers formalise their production into farmers' organisations, thereby giving them greater voice and bargaining power as evidenced in Kenya's potato farmers, coffee farmers on Kilimanjaro, and Uganda's cassava growers. Members of these organisations can access facilities like loans and better-negotiate with traders and private sectors to enhance sales of outputs and purchase of inputs. Beneficiaries of the Orange Flesh Sweet Potato project in Kenya bargained with Farm Concern International and the One Acre Project for the sale of their produce with each farmer receiving US$ 340 during that season compared to US$ 140 in the previous season (Warinda et al., 2020).

Farmers still require more information on these technologies and there is a need for concerted and joined up efforts between researchers, the private sector, policymakers, extension officers, traders, and farmers

(Ng'ang'a et al., 2020). As with the previous spread of agriculture, there has been a rapid expansion of exotic crops (like Maize and Avocado) with traditional crops, cropping systems, and land management practices all being replaced. Traditional water management, terracing, and furrow systems have been largely replaced by irrigated and mechanised systems. While more intensive, industrial agriculture has its place this is not the case everywhere and again, local bespoke knowledge is key to success. Many resource-constrained smallholder farmers are incorporating cereal crops with greater drought resistance (e.g. sorghum and millet) and replacing cattle with small ruminants for dairy production (Khan et al., 2017; Pretty et al., 2011) to deal with contemporary land degradation and climate change combinations. With many emerging agricultural challenges, a different mind-set is needed that combines a science-based approach that embraces, and works with, traditional ecological knowledge informed from historical ecology. National agriculture research systems have local mandates and are best suited to implement specific projects addressing low-impact local issues. Common issues affecting multiple countries such as limitations in agriculture technology, human capacity, cross-border biosafety, market access, agriculture policy, and response to climate change can be best tackled by deploying a regional approach for enhanced shared prosperity (Warinda et al., 2020).

Benefits and synergies are possible through a regional focus on agricultural research for development—shifting from country-wide projects to more scalable projects that can be implemented in multiple neighbouring countries, thereby generating wider benefits, and tackling common regional challenges (Warinda et al., 2020). However, these regional initiatives come at a cost and understanding the local context is key to minimising these costs where the past failings of trying to adopt a one-size-fits all approach must be learnt. Certainly, the governance systems are developing to allow for joined up benefits such as the East Africa Community with harmonised policy and the emergence of the African-wide free trade agreement that this will allow for. Prioritising context-specific agriculture technologies that work with heterogeneity and the role of different genders will remain a challenge (Brandt et al., 2017; Warinda et al., 2020).

Reduced intra- and inter-country and regional trade barriers have contributed to farmers' willingness to expand their cropped areas for adoption of productivity-enhancing technologies such as maize, cassava, banana, and sorghum varieties. Regional initiatives such as the East African Food Prices data portal, a web-based information system for food, and input prices for these countries contain updated food prices of several commodities (Warinda et al., 2020). Regional agricultural projects generate economic and capacity-building benefits to farmers and other stakeholders, such as gains in crop yield, livestock productivity, farming- and non-farming income, and food security. Wider benefits include enhanced farmer participation and access to improved technologies, early harvest, better soil fertility, effective water management, greater value addition, and participation in markets and increased awareness, confidence, and a better outlook towards the adoption of these improved agricultural technologies (Warinda et al., 2020). Regional agricultural development interventions can also deliver notable benefits through coordinating inter-country trade policies and the harmonisation of standards, regulations, and streamlined cross-border procedures for transit of food commodities and food aid. Instead of operating in knowledge silos, the regional approach to agricultural development provides a shared platform to discuss emerging challenges including drought mitigation plans, cross-border disease management, biosecurity, and food security challenges (Warinda et al., 2020).

7.4.1.1 Pastoral to Agricultural Transitions

Keeping livestock in the traditional pastoralist transhumance systems has become increasingly difficult (Chapter 6). Modern economies require pastoralists to have cash to pay for health care, education, and food, increasing the need for livelihood diversification to generate income (Homewood et al., 2012). From wage labour in cities or the tourist industry (Fig. 7.13), to involvement in gemstone mining and selling of beadwork and livestock products (Smith, 2014), economic shifts are coping strategies to protect households from having to sell livestock (Lankester & Davis, 2016). One livelihood change that is rapidly

Fig. 7.13 Alternative livelihoods for many communities are being developed such as these Masai pastoralists working in the tourist industry or selling bone products (All photographs: Rob Marchant)

increasing and has the potential to devastate pastoralism and wildlife relationships, is the conversion of land for cultivation, particularly large-scale cultivation (Homewood et al., 2001; Nyumba et al., 2021), which is often funded by outside investors. Despite growing pressures to cultivate, much of East Africa rangeland is too arid and soil fertility too marginal for cultivated food production without large investment into irrigation. In places that can irrigate there are clearly short to medium-term benefits although the longer-term impacts are uncertain. As elsewhere in the world where irrigating has been practised, it is likely there will be impacts such as salination of soils or impacts on adjacent properties as water levels in aquifers are reduced. Many pastoral communities, like the Maasai, are undergoing rapid social and economic change. The growing human population coupled with a fluctuating livestock population is changing the nature of relationships between societies and land. The process of change has included increased sedentarisation, the desire for education, and the understanding that wage labour may be a necessary component in a future diversified livelihood strategy.

7.4.2 Climate Change

East Africa is one of the regions of the world most severely affected by current global climate change as documented by multiple studies—from IPCC to IPBES and a myriad of local evidence. Under global climate change, like the rest of the world, East Africa is becoming warmer, but it is the increased unpredictability of rainfall, reduction in plant-available moisture, and an increased frequency of extreme climatic events that are bringing the main challenges to the region. The economy of East Africa is heavily dependent on rain-fed agriculture, particularly in the densely settled mountain areas that are characterised by very steep moisture-balanced gradients resulting in a mosaic of landscapes that provide a range of ecological and economic goods and services that support the national economies. Drought is a recurrent problem for mountain communities and downstream pastoral herders, further reducing carrying capacities and undermining the subsistence basis of pastoral livelihoods. Highly variable rainfall, both amount and distribution, provides more uncertainty as droughts are becoming more frequent and intense (Fig. 7.14), and increasingly leading to prolonged famine, particularly in the already water-stressed areas of Northern Uganda and Northern Kenya. This is not a bidirectional impact as increasing climatic impacts will ultimately lead to removal of vegetation by burning, overgrazing, cutting, and trampling that can expose soils to erosion, particularly during ensuing periods of intense rains (Blake et al., 2018). As seen for the areas in Northern Kenya around Mount Marsabit, droughts and ensuing impacts in lowland areas lead to increased pressure on the Mountain oasis; cutting of trees for fodder as a short-term solution is ultimately curtailing the longer-term functioning of the Mountain to maintain hydrological operation (Cuní-Sanchez et al., 2018).

Increasing climate variability has undermined people's confidence in any single livelihood strategy be it growing a certain crop or keeping a certain type of livestock. Strategies to cope with the climate change

Fig. 7.14 One of the key climate change challenges ahead is going to be associated with the timing and quantity of rainfall. Although the climate is predicted to be wetter, in common with the result of the world, increased extremes are going to have the biggest impact. These flash floods near Suswa (Kenya) have been exacerbated by upland forest clearance (All photographs: Rob Marchant)

impacts, particularly focusing on managing the water tower resource are clearly going to be crucial for managing a future that is going to be characterised by increasing hydrological variability—both droughts and floods (Fig. 7.15). In addition to managing upland 'catchment forests' there is a need to manage water and lowland user communities (Fig. 7.15) to diversify livelihoods that can include a combination of adaptive management (e.g. reforestation for carbon sequestration projects, biofuel production, soil management) and risk management (e.g. choice of alternate crop types or varieties) options. It is crucial that such management decisions are informed by knowledge on the response of ecosystems to climatic change, local environmental knowledge, and direct evidence of how communities respond to climate change: the regional historical ecology. Applying knowledge that integrates such insights into future planning and policy is a central goal of the scientific community and builds on identifying paths to solutions as well as documenting the evidence of climate change impacts and how this is manifested across societies and initiatives. This application is particularly key for East Africa where people's livelihoods connect strongly with the environment and the dynamic nature of this past, present, and future.

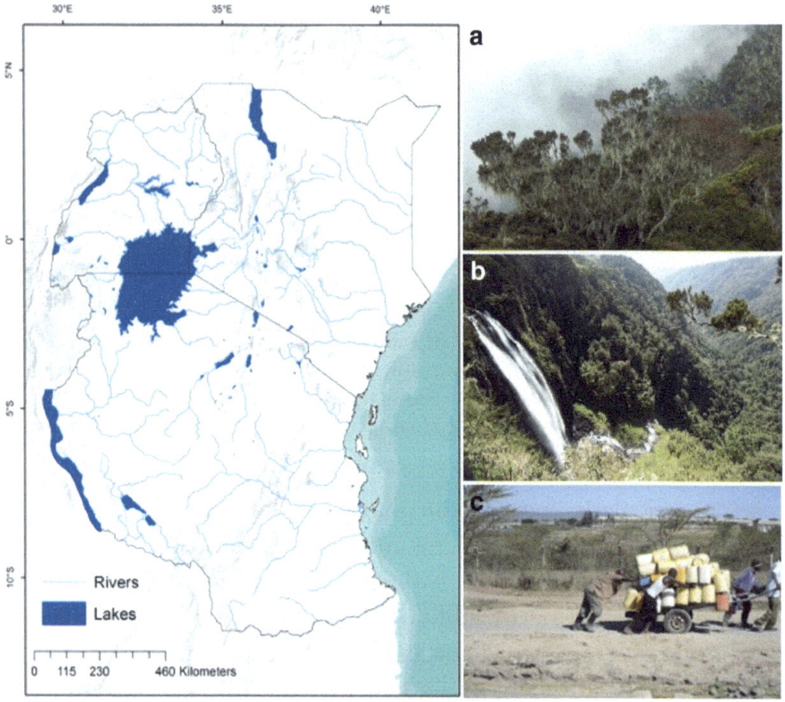

Fig. 7.15 With the main climate change challenge linked with rainfall, key to mitigating the impact will be expansion of catchment-based approaches where water towers (**a**, **b**) are managed to capture water, deliver this is a managed way to avoid unintended downstream impacts (**c**). East African mountains are key ecosystems that capture water from low-level clouds drifting through the forest to allow these water towers to regulate supply to down-stream users (**c**) (All photographs: Rob Marchant)

7.4.3 Population Growth in East Africa and Expansion of Urban Centres

The African continent has the fastest growing population in the world, projected to increase from 1.1 billion in 2015 to 2.5 billion people by 2050 (Nieves et al., 2017). Zooming into the population of East Africa we can see that this has experienced exponential growth from an estimated 6–12 million in the 20s (Anderson, 1984; Trewartha & Zelinsky,

1954), 24 million in 1950, 56 million in 1980s, to 173 million in 2017 (UNDESA, 2017). This rapidly increasing population will have huge implications for the use of natural resources, agriculture, food security, investments, and public policy. Governments and development partners across East Africa have launched many initiatives to transform agriculture and reduce poverty. However, scapegoating the problems facing the region to overpopulation and overexploitation of natural resources lacks understanding of the complex human-environment interactions and how these have evolved, and can thus fuel potentially detrimental policies (Blake et al., 2018).

An increasing population means more mouths to feed, but also more livelihoods to find (Korotayev & Zinkina, 2015). With an estimated 75% of East Africans dependent on agriculture or pastoralism; without livelihood diversification, the next generation will also be forced to find their livelihood in these sectors (Jayne et al., 2014). There is ample evidence showing that rising populations have resulted in increased competition for land that pushes farmers to smaller and/or unsuitable farming areas and increases grazing pressures on rangelands. Sustainable intensification can reduce population pressure, diversify livelihood, promote technological and organisational innovation, and can increase productivity while simultaneously preserving other ecosystem services. Examples from sustainable intensification responses in African systems demonstrate communities' potential to adapt to increasing population pressure by investing in soil conservation methods and allowing a sustainable increase in productivity and revenues (Barbier, 1998). This call for external intervention to escape the poverty and population trap leads us to the role of governance, whereby the rate of institutional adaptability relative to environmental dynamics is crucial regarding land and soil management (López, 1997).

Population growth, together with socio-economic development, unavoidably results in resource competition between different land use requirements (e.g. agriculture, forest, and biodiversity conservation, pastoralism, water provision, and carbon sequestration). When assessed across the region it is apparent that accelerated human population growth and intensive land use transformations (e.g. the expansion of intensive agricultural production, and urbanisation) have a spatial pattern. One area where there is relatively high population growth is focused on the boundaries around many Protected Areas (Pfeifer et al., 2012), thus a paradigm of conservation inside Protected Areas can result in enhanced development and livelihood challenges outside of them. Intensive human impacts on the lands surrounding Protected Areas may bleed inside the protected land, resulting in biodiversity decline and ecological degradation inside the Protected Areas, particularly around the park boundaries. As a result of these challenges, conservation policies in East Africa continue to call for benefits to incentivise local residents to support conservation while alleviating poverty (Brockington & Wilkie, 2015) that can reduce negative impact within the Protected Areas.

Another area of high population growth in East Africa is experienced by the rapidly escalating developmental and ecological challenges around the sprawling peri-urban and urban areas where rapid infrastructure growth is resulting in overstretched transport and waste management infrastructure. This rapid change results in the formation of some of the fastest-growing cities like Nanyuki and Dar es Salaam, mostly unplanned, large peri-urban areas. The urban poor, who rely heavily on biodiversity (food, fuel), will be among the most severely affected by these rapidly developing cityscapes (Thorn et al., 2021). Peri-urban areas tend to be located on previously undeveloped lands (e.g. wetlands, forests, farmlands)—leading to fragmentation, encroachment, over-extraction of natural resources, and food security. Two of the deadliest hydro-meteorological hazards for the urban poor are flooding and heat exposure. Riparian areas are often flooded and contaminated from overflowing pit latrines, industrial waste, polythene, and blocked stormwater drains—with wider water-related health impacts (e.g. malaria, cholera). Second, urbanisation increases the area of impervious surfaces and alters albedo and geometry leading to the urban heat island effect, which is amplified by galvanised sheet housing, little vegetation, narrow paths, and inadequate ventilation. Such conditions can increase heat-related illnesses and cooling requirements. As with previous change areas, understanding of the multidimensional nature of interlinkages across landscapes and time, particularly in terms of multi-directional resource flows, people's movements, governance regimes, and climate services in both peri-urban destinations and those with rural origins.

7.5 Past to the Present and Towards the Future: Underpinning Sustainability Science with the Long Term

East Africa can be viewed as a complex adaptive system, shaped by intertwined, coevolving interactions between nature and society (Fig. 7.16). Because that system is complex and heterogeneous, local insights and experiences from one location cannot be translated to another, particularly as people have their own agency and goals, and power struggles will play a central role in shaping pathways of development (Clark & Harley, 2020).

Given these properties, we can understand that there is a limited ability to predict how development pathways will unfold, or how particular interventions will work out (Clark & Harley, 2020). Sustainability

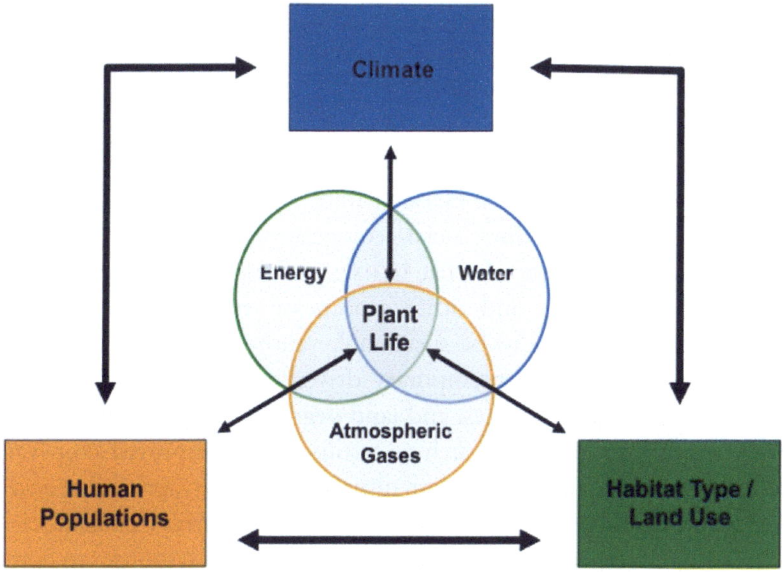

Fig. 7.16 East Africa is characterised by inherently strong interactions between nature, human populations, and the environment. The drivers behind land use transformation ultimately impact on the foundation for all life—that of plant life

science, like historical ecology, draws from a great variety of perspectives including tacit (traditional and practical) knowledge, ecology and economics, engineering and medicine, political science, and law. These multiple perspectives are generally a source of strength, bringing potentially complementary bodies of theory, data, and methods to bear on the challenges of sustainable development. Due to the interconnected nature of the social-ecological system, the integration remains incomplete, with our understanding remaining substantially less than the sum of its impressive parts. Transforming unsustainable development pathways into more sustainable ones is perhaps the grand challenge ahead. Meeting that challenge will require, above all, a greater capacity to foster innovation, but also a capacity to shape the collective visions of sustainable futures required to encourage innovation. Again, if these capacities are to be brought together an understanding past sustainable and unsustainable actions is required. The pursuit of sustainability can be strengthened by fostering a set of six essential capacities: (*a*) the capacity to measure sustainable development, (*b*) the capacity to promote equity, (*c*) the capacity to adapt to shocks and surprises, (*d*) the capacity to transform the system onto more sustainable development pathways, (*e*) the capacity to link knowledge with action, and (*f*) the capacity to devise governance arrangements that allow people to work together in exercising the other capacities (Clark & Harley, 2020).

Ultimately, understanding social-ecological history is fundamental to permit an understanding of what has shaped and influenced attitudes, perceptions of culture, and other aspects of society. For sustainable land management plans to be sustainable, authorities need to take the complex historical and contemporary drivers of increased soil erosion, population, changing climates, and land degradation into consideration. While modern technologies such as fertilisers and improved crop varieties are essential for increasing agricultural productivity, providing access to modern technologies alone will not solve the current challenges in East Africa. Examples from sustainable intensification responses to population growth highlight the adaptive capacity of communities with a well-developed social domain to increase economic production without compromising the continued usage of natural resources (Wynants et al., 2019). Locally adapted management practices need to be integrated into regional, national, and supra-national institutions. A nested political and economic framework allowing local communities to access market and governance services can enable regional development of sustainable agropastoral systems that safeguard ecosystem health, food security, and livelihood security (Wynants et al., 2019). To fully utilise these insights and incorporate them into future trajectories, we need to understand and integrate long-term perspectives (Crumley, 1994; Gillson & Marchant, 2014). Clearly, archaeological, historical, anthropological, ecological, and sedimentary sequences can contribute to our wider understanding of environmental history in East Africa, proving especially important for a part of the world where the instrumental records are relatively short and incomplete.

7.6 Commodification of Nature and the Rise of Ecosystem Services: Nature-Based Solutions to Challenges

Ecosystems and the services they provide (e.g. food, timber, water, carbon storage, nutrient cycling, soil formation, etc.) are increasingly being recognised for their fundamental importance for future livelihoods (Rockström et al., 2009), particularly in places like East Africa where this is a strong reliance on natural resources and where people's livelihoods connect strongly with their environment. So-called 'nature-based solutions ' are increasingly being brought to bear to address some of the future challenges around climate change, conservation, and livelihoods. It is critically important to understand and manage the delivery and use of ecosystem services (Fig. 7.17) but to do this effectively we must understand how they have varied temporally and spatially (Dearing et al., 2014). Compounding these biophysical challenges are further challenges around developing and implementing workable Payment for Ecosystem Services (PES) schemes such as seen through the Reducing Emissions from Deforestation and Forest Degradation (REDD+) policy (Chapter 6). As this is one of the key areas that will shape the future relationship between people and a highly dynamic suite of ecosystems and

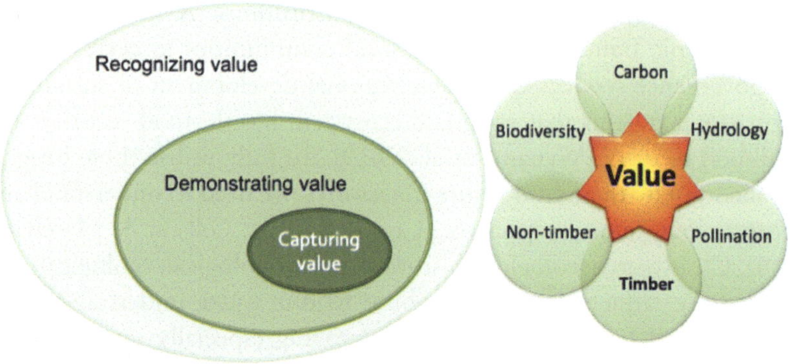

Fig. 7.17 Ecosystem services, or Natures contributions to people, are increasingly being recognised as vital to managing some of the Sustainable Development challenges as well as underpinning future economic development. Be it carbon storage or water cycling, tourism or pollination, key to mainstreaming ecosystem services is being able to adequately recognize, demonstrate and capture the value of the myriad of Ecosystem services

how ecosystems provide services, it is vital that such schemes do not lead to unintended consequences.

The integration of long-term perspectives into planning and ecosystem management tools is vital for the future delivery of ecosystem services (Jeffers et al., 2015). An understanding of land use change can support policies aimed at mitigating both the driving factors behind such change, and the behaviour of complex agro-ecosystems under changing conditions (de Koning et al., 1999). For example, in East Africa, it is clear from palaeoecological and historical evidence (Chapter 3) that precipitation regimes can change suddenly, and that change is not synchronous across the region, therefore future adaptation planning must include provision for such variation (Shanahan et al., 2015). The Ecosystem Services approach demonstrates the spatial connections between the service-providing areas and the service-benefitting areas, thus promoting regional conservation planning (Opdam, 2013). Payment schemes for ecosystem services (PES) could be a means to adequately value nature and natural capital that has hitherto been largely out of the planning decisions. For example, promoting biodiversity conservation could support rural development (Wei, Wang, Fu, Zhang, et al., 2018). In

PES, incentives are typically provided to local communities in exchange for the community's support for conservation, while the costs are borne by beneficiaries (Calvet-Mir et al., 2015). Indeed, many human activities have been prohibited in Protected Areas, which not only restricts access to several ecosystem services (mostly provisioning, such as gathering, hunting, or wood collecting) but can also create poverty and social conflicts (Schmidt-Soltau & Brockington, 2013) and ignore the important role of local communities in managing ecosystems and biodiversity. Ecosystems and the services they provide are numerous and continue to be added to as the cultural value and value for regulating nutrient flows and regulation are becoming increasingly clear. For the purposes of this book, Ecosystem services will be considered through the lens of three main commodities: carbon, water, and soil.

7.6.1 Carbon: Reforest Africa—The New Green Revolution?

Forests throughout East Africa are threatened by agricultural expansion either in the form of expanding food crops, exotic tree plantations, and ever-increasing demands for food, firewood, and timber for construction. A persistently high reliance on wood as a fuel source, sustained forest conversion to agriculture and development, and selective extraction of valuable medicinal and timber species, continues to put pressure on East Africa's forests and forest resources (Shaw et al., 2016). Although the use of forest and forest resource is not new, increasingly the value of relatively undisturbed forests cannot be anticipated in monetary terms and the gains and costs associated with forest clearance has usually been underestimated. Forests are much more than timber or potential land for agriculture, they continually act as watersheds (both production and regulation), stores of genetic resources (e.g. for medical and industrial purposes), protectors against soil erosion and floods, climatic stabilisers, biological standards to compare with adjacent land used for different purposes, and tourist attractions (Struhsaker, 1981). A rapidly growing focus, that is likely to increase in the future due to the global climate emergency and need to regulate carbon cycle, is how forests can be used to manage carbon budgets. Large areas of forest have been lost, particularly during the past 2000 years, with a massive increase in forest resource use during the colonial period resulting in carbon emissions, reduced habitat for forest-dependent biodiversity, and reduced provision and flow of essential ecosystem services.

Recent forest restoration commitments from African countries, including the AFR100 (a pan-African, country-led effort to restore 100 million hectares of degraded and deforested landscapes by 2030) and Bonn Challenge that pledges to restore 150 million hectares of the world's deforested and degraded lands by 2020 and 350 million hectares by 2030, are commitments feeding into the United Nations Framework Convention on Climate Change (UNFCCC). One specific scheme entitled Reducing Emissions from Deforestation and Forest Degradation (REDD+) and tied to the UNFCCC process the Bonn Challenge of Af100 aims to reforest vast tracts of Sub-Saharan Africa while combating the global trade in illegal forest products and securing property rights to conserved forests (Ahrends et al., 2021). Schemes like REDD+ rely on being able to assess carbon drawdown and sequestration (Asner et al., 2010), which are then equated to a monetary value. Baseline information is required on the potential of the forests to sequester and store carbon, particularly where many of the previously extensive forests of East Africa have been largely converted to agricultural land, especially in montane areas (Cuni-Sanchez et al., 2021). While we can see the impact of relatively recent (decadal) land cover changes through remote sensing, again a deeper perspective is really required to provide the full picture of the potential, and to enable this potential to be realised.

The expansion of interests in forest conservation is not without multiple challenges. For example, some have suggested the incentives may provide for states the tool to further marginalise indigenous forest-dwelling populations, charting an ominous trend towards the militarisation of forest and biodiversity conservation. Long histories of colonial and post-colonial marginalisation of forest-dwelling populations have more recently been exacerbated by evictions reportedly associated with REDD+ and other carbon offset forestry initiatives, which provide states with new financial incentives to pursue exclusionary forest conservation and recentralising control over natural resources. Consequently, in the context of both REDD+ and new forest management legislation in East Africa, it may be groups such as the Ogiek or Sengwer—rather than Al-Shabaab or Congolese militants—that find themselves at the brunt of multi-national efforts to fight the climate change challenge (Lunstrum, 2014). Duffy (2014) considers how the familiar tensions between conservation and communities seem to have regressed into a 'war for biodiversity'—one exacerbated by global markets for illegal wildlife products, as well as the grinding poverty that continues to afflict many protected area-adjacent populations. Indeed, it might be that REDD+ scheme has the potential to provide alternative income streams although there are multiple challenges around elite capture, and ensuing the benefits flow across to the local level.

Clearly, avoiding deforestation is key although it is the more diffuse carbon emissions from forest degradation that are particularly likely to be underestimated and could account for as much as 25–69% of the combined gross carbon losses due to deforestation in the tropics (Baccini et al., 2017). Significant progress has been made with measuring forest degradation from space (Woodcock et al., 2020) as changes in tree cover and biomass can now be monitored at high spatial and temporal resolution; providing policy makers and conservation planners with an unprecedented wealth of data to guide interventions (Ahrends et al., 2021). Although there are still challenges in our assessment tools and ground-truthing, remotely sensed data is a key component to produce accurate assessments (Ahrends et al., 2021). Collecting data on species, stem diameter, height, crown cover, and various biotic and abiotic parameters is a vital tool in biodiversity and environmental research (Baker et al., 2017). Consequently, while countries increasingly monitor wall-to-wall forest cover change using remote sensing, and they also have some inventory data, they still lack representative quantitative data on forest degradation (Romijn et al., 2015). The consideration of degradation in forest reporting is important—particularly in Eastern Africa where overall carbon emissions from forest degradation are likely to exceed those from deforestation (McNicol et al., 2018).

7.6.2 Water: Adapting to the Climate Variability Challenge

Water is life. Impending changes in the environment, population, land use, and economic development will dictate future patterns of water supply and demand (Vörösmarty et al., 2000). As population continues to grow, and as the regional climate seasonality enhances, there is increasing pressure on water supplies; this pressure is accentuated further as freshwater supply becomes even more restricted due to pollution (Heathwaite, 2010). Water is one of the key elements that is being challenged by an increasingly changing climate and is thus causing impacts on people's livelihoods and agricultural systems, particularly in water-limited ecosystems that are particularly vulnerable to decreasing or more variable rainfall regimes (Wei, Wang, Fu, Pan, et al., 2018). Jeffers et al. (2015) and Ekblom et al. (2017) identify palaeoecology as being ideally situated to address questions around vulnerability in water supplies under differing climate change scenarios, and for potentially providing baseline information on water supply and quality. These issues are gaining increasing importance as government agencies around the world, but particularly in East Africa, become responsible for implementing legislation to ensure that reliable and clean water supplies exist and are maintained.

One of the key impacts of land cover change is on hydrological function (Fig. 7.15). Mwangi et al. (2018) estimated that about 97% of change in the streamflow of a tributary of the Mara River has been caused by land use change, particularly deforestation and expansion of agriculture. It is increasingly obvious that the observed land use change in the Mara River Basin has affected the watershed's capacity to provide some ecosystem services (e.g. good quality water, good habitat for some wild animals and people). It is important therefore that the patterns, trends, and dominant processes of land use change and impacts on water be used to sustainably control expansion of agriculture (Mwangi et al., 2018) that does not lead to unintended downstream impacts. Such a catchment-based approach is key as cloud forests across the East African mountains are important ecosystems that can capture water from fog or low-level clouds drifting through the forest (Fig. 7.15). The area around Mt Marsabit is affected to a large degree by humans, e.g. by land cover change, groundwater withdrawal, and selective harvesting for use as fodder for livestock (Cuní-Sanchez et al., 2018). Appropriate management of the forest is needed to mitigate the adverse effects of global warming on hydrological resources; this is not only important for Mt Marsabit and but also for all 'water towers' across the region. There are unprecedented threats to the functioning of these water towers in the future. Although there is great uncertainty, it has been suggested that an increase of 250 m in the cloud-base height is expected with a future 2 °C global warming that could reduce forest extent substantially and curtail its ability to generate moisture for lowland communities (Los et al., 2019). Unfortunately, the amount of occult precipitation a forest captures is largely unknown and hence not factored into land use decisions or modelling the response of cloud forests to climate change, but it is clearly an area where advance is needed if we are to manage the impacts in the future.

One of the key tools used to address challenges of water availability and use is irrigation of dams and hard infrastructure so that water can be accessed during times of drought or agriculture possible in hitherto too arid a region. However, this is clearly not new technology and there are many examples of irrigation or terracing system landscapes that are relict and could be one of the salutary lessons from the past on land management that offer hope for the future. The earliest firmly dated irrigation features on Kilimanjaro were built in the eighteenth century; these customary management systems have been undermined by both colonial and post-colonial forest management authorities but do offer restoration potential if they can be brought back to function; not just in their ability to manage water and offset water shortages, but also how they can provide social cohesion through the collective management of the resource.

Wetland environments have clearly been a locus for East African populations and continue to be important to many societies, wildlife, and livelihoods: such environments are, and always have been, attractive areas within a landscape due to their high diversity in resources, high productivity, and their reliable water supply (Nicholas, 1998). While we do not know the detail on how most swamps are hydrologically sustained, they currently serve as a permanent freshwater source for animals and people, and as such are critical for maintaining biodiversity (Homewood & Lewis, 1987). These spring-fed wetlands are vitally important resources and connectivity nodes in arid regions of East Africa, not only for providing cooking and drinking water, but also for providing materials for fodder, thatching roofs, and for mat making (Ashley et al., 2002; Cuthbert et al., 2017). However, spring-fed wetlands are increasingly being put under pressure as population increases (e.g. Thenya, 2001), and the expansion of irrigated agriculture that taps the groundwater will ultimately divert water away from swamp systems to support short-term agricultural gains. The wider regulatory role of these swamps and wetlands is slowly being quantified in terms of their role in storing carbon in the sediments and purifying water through filtering effects of plant growth.

7.6.3 Soil: The Foundation for Agricultural Production

Human societies and their expanding use have become an integral fulcrum on the balance between soil erosion and soil production. East Africa's agropastoral and agricultural systems have been shaped by millennia of co-adaptation, reciprocal influencing, and feedback mechanisms between communities and ecosystems (Wynants et al., 2019). Soil fertility and structure underpins agropastoral and agricultural production as well as several ecosystem services such as nutrient cycling, below-ground carbon storage, and water supply (Jeffers et al., 2015). Paleoenvironmental and archaeological insights provide information that can inform and validate soil formation and erosion models (Kabora et al., 2020) as well as salutary lessons for successful or maladapted land use practices. For example, changes in vegetation cover and land use cover that result in pedogenic changes, as observed through detailed soil analyses at Engaruka in Northern Tanzania, are driven by agriculture and irrigation (Lang & Stump, 2017).

Human activities and climate change can accelerate soil erosion processes, which, in turn, are likely to produce an economic effect as productive agricultural land becomes degraded and water quality declines. Soil resources in many agricultural and pastoral landscapes of East Africa are depleted by increased erosion, contributing to widespread land degradation, which threatens food security, water security, and livelihood security (Blaikie & Brookfield, 2015). Multiple studies have charted the acceleration of surface and gully erosion leading to decreased agricultural productivity, pollution of water bodies, and ecosystem degradation (Fleitmann et al., 2007; Wynants et al., 2019). Many soils in semi-arid East Africa are particularly vulnerable due to low organic matter content and instability, and thus have a high prevalence of crusting and overall weak structural development (Blake et al., 2018). Although soil erosion is a physical process, its underlying causes are also firmly rooted in the social, economic, and institutional environment in which land users make decisions. Increased land degradation is rooted in historic disruptions to co-adapted agropastoral systems (Wynants et al., 2019). Coercive policies of land use, privatisation, sedentarisation, exclusion, and marginalisation through the colonial and postcolonial periods led to a gradual erosion of the indigenous social and economic structures (Wynants et al., 2019).

Due to the complex and dynamic East African environment, there are some mismatches in what grows well where, some of which has been driven by misguided policies such as the rapid planting of *Eucalyptus* that while growing quickly has a massive water demand and will tap into deeper aquifers taking water away from other crops. Similarly, recent expansion in avocados, which have a very high nutrient demand and extensive lateral root systems, can reduce nutrient availability for surrounding crops. One of the staples of East African agriculture, maize is again quite water demanding, particularly in more marginal areas where it is being replaced by indigenous crops of sorghum and millet. Such 'wrong crop in the wrong place' decisions again clearly underline the need to be guided by history on where best to avoid soil exhaustion, declining fertility, and increased soil erosion. Communities can increase productivity and diversify their economy by building upon, not abandoning, existing linkages between the social, economic, and natural domains. Locally adapted management practices need to be integrated into regional, national, and supra-national institutions. A nested political and economic framework, guided by history wherein local communities can access agricultural technologies and state services, is a key prerequisite towards regional development of sustainable agropastoral systems that safeguard soil health, food, and livelihood security. For the development and implementation of sustainable land management plans to be maintainable, authorities need to take the complex drivers of increased soil erosion into consideration with an emphasis on locally adapted management practices (Wynants et al., 2019). Drawing on examples from sustainable intensification responses to the demands of population increase, these demonstrate that the integrity of locally adapted systems needs to be protected, but not isolated, from external pressures. Communities can increase productivity and diversify their economy by building upon, not abandoning, deep-rooted linkages between the social, economic, and natural domains.

7.7 Current Conservation Challenge: Future-Proofing Protected Areas

In East Africa, there can be little doubt of the value of Protected Areas in terms of the world's natural heritage. National Parks and other strictly protected areas have conserved much (Sinclair et al., 2002; Struhsaker, 1981), and have offered the best option available for safeguarding wild nature (Gardner et al., 2007; Terborgh et al., 2002). Some 50 years ago the futures of the Protected Areas were questioned by some of the leading conservations/ecologists of the time (Myers, 1972). Although the rapid social and environmental changes that East Africa is experiencing is resulting in the loss of biodiversity and impacting on livelihoods that depend on natural resources (Calvet-Mir et al., 2015), the protected area system is intact, albeit the functioning of this has evolved and is under increasing pressure from a whole series of challenges. Some of these challenges have their causes in the origins: the National Park system was initially set up as hunting reserves, as clearly documented by early 'Great' white hunters, to the conservation of the 'Big 5 (Elephant, Lion, Buffalo, Rhino, and Leopard)' for tourism, then to biodiversity conservation of the wider ecosystem and migration corridors, and more recently to livelihoods, protection of ecosystem services and to underpin the Sustainable Development Goal agenda. Despite these massive transformations in rationale, the boundaries of the National Park are largely unchanged since the 1940s. How these can make the most of the natural capital agenda and support community-based conservation are just some of the contemporary challenges. One of the key challenges comes from the background of fortress conservation—people are excluded from National Parks. Whereas in some cases this is warranted to prevent agricultural expansion, in other cases indigenous populations have lived within the protected area a long time before being demarcated, and in some cases have been instrumental in their persistence while land outside the protected area has been converted to agriculture (Hamilton et al., 2000). As we have seen, protected areas are part of the wider social-ecological system and respond to, and are influenced by, a wide range of ecological, social, and political processes (Wei, Wang, Fu, Zhang, et al., 2018). Therefore, it is critical to detect certain key drivers and feedbacks of the coupled social-ecological system and understand how this has transitioned through time.

Virtually all nature reserves are, or soon will be, islands of natural habitat in an area of inhospitable terrain (Soulé et al., 1979). Building protected areas that can withstand the Anthropocene challenges is a challenge, particularly as the boundaries are inherited and increasingly fixed by population expansion and transformation. Following years of contradictory management regimes, communicating clear and convincing conservation storylines is fundamental to the future success of protected areas (Folke et al., 2005). At the heart of this challenge is general distrust towards the government, who by and large are still seen as corrupt and trying to steal land. New devolved ownership and decentralised governance systems in Kenya appear to be as susceptible to the same power asymmetries as their higher order origins. There are no shortcuts in rebuilding meaningful relationships, and livelihoods and equitable benefits must be at the heart of conservation. Future measures that deliver the best outcomes for both conservation and equity are founded upon productive co-management regimes (Oldekop et al., 2015); working with local communities. Again, historical perspective is key here: local people are not homogenous entities; they differ in their histories, social structures, and belief systems that not only affect their ability to actively partake in conservation-related activities but their desire to do so. Local context is key with one-size-fits-all policy intervention being doomed to failure.

Protected areas provide important global, national, and local benefits by conserving biodiversity and maintaining ecosystem services. Yet such benefits may come at a cost to indigenous and local communities (Schreckenberg et al., 2016). In addition to the normative (or moral) argument for equitable conservation, there is growing acknowledgement that resentment and a sense of injustice among those affected by protected areas can drive threats to protected area conservation. A growing body of research provides evidence that empowerment of local people and more equitable sharing of benefits increase the likelihood of effective conservation (Oldekop et al., 2015). In response to calls from various decisions coming from the World Parks Congresses, specifically expressed in the Aichi Targets, there has been rapid progress in developing tools for assessing the effectiveness of protected areas. There is a shift of the conservation narrative from an overly narrow focus on livelihoods to a broader focus on equity that fully integrates the issue of protected area costs and benefits with protected area governance at the level of a whole protected area system (Schreckenberg et al., 2016).

Management decisions and human actions should integrate both local livelihoods and biodiversity to achieve a win-win or 'small loss, big gain' goal (Reid et al., 2016). Clearly, there are many conservation challenges ahead and one way to rise to these challenges is to take a more holistic approach. Traditional conservation strategies (e.g. strictly protected areas) often not only generate high opportunity costs for local communities, but also lack sustainable funding for management and safeguarding the protection and integrity of biodiversity. National Park boundaries are expensive to delimit and maintain (Pullan, 1988). Therefore, a major challenge for contemporary conservation policies and practices is formulating workable compromises between biodiversity conservation and community needs. Community conservation initiatives often create mixed-use landscapes containing both wildlife and livestock (Tyrrell et al., 2017) that result in trade-offs of costs and benefits between wildlife and people. Countrywide declines in wildlife have highlighted the need for protection of wildlife outside of protected areas and across the rangelands: in Kenya, these areas contain c. 60% of wildlife (Western et al., 2009).

Conservancies are a vital tool for the long-term survival of wildlife, both outside and inside protected areas (Western et al., 2015). Many of these conservancies are community conservancies, involving multiple indigenous landowners. Wildlife conservancies can create substantial local revenue through wildlife-based tourism (Naidoo et al., 2016). In Kenya, the number of conservancies has grown rapidly, from less than five in the early 1990s to over 140 in 2016 covering 30,000 km^2 (KWCA, 2016); growth that is expected to continue. A big challenge for the sustainable management of Protected Areas in East Africa is to balance conservation and development objectives. Wildlife tourism is an essential part of national income in East Africa, providing job opportunities and making people rely less on natural resources, thus optimising the tourism-benefit-sharing mechanism and assuring the inclusion of benefits to local people (Wei, Wang, Fu, Zhang, et al., 2018). New opportunities to reconcile biodiversity conservation and rural livelihoods could be derived from changes in conservation philosophies to incorporate benefits to local people, while also necessitating a revolution of human value towards a sustainable planet where people live in harmony with nature (Wei, Wang, Fu, Zhang, et al., 2018). Notwithstanding the massive shock that these types of revenue-sharing schemes have encountered through the impacts of COVID-19 where the tourist revenue has dried up almost overnight. Conservation of wide-ranging endangered species, such as wild dogs, poses significant challenges. However, utilisation distributions and resource selection models derived from tracking data can help to identify core-use areas and ecological factors influencing space use, ultimately helping to prioritise conservation actions (Buechley et al., 2018). To develop effective landscape-scale conservation strategies, it is necessary to evaluate the contribution of protected areas towards the protection of regional diversity, and the biodiversity value of land outside strictly protected areas (Gardner et al., 2007); including lightly protected conservation areas that allow limited forms of human use, game reserves (Brooks et al., 2006), and agricultural areas (Daily et al.,

2003; Naidoo, 2004). This lack of appreciation towards parks outside the park can result in areas outside strictly protected areas being abandoned in the wake of rising population pressure and economic development and succumbing to increasingly severe levels of exploitation and degradation. This process rapidly leads to National Parks becoming isolated islands within highly developed landscapes (Gardner et al., 2007). With a wider appreciation of a Nature-based management future hopefully the balance can be tipped back in favour of ecosystems and animals that have been, and continue to be, squeezed at the expense of agricultural expansion (Fig. 7.18).

Fig. 7.18 Conservation vs agriculture in the balance. Around many Protected Areas increasingly intensive and irrigated farms, such as this one near Amboseli, is reducing space and functionality of the landscape to support wildlife populations and increasing human–wildlife conflict that is shifting the balance towards agriculture and reducing space for wildlife, particularly migratory species like elephants. Key habitats, such as these swamps (**a**) are drained for agricultural production (**b**) further increasing human-wildlife conflict, particularly during periods of drought (All photographs: Rob Marchant)

While there have been conservation success stories in the lowland rangelands, there remains a pressing need to protect and conserve many more areas of lowland and montane humid forest in East Africa if these ecosystems and their flora and fauna are to survive into the next century. Their designation as forest reserves and game reserves has not, in most cases, protected them in the past from modification and destruction by human activities. As these pressures continue to intensify and modify areas, there is a need for policies and practice to promote successful management of key forest areas, as many of these are the pump of life and livelihoods through their role as water towers. Indeed, to withstand future climatic changes, the rich panoply of plants and animals in East Africa must have the latitude to move up and down the altitudinal gradients and across large open landscapes. Two human-induced changes are underway that in tandem threaten the biodiversity of the region. Land subdivision and settlement are breaking up the open landscape and the continuity of habitats strung along altitudinal gradients with fundamental relationships, as we understand them, are being repositioned by the growing impacts of climate change. These two forces threaten the region's natural buffering system and put at risk species evolved over millions of years and accumulated over climatic oscillations (Western, 2008). A more balanced view that biodiversity is part and product of complex and linked natural and anthropogenic interactions (Pinedo-Vasquez et al., 2002) is crucial if we are to manage ecosystems in a future of uncertain change. For example, the diversity in resource use, cultural practices, and the capacity of adaptations for social change are only rarely considered for conservation and development schemes (Pinedo-Vasquez et al., 2002). There is nevertheless a growing realisation that the idea of virgin rainforests, initially seen as largely 'untouched' by 'native' people, is a misnomer (Willis et al., 2004). Instead the extent of, and potential for relationships between, early populations and ecosystems, is supporting the realisation that humans have sculpted closed forests for millennia (Pearce, 1999; Roberts et al., 2017). With long-term perspectives on human-ecosystem interaction, we can modify our conservation and protectionist ethos to adopt more dynamic strategies for conserving biodiversity to accommodate the change that inevitably occurs in species ranges and environmental regimes over time (Marchant, 2010; Tallis, 1991).

Salutary lessons can be learned from a historical perspective. A longer-term evaluation on land use change that also brings together conservation measures to assess the challenges to biodiversity conservation and to provide status assessments would be a helpful step forward. A good example of the utility of this long-term perspective is clear from our understanding of fluctuating elephant populations. Following the ban on international trade in ivory from 1989 (Lemieux & Clarke, 2009), poaching dramatically declined. However, we still know relatively little about the effects of changing herbivore numbers, and what the original carry capacity of elephant populations may have been. There remains an urgent need for more data on long-term vegetation variability in response to changing herbivore densities. Gaining insight into this relationship would have the effect of improving our understanding of ecological processes in savannahs; past distribution of elephant populations (Coutu et al., 2016); and their mobility patterns and geographical ranges (Coutu, 2015). And crucially, it could also provide a basis for future management decisions. We know today the significant impact that elephant populations have in maintaining both open and grassland (Morrison, Gaillard et al., 2016), and that the impact of these important 'ecosystem engineers' has been significantly curtailed through the wholesale removal of the elephant population from some areas, initially through the European and American-driven demands for ivory during the nineteenth and early twentieth centuries, and now, owing to the recent upsurge in poaching and illicit trade in ivory, largely driven by Asian markets (Sommerville, 2016). An appreciation that integrates the historical and political ecologies of these trades, that connects to previous phases of forest and savannah elephant ivory extraction, offers only one model for addressing these challenges (Håkansson, 2004; Lane, 2010).

Conserving ecosystems and communities, as they presently exist or intervening to restore them to a previous state, has become redundant. Instead, dynamic strategies are supported as these can allow for a type of conservation that accommodates climate change by embracing concepts that encapsulate networks and connectivity as these can support fluidity and change (Hannah, Midgley, & Millar, 2002). Given current climate projections (Platts et al., 2015), conservation institutions and approaches will fail if they do not factor in change as a major component of the system. Equally, individuals, communities, and societies will also need to be able to adapt and change to reduce their vulnerability to climate change (Rayner & Malone, 1997). It is also the responsibility of the academic community to rethink how we research and communicate our understandings of climate change, and this includes challenging our current theories and models for working on issues surrounding climate change and how people and the environment relate (Marchant & Lane, 2014; Rayner & Malone, 1997). An example of this is the debate about paved roadways through the Serengeti ecosystem (Dobson et al., 2010; Fyumagwa et al., 2013; Homewood et al., 2010) and further development of new railway lines; both of which are long linear features that fragment landscapes and have varied socio-ecological benefits and impacts (Hopcraft et al., 2015). One of the many ways that we can engage with the call to radically alter how we think about these issues is to again reiterate the importance of relevant and scientifically robust palaeoecological and archaeological records.

Some would argue that historically, the role of protected areas in maintaining ecological functions of ecosystems is built on an antagonism between the alleged predatory activities of pastoralist societies and conservation practices (Hazard & Adongo, 2012). Today there is changing attitude and perception of protection that has changed from top-down to bottom-up with environmental education playing a significant role in conservation. This has ensured success of conservation models like conservancies. Protection can be initiated and implemented by the government, non-state actors, or the community themselves. This might also occur in three perspectives: enhance biological perspective, collaborative management approach, or pure community-based conservation rendering control to the communities only perhaps with technical support from the Government or Non-Governmental Organisations. A balance between protected areas/conservation and livelihoods is important to the local communities as this elaborates how the community will benefit from conservation. Coexistence of wildlife, conservation, and pastoral production is filled with problems, but trying to combine conservation and pastoral development is the only solution if both pastoral societies of tomorrow and large-scale conservation are to continue to exist. Fundamental to future success is ensuring that rights and responsibilities over natural resources are devolved to the lowest appropriate level (Western & Nightingale, 2003). As there are avoided opportunity costs by releasing grazing resource there should be reciprocal grazing arrangements in pace to allow mobility and access to pastures that considers seasonality, drought, and the benefits of mobility. To ensure pressure is taken off the grazing resource there should be a diversification of opportunities including agriculture, irrigation, wildlife-based income, cultural tourism, and value-added livestock industries such as tanning and establishing of trade associations. These diverse opportunities need additional support with improved health and educational facilities.

The challenge is not just for communities to adapt. International organisations also need to take responsibility and push for change as well as the national governments and corporate sectors. There is massive momentum in the area around meeting the net zero challenges, links between climate change, conservation, development, and Biodiversity are going to be key here with the most obvious link being Nature Based Solutions to address some of the multiple challenges coming down the line. The current network of land use and protected area institutions in East Africa needs to be strengthened if we are to conserve regional biodiversity effectively and prevent the insularisation of strictly protected areas (Gardner et al., 2007). The design of nature reserves can be placed on a secure scientific footing with the application of numerous analytic methods for delineating the size and boundaries of proposed reserves (Caro, 2003) that must draw on history, and not one that just started with the remotely sensed view of the region. In a simple world, where best to place reserves can be determined by computational algorithms applied to data sets on the distribution of different taxa across a large area to conserve areas where the greatest number of species' geographic ranges overlap (Williams et al., 1996). However, hotspots of species richness in one taxonomic group (e.g. butterflies) are usually weak predictors of hotspots of other groups (e.g. mammals or liverworts) availability (Caro, 2003). This 'umbrella-species' concept was originally defined by Noss (1990) as 'species with large area requirements, which if given sufficient protected habitat area, will bring many other species under protection'. It is important that the umbrella population remains viable for political reasons (Berger, 1997) as local extinction could open the door to developers or agricultural expansion (Caro, 2003). However, it raises the question: how does the implementation of models of management of natural resources interact with the future of pastoralist economies and their socio-cultural heritage, in a context of socio-ecological transition due to a specific history, geopolitical uncertainty, and environmental degradation, exacerbated by climate variability/change and limited adaptability of social and economic organisation?

7.8 Next Steps

The road ahead will be characterised by obstacles and problems that can be negated by learning from landscapes, listening to different land users, and embracing the opportunities, particularly those stemming from a young and entrepreneurial population. For example, the young population commonly using mobile technology should be able to capitalise on low carbon futures and build on the development lessons learnt from other countries around the world. Clearly the frequency of atypical weather extremes linked to anthropogenically induced climate change is disrupting traditionally predictable seasons with this increased frequency of extremes (droughts and floods) having major implications on social structures and livelihoods of pastoralists and farmers. Creating knowledge and integrating past understandings into future planning and policy to enhance community cohesion, ecosystem resilience, and adaptive capacities through future scenarios is crucial. A better understanding of how people have responded to past environmental change, and how ongoing adaptation measures have evolved, are vital if we are to craft visions, policies, and programmes aimed at promoting successful adaptation to future environmental shifts and cultural changes.

Datasets and synthesis provide baselines to build future scenarios based on the range of variability and potential threshold responses, and as valuable information for decision-making in land management, food production, and security. Planning for the long-term sustainable use of natural resources requires a long-term, nested spatial and temporal perspective on human-ecosystem-environment interactions and engagement with the biocultural heritage and societal evaluations of these spaces to achieve an increasingly diverse set of conservation, social, and economic objectives. While critical earth system boundaries have been identified and exceeded at the global level, most sustainability targets (e.g. Sustainable Development Goals, African Union Agenda 2063) appear highly challenging to achieve; partly because there is little synthesised knowledge of environmental, social, and heritage interactions at the landscape scale. The socio-environmental examination of how East African social-ecological systems have responded to environmental and social shifts in the past will hopefully contribute to informing how ecosystems, people, and wildlife will respond in the future. Maintaining and restoring social-ecological connectivity is a fundamental piece of the puzzle. We need a unified vision for the future with infrastructure and solutions that work for the local places, local contexts, and future challenges that have been shaped by history. Conceptualising East Africa in a 'global history' can do justice to the views of the local, particularly when the local has been intimately connected to the global for millennia. Suffice to conclude, it has never been more important to provide a more robust, supportive environment, and an appreciation of this history to build a socially equitable, resilient, and bright future that maintains the diversity of people, cultures, and ecosystems for the millennia to come.

References

Ahrends, A., Bulling, M. T., Platts, P. J., et al. (2021). Detecting and Predicting Forest Degradation: A Comparison of Ground Surveys and Remote Sensing in Tanzanian Forests. *Plants, People, Planet, 3*, 268–281. https://doi.org/10.1002/ppp3.10189

Alexandre, A., Meunier, J. D., Lézine, A. M., Vincens, A., & Schwartz, D. (1997). Phytoliths: Indicators of Grassland Dynamics During the Holocene in Intertropical Africa. *Palaeogeography, Palaeoclimatology, Palaeoecology, 136*, 213–229.

Ambrose, S. H. (1984). The Introduction of Pastoral Adaptations to the Highlands of Eastern Africa. In J. D. Clark & S. A. Brandt (Eds.), *From Hunters to Farmers: The Causes and Consequences of Food Production in Africa* (pp. 212–239). University of California Press.

Anderson, D. (1984). Depression, Dust Bowl, Demography, and Drought: The Colonial State and Soil Conservation in East Africa During the 1930s. *African Affairs* [online], *83*, 321–343. https://www.jstor.org/stable/722351. Accessed 11 June 2018.

Araújo, M. B., & Peterson, A. T. (2012). Uses and Misuses of Bioclimatic Envelope Modeling. *Ecology, 93*, 1527–1539.

Ashley, G. M., Goman, M., Hover, V. C., Owen, R. B., Renaut, R. W., & Muasya, A. M. (2002). Artesian Blister Wetlands, a Perennial Water Resource in the Semi-Arid Rift Valley of East Africa. *Wetlands, 22*(4), 686–695.

Asner, G. P., Powell, G. V. N., Mascaro, J., Knapp, D. E., Clark, J. K., Jacobson, J., Kennedy-Bowdoin, T., Balaji, A., Paez-Acosta, G., Victoria, E., Secada, L., Valqui, M., & Hughes, R. F. (2010). High-Resolution Forest Carbon Stocks and Emissions in the Amazon. *Proceedings National Academy of Sciences, 107*, 16738–16742.

Baccini, A., Walker, W., Carvalho, L., Farina, M., Sulla-Menashe, D., & Houghton, R. A. (2017). Tropical Forests Are a Net Carbon Source Based on Aboveground Measurements of Gain and Loss. *Science, 358*(6360), 230–234.

Baker, T. R., Pennington, R. T., Dexter, K. G., Fine, P. V. A., Fortune-Hopkins, H., Honorio, E. N., & Vasquez, R. (2017). Maximising Synergy Among Tropical Plant Systematists, Ecologists, and Evolutionary Biologists. *Trends in Ecology and Evolution, 32*(4), 258–267.

Barbier, B. (1998). Induced Innovation and Land Degradation: Results from a Bioeconomic Model of a Village in West Africa. *Agric Econ* [online], *19*, 15–25. https://doi.org/10.1016/S0169-5150(98)00052-8

Berger, J. (1997). Population Constraints Associated with the Use of Black Rhinos as an Umbrella Species for Desert Herbivores. *Conservation Biology, 11*, 69–78.

Birks, H. H., & Birks, H. J. B. (2000). Future Uses of Pollen Analysis Must Include Plant Macrofossils. *Journal of Biogeography, 27*(1), 31–35.

Blaauw, M., van Geel, B., Kristen, I., Plessen, B., Lyaruu, A., Engstrom, D. R., van der Plicht, J., & Verschuren, D. (2011). High-Resolution ^{14}C Dating of a 25,000-Year Lake-Sediment Record from Equatorial East Africa. *Quaternary Science Reviews, 30*(21), 3043–3059.

Blaikie, P., & Brookfield, H. (2015). *Land Degradation and Society*. Routledge.
Blake, W. H., Rabinovich, A., Wynants, M., Kelly, C., Nasseri, M., Ngondya, I., Patrick, A., Mtei, K., Munishi, L., Boeckx, P., Navas, A., Smith, H. G., Gilvear, D., Wilson, G., Roberts, N., & Ndakidemi, P. (2018). Soil Erosion in East Africa: An Interdisciplinary Approach to Realising Pastoral Land Management Change. *Environ Res Lett* [online], 124014. https://doi.org/10.1088/1748-9326/aaea8b
Bollig, M. (2014). Resilience—Analytical Tool, Bridging Concept or Development Goal? Anthropological Perspectives on the Use of a Border Object. *Zeitschrift Für Ethnologie, 139*(2), 253–279.
Bond, W. J., & Midgley, G. F. (2000). A Proposed CO_2-Controlled Mechanism of Woody Plant Invasion in Grasslands and Savannas. *Global Change Biology, 6*(8), 865–869.
Braconnot, P., Harrison, S. P., Kageyama, M., Bartlein, P. J., Masson-Delmotte, V., Vabe-Ouchi, A., Otto-Bliesner, B. L., & Zhao, Y. (2012). Evaluation of Climate Models Using Palaeoclimatic Data. *Nature Climate Change, 2*, 417–424.
Brandt, P., Kvaki, C. M., Butterbach-Bahl, K., & Rufino, M. C. (2017). How to Target Climate-Smart Agriculture? Concept and Application of the Consensus-Driven Decision Support Framework "targetCSA." *Agricultural Systems, 151*, 234–245.
Brockington, D., & Wilkie, D. (2015). Protected Areas and Poverty. *Philosophical Transactions of the Royal Society B, 370*, 20140271.
Brooks, J. S., Franzen, M. A., Holmes, C. M., Grote, M. N., & Borgerhoff Mulder, M. (2006). Testing Hypotheses for the Success of Different Conservation Strategies. *Conservation Biology, 20*, 1528–1538.
Buechley, E. R., McGrady, M. J., Çoban, E., & Şekercioğlu, Ç. H. (2018). Satellite Tracking a Wide-Ranging Endangered Vulture Species to Target Conservation Actions in the Middle East and East Africa. *Biodiversity and Conservation, 27*(9), 2293–2310.
Calvet-Mir, L., Corbera, E., Martin, A., Fisher, J., & Gross-Camp, N. (2015). Payments for Ecosystem Services in the Tropics: A Closer Look at Effectiveness and Equity. *Current Opinion in Environmental Sustainability, 14*, 150–162.
Canning, D., Raja, S., & Yazbeck, A. S. (2015). *Africa's Demographic Transition: Dividend or Disaster? Africa Development Forum Series*. World Bank. https://doi.org/10.1596/978-1-4648-0489-2
Capitani, C., Mukama, K., Mbilinyi, B., Malugu, I., Munishi, P. K. T., Burgess, N. D., Platts, P. J., Sallu, S., & Marchant, R. (2016). From Local Scenarios

to National Maps: A Participatory Framework for Envisioning the Future of Tanzania. *Ecology and Society, 21*(3), 4.

Capitani, C., Van Soesbergen, A., Mukama, K., Malugu, I., Mbilinyi, B., Chamuya, N., … Marchant, R. (2019). Scenarios of land use and land cover change and their multiple impacts on natural capital in Tanzania. *Environmental Conservation, 46*(1), 17–24.

Capitani, C., Garedew, W., Mitiku, A., Berecha, G., Hailu, B. T., Heiskanen, J., ... Marchant, R. (2019b). Views from two mountains: Exploring climate change impacts on traditional farming communities of Eastern Africa highlands through participatory scenarios. *Sustainability Science, 14*(1), 191–203.

Caro, T. M. (2003). Umbrella Species: Critique and Lessons from East Africa. *Animal Conservation, 6*(2), 171–181.

Chevalier, M. A., Brewer, S., & Chase, B. M. (2017). Qualitative Assessment of PMIP3 Rainfall Simulations Across the Eastern African Monsoon Domains During the Mid-Holocene and the Last Glacial Maximum. *Quaternary Science Review, 156*, 107–120.

Clark, W. C., & Harley, A. G. (2020). Sustainability Science: Toward a Synthesis. *Annual Review of Environment and Resources, 45*, 331–386.

Clark, P. U., Pisias, N. G., Stocker, T. F., & Weaver, A. J. (2002). The Role of the Thermohaline Circulation in Abrupt Climate Change. *Nature, 415*, 863–869.

Claussen, M., Kubatzki, C., Brovkin, V., Ganopolski, A., Hoelzmann, P., & Pachur, H. J. (1999). Simulation of an Abrupt Change in Saharan Vegetation in the Mid-Holocene. *Geophysical Research Letters, 26*(14), 2037–2040.

Courtney-Mustaphi, C. J., & Marchant, R. (2016). A Database of Radiocarbon Dates for Palaeoenvironmental Research in Eastern Africa. *Open Quaternary, 2*(3), 1–7.

Courtney-Mustaphi, C. J., Rucina, S. M., & Marchant, R. (2014). Training in Emerging Palaeoenvironmental Approaches to Researchers on the Dynamics of East African Ecosystems. *Frontiers of Biogeography, 6*, 169–172.

Coutu, A. N. (2015). The Elephant in the Room: Mapping the Footsteps of Historic Elephants with Big Game Hunting Collections. *World Archaeology, 47*(3), 486–503.

Coutu, A. N., Lee-Thorp, J., Collins, M. J., & Lane, P. J. (2016). Mapping the Elephants of the 19th Century East African Ivory Trade with a Multi-Isotope Approach. *PloS One, 11*(10), e0163606.

Crowther, A., Prendergast, M. E., Fuller, D. Q., & Boivin, N. (2017). Subsistence Mosaics, Forager-Farmer Interactions, and the Transition to Food

Production in Eastern Africa. *Quaternary International* [online]. https://doi.org/10.1016/j.quaint.2017.01.014

Crumley, C. L. (1994). *Historical Ecology: Cultural Knowledge and Changing Landscapes*. School of American Research Press.

Cuní-Sanchez, A., Omeney, P., Pfeifer, M., Olaka, L., Boru Mamo, M., Marchant, R., & Burgess, N. D. (2018). Climate Change and Pastoralists: Perceptions and Adaptation in Montane Kenya. *Climate and Development* [online], 1–12. https://doi.org/10.1080/17565529.2018.1454880

Cuni-Sanchez, A., Sullivan, M. J., Platts, P. J., Lewis, S. L., Marchant, R., Imani, G., ... Zibera, E. (2021). High aboveground carbon stock of African tropical montane forests. *Nature, 596*(7873), 536–542.

Cuthbert, M. O., Gleeson, T., Reynolds, S. C., Bennett, M. R., Newton, A. C., McCormack, C. J., & Ashley, G. M. (2017). Modelling the Role of Groundwater Hydro-Refugia in East African Hominin Evolution and Dispersal. *Nature Communications, 8*(1), 1–11.

Daily, G. C., Ceballos, G., Pacheco, J., Suzan, G., & Sanchez-Azofeifa, A. (2003). Countryside Biogeography of Neotropical Mammals: Conservation Opportunities in Agricultural Landscapes of Costa Rica. *Conservation Biology, 17*, 1814–1826.

de Jong Cleyndert, G., Cuni-Sanchez, A., Seki, H. A., Shirima, D. D., Munishi, P. K., Burgess, N., Calders, K., & Marchant, R. (2020). The Effects of Seaward Distance on Above and Below Ground Carbon Stocks in Estuarine Mangrove Ecosystems. *Carbon Balance and Management, 15*(1), 1–15.

de Koning, G. H. J., Veldkamp, A., & Fresco, L. O. (1999). Exploring Changes in Ecuadorian Land Use for Food Production and Their Effects on Natural Resources. *Journal of Environmental Management, 57*, 221–237.

Dearing, J. A., Wang, R., Zhang, K., Dyke, J. G., Haberl, H., Hossain, M. S., Langdon, P. G., Lenton, T. M., Raworth, K., Brown, S., & Carstensen, J. (2014). Safe and Just Operating Spaces for Regional Social-Ecological Systems. *Global Environmental Change, 28*, 227–238.

Demenocal, P., Ortiz, J., Guilderson, T., Adkins, J., Sarnthein, M., Baker, L., & Yarusinsky, M. (2000). Abrupt onset and termination of the African Humid Period: Rapid climate responses to gradual insolation forcing. *Quaternary science reviews, 19*(1–5), 347–361.

Dobson, A. P., Borner, M., Sinclair, A. R., Hudson, P. J., Anderson, T. M., Bigurube, G., Davenport, T. B., Deutsch, J., Durant, S. M., Estes, R. D., & Estes, A. B. (2010). Road Will Ruin Serengeti. *Nature, 467*(7313), 272–273.

Dublin, H. T. (1995). Vegetation Dynamics in the Serengeti-Mara Ecosystem: The Role of Elephants, Fire, and Other Factors. In A. R. E. Sinclair & P. Arcese (Eds.), *Serengeti II: Dynamics, Management, and Conservation of an Ecosystem* (pp. 71–90). University of Chicago Press.

Duffy, R. (2014). Waging a War to Save Biodiversity: The Rise of Militarized Conservation. *International Affairs, 90*(4), 819–834.

Eggert, M. K. H. (2005). The Bantu Problem and African Archaeology. In A. B. Stahl (Ed.), *African Archaeology: A Critical Introduction* (pp. 301–326). Blackwell Publishing.

Ejarque, A., Miras, Y., & Riera, S. (2011). Pollen and Non-Pollen Palynomorph Indicators of Vegetation and Highland Grazing Activities Obtained from Modern Surface and Dung Datasets in the Eastern Pyrenees. *Review of Palaeobotany and Palynology, 167*, 123–139.

Ekblom, A., & Gillson, L. (2010). Dung Fungi as Indicators of Past Herbivore Abundance, Kruger and Limpopo National Park. *Palaeogeography, Palaeoclimatology, Palaeoecology, 296*, 14–27.

Ekblom, A., Gillson, L., & Notelid, M. (2017). Water Flow, Ecological Dynamics, and Management in the Lower Limpopo Valley: A Long Term View. *Wiley Interdisciplinary Reviews: Water, 4*(5), e1228.

Finch, J., Marchant, R., & Courtney Mustaphi, C. J. (2017). Ecosystem Change in the South Pare Mountain Bloc, Eastern Arc Mountains of Tanzania. *The Holocene, 27*(6), 796–810.

Fleitmann, D., Dunbar, R. B., McCulloch, M., Mudelsee, M., Vuille, M., McClanahan, T. R., Cole, J. E., & Eggins, S. (2007). East African Soil Erosion Recorded in a 300 Year Old Coral Colony from Kenya. *Geophysical Research Letters, 34*, L04401.

Folke, C., Hahn, T., Olsson, P., & Norberg, J. (2005). Adaptive Governance of Social-Ecological Systems. *Annual Review of Environment and Resources, 30*, 441–473.

Friedlingstein, P., Cox, P., Betts, R., Bopp, L., Von Bloh, W., Brovkin, V., Cadule, P., Doney, S., Eby, M., Fung, I., & Bala, G. (2006). Climate-Carbon Cycle Feedback Analysis: Results from the C4MIP Model Intercomparison. *Journal of Climate, 19*(14), 3337–3353.

Fyumagwa, R., Gereta, E., Hassan, S., Kideghesho, J. R., Kohi, E. M., Keyyu, J., Magige, F., Mfunda, I. M., Mwakatobe, A., Ntalwila, J., & Nyahongo, J. W. (2013). Roads as a Threat to the Serengeti Ecosystem. *Conservation Biology, 27*(5), 1122–1125.

Gaillard, M. J., Lezine, A. M., Morrison, K. D., & LandCover6k Steering Group (2015). Launching Workshop of PAGES' Working Group LandCover6k. *PAGES Magazine, 23*(2), 81.

Gardner, T. A., Caro, T. I. M., Fitzherbert, E. B., Banda, T., & Lalbhai, P. (2007). Conservation Value of Multiple-Use Areas in East Africa. *Conservation Biology, 21*(6), 1516–1525.

Gillson, L., & Marchant, R. (2014). From Myopia to Clarity: Sharpening Ecosystem Management. *Trends in Ecology and Evolution, 29*, 317–325.

Githumbi, E., Courtney Mustaphi, C., & Marchant, R. (2021). Late Pleistocene and Holocene Afromontane Vegetation and Headwater Wetland Dynamics Within the Eastern Mau Forest, Kenya. *Journal of Quaternary Science, 36*, 239–254. https://onlinelibrary.wiley.com/doi/full/10.1002/jqs.3267

Håkansson, N. T. (2004). The Human Ecology of World Systems in East Africa: The Impact of the Ivory Trade. *Human Ecology, 32*(5), 561–591.

Hamilton, A., Cunningham, A., Byarugaba, D., & Kayanja, F. (2000). Conservation in a region of political instability: Bwindi Impenetrable Forest, Uganda. *Conservation Biology, 14*(6), 1722–1725.

Hannah, L., Midgley, G. F., Lovejoy, T., Bond, W. J., Bush, M., Lovett, J. C., Scott, D., & Woodward, F. I. (2002). Conservation of Biodiversity in a Changing Climate. *Conservation Biology, 16*, 264–268.

Hannah, L., Midgley, G. F., & Millar, D. (2002). Climate Change-Integrated Conservation Strategies. *Global Ecology and Biogeography, 11*(6), 485–495.

Hazard, B., & Adongo, C. (2012). *Natural Protected Areas and Dynamics of Pastoralist Heritage in Marsabit Greater Ecosystem (Northern Kenya)*: Proceedings. Intermediary report—IFRA Workshop (7–9 November 2012, Nairobi): Proceedings of IFRA Workshop (7–9 November 2012, Nairobi, Kenya). 2012. https://hal.archives-ouvertes.fr/hal-01688975

Heathwaite, A. L. (2010). Multiple Stressors on Water Availability at Global to Catchment Scales: Understanding Human Impact on Nutrient Cycles to Protect Water Quality and Water Availability in the Long Term. *Freshwater Biology, 55*, 241–257.

Homewood, K., Brockington, D., & Sullivan, S. (2010). Alternative View of Serengeti Road. *Nature, 467*(7317), 788–789.

Homewood, K., Lambin, E. F., Coast, E., Kariuki, A., Kikula, I., Kivelia, J., Said, M., Serneels, S., & Thompson, M. (2001). Long-Term Changes in Serengeti-Mara Wildebeest and Land Cover: Pastoralism, Population, or Policies? *Proceedings of the National Academy of Sciences of the United States of*

America [online], *98*(22), 12544–12549. https://doi.org/10.1073/pnas.221053998
Homewood, K., & Lewis, J. (1987). Impact of Drought on Pastoral Livestock in Baringo, Kenya 1983–85. *Journal of Applied Ecology, 24*(2), 615–631.
Homewood, K. M., Trench, P., & Brockington, D. (2012). Pastoralist Livelihoods and Wildlife Revenues in East Africa: A Case for Coexistence? *Pastoralism: Research Policy and Practice* [online], *2*(1), 19. https://doi.org/10.1186/2041-7136-2-19
Hopcraft, J. G. C., Mduma, S. A., Borner, M., Bigurube, G., Kijazi, A., Haydon, D. T., Wakilema, W., Rentsch, D., Sinclair, A. R. E., Dobson, A., & Lembeli, J. D. (2015). Conservation and Economic Benefits of a Road Around the Serengeti. *Conservation Biology, 29*(3), 932–936.
IPCC. (2021). 7th Assessment Report.
Jayne, T. S., Chapoto, A., Sitko, N., Nkonde, C., Muyanga, M., & Chamberlin, J. (2014). Is the Scramble for Land in Africa Foreclosing a Smallholder Agricultural Expansion Strategy. *Journal of International Affairs* [online], *67*, 35–53. https://www.jstor.org/stable/24461734 Accessed 13 June 2018.
Jeffers, E., Norgué, S., & Willis, K. (2015). The Role of Palaeoecological Records in Assessing Ecosystem Services. *Quaternary Science Reviews, 112*, 17–32.
Jolly, D., Harrison, S. P., Damnati, B., & Bonnefille, R. (1998). Simulated Climate and Biomes of Africa During the Late Quaternary: Comparisons with Pollen and Lake Status Data. *Quaternary Science Reviews, 17*, 629–657.
Jolly, D., & Haxeltine, A. (1997). Effect of Low Glacial Atmospheric CO_2 on Tropical African Montane Vegetation. *Science, 276*, 786–788.
Jung, M., Hill, S. L. L., Platts, P. J., Marchant, R., Siebert, S., Fournier, A., Munyekenye, F. B., Purvis, A., Burgess, N. D., & Newbold, T. (2017). Local Factors Mediate the Response of Biodiversity to Land Use on Two African Mountains. *Animal Conservation, 20*(4), 370–381.
Kabora, T. K., Stump, D., & Wainwright, J. (2020). How Did That Get There? Understanding Sediment Transport and Accumulation Rates in Agricultural Landscapes Using the ESTTraP Agent-Based Model. *Journal of Archaeological Science: Reports, 29*, 102115.
Kaplan, J. O., & Krumhardt, K. M. (2011). The KK10 Anthropogenic Land Cover Change Scenario for the Preindustrial Holocene, link to data in NetCDF format. *PANGAEA*. https://doi.org/10.1594/PANGAEA.871369

Kariuki, R. W., Munishi, L., Courtney Mustaphi, C. J., Capitani, C., Shoemaker, A., Lane, P., Marchant, R. (2021). Integrating Stakeholders' Perspectives and Spatial Modelling to Develop Scenarios of Future Land Use and Land Cover Change in Northern Tanzania. *PLoS ONE* [Open Access]. https://journals.plos.org/plosone/article?id=10.1371/journal.pone.0245516

Kariuki, R., Western, D., Willcock, S., & Marchant, R. (2021a). Exploring the Relationships Between Land Use and Ecosystem Services. *Land, 10*(1), 46. https://www.mdpi.com/journal/land/special_issues/landuse_ES

Kariuki, R. W., Western, D., Willcock, S., & Marchant, R. (2021b). Assessing interactions between agriculture, livestock grazing and wildlife conservation land uses: A historical example from East Africa. *Land, 10*(1), 46.

Kariuki, R., Willcock, S., & Marchant, R. (2018). Rangeland Livelihood Strategies under Varying Climate Regimes: Model Insights from Southern Kenya. *Land, 7*(2), 47. https://doi.org/10.3390/land7020047

Kay, A. U., & Kaplan, J. O. (2015). Human Subsistence and Land Use in Sub-Saharan Africa, 1000BC to AD1500: A Review, Quantification, and Classification. *Anthropocene, 9*, 14–32.

Kay, A. U., Phelps, L. N., & Kaplan, J. O. (2016). Mapping Human Subsistence in West Africa (1000 BC to AD 1500). *PAGES Magazine* [online], *24*(1), 38. https://doi.org/10.22498/pages.24.1

Keeso, A. (2014). *Big Data and Environmental Sustainability: A Conversation Starter*. Smith School of Enterprise and the Environment. Working Paper Series Number 14–04. University of Oxford.

Kenya Wildlife Conservancies Association. (2016). *State of wildlife conservancies in Kenya.*

Khan, Z. R., Midega, C. A., Pittachar, J. O., Murage, A. L. I. C. E., & Pickett, J. O. H. N. (2017). Climate-Smart Push-Pull: A Conservation Agriculture Technology for Food Security and Environmental Sustainability in Africa. *Conservation Agriculture for Africa: Building Resilient Farming Systems in a Changing Climate. BABI, Wallingford* (151–166).

Klein Goldewijk, K. (2017). Anthropogenic Land-Use Estimates for the Holocene; HYDE 3.2. DANS. https://doi.org/10.17026/dans-25g-gez3

Klein Goldewijk, K., Belsen, A., & Janssen, P. (2010). Long-Term Dynamic Modelling of Global Population and Built-Up Area in a Spatially Explicit Way: HYDE 3.1. *The Holocene, 20*(4), 565–573.

Klein Goldewijk, K., Beusen, A., van Drecht, G., & de Vos, M. (2011). The HYDE 3.1 Spatially Explicit Database of Human-Induced Global Land-Use Change over the Past 12,000 Years. *Global Ecology and Biogeography* [online], *20*, 73–86. https://doi.org/10.1111/j.1466-8238.2010.00587.x

Knorr, W., Schnitzler, K. G., & Govaerts, Y. (2001). The Role of Bright Desert Regions in Shaping North African Climate. *Geophysical Research Letters, 28*(18), 3489–3492.

Korotayev, A., & Zinkina, J. (2015). East Africa in the Malthusian Trap? *Journal of Developing Societies* [online], *31*, 385–420. https://doi.org/10.1177/0169796x15590322

Lal, R. (2004). Soil Carbon Sequestration Impacts on Global Climate Change and Food Security. *Science (New York, N.Y.)* [online], *304*(5677), 1623–1627. https://doi.org/10.1126/science.1097396

Lane, P. J. (2010). Developing Landscape Historical Ecologies in Eastern Africa: An Outline of Current Research and Potential Future Directions. *African Studies, 69*(2), 299–322.

Lane, P. J. (2015). Just How Long Does 'Long-Term' Have to Be? Matters of Temporal Scale as Impediments to Interdisciplinary Understanding in Historical Ecology. In C. Isendahl & D. Stump (Ed.), *The Oxford Handbook of Historical Ecology and Applied Archaeology.* Oxford University Press. https://doi.org/10.1093/oxfordhb/9780199672691.013.5

Lang, C., & Stump, D. (2017). Geoarchaeological Evidence for the Construction, Irrigation, Cultivation, and Resilience of 15th–18th Century AD Terraced Landscape at Engaruka, Tanzania. *Quat Res* [online], *88*, 382–399. https://doi.org/10.1017/qua.2017.54

Lankester, F., & Davis, A. (2016). Pastoralism and Wildlife: Historical and Current Perspectives in the East African Rangelands of Kenya and Tanzania. *Revue Scientifique Et Technique (international Office of Epizootics), 35*(2), 473–484.

Leakey, L. S. B. (1936). Preliminary Report on Examination of the Engaruka Ruins. *Tanganyika Notes and Records, 1*, 57–60.

Lemieux, A. M., & Clarke, R. V. (2009). The International Ban on Ivory Sales and Its Effects on Elephant Poaching in Africa. *The British Journal of Criminology, 49*(4), 451–471.

Li, B. L. (2002). A Theoretical Framework of Ecological Phase Transitions for Characterizing Tree-Grass Dynamics. *Acta Biotheoretica, 50*(3), 141–154.

López, R. (1997). *Where Development Can or Cannot Go: The Role of Poverty-Environment Linkages*. The World Bank.

Los, S. O., Street-Perrott, F. A., Loader, N. J., Froyd, C. A., Cuní-Sanchez, A., & Marchant, R. A. (2019). Sensitivity of a Tropical Montane Cloud Forest to Climate Change, Present, Past and Future: Mt. Marsabit N. Kenya. *Quaternary Science Reviews, 218*, 34–48.

Lunstrum, E. (2014). Green Militarization: Anti-poaching Efforts and the Spatial Contours of Kruger National Park. *Annals of the American Association of Geographers, 104*(4), 816–832.

Maiorano, L., Cheddadi, R., Zimmermann, N. E., Pellissier, B., Pottier, J., Laborde, H., Hurdu, B. I., Pearman, P. B., Psomas, A., Singarayer, J. S., Broennimann, O., Vittoz, P., Dubuis, A., Edwards, M. E., Binney, H. A., & Guisan, A. (2013). Building the Niche Through Time: Using 13,000 Years of Data to Predict the Effects of Climate Change on Three Tree Species in Europe. *Global Ecology and Biogeography, 22*, 302–317.

Maitima, J. M., Mugatha, S. M., Reid, R. S., Gachimbi, L. N., Majule, A., Lyaruu, H., Pomery, D., Mathai, S., & Mugisha, S. (2009). The Linkages Between Land Use Change, Land Degradation and Biodiversity Across East Africa. *African Journal of Environmental Science and Technology, 3*(10), 310–325.

Marchant, R. A. (2010). Understanding Complexity in Savannas: Climate, Biodiversity and People. *Current Opinion in Environmental Sustainability, 2*, 101–108.

Marchant, R. A., & Lane, P. (2014). Past Perspectives for the Future: Foundations for Sustainable Development in E. Africa. *Journal of Archaeological Science, 51*, 12–21.

Marchant, R., Richer, S., Boles, O., Capitani, C., Courtney-Mustaphi, C. J., Lane, P., Prendergast, M. E., Stump, D., de Cort, G., Kaplan, J. O., Phelps, L., Kay, A., Olago, D., Petek, N., Platts, P., Punwong, P., Widgren, M., Wynne-Jones, S., Ferro-Vázquez, C., … Wright, D. (2018). Drivers and Trajectories of Land Cover Change in East Africa: Human and environmental Interactions from 6000 Years Ago to Present. *Earth-Science Reviews, 178*, 322–378.

Marlon, J. R., Kelly, R., Daniau, A. L., Vannière, B., Power, M. J., Bartlein, P., Higuera, P., Blarquez, O., Brewer, S., Brücher, T., Feurdean, A., Gil-Romera, G., Iglesias, V., Maezumi, S. Y., Magi, B., Courtney Mustaphi, C. J., & Zhihai, T. (2016). Reconstructions of Biomass Burning from Sediment Charcoal Records to Improve Data-Model Comparisons. *Biogeosciences, 13*, 3225–3244.

McNicol, I. M., Ryan, C. M., & Mitchard, E. T. A. (2018), Carbon Losses from Deforestation and Widespread Degradation Offset by Extensive Growth in African Woodlands. *Nature Communications* [online], *9*(1), 3045. https://doi.org/10.1038/s41467-018-05386-z

Mgomezulu, G. G. (1981). Recent Archaeological Research and Radiocarbon Dates from Eastern Africa. *The Journal of African History, 22*(04), 435–456.

Midgley, G. F., & Bond, W. J. (2015). Future of African Terrestrial Biodiversity and Ecosystems Under Anthropogenic Climate Change. *Nature Climate Change, 5*(9), 823–829.

Morrison, K. D., Gaillard, M. J., Madella, M., Whitehouse, N., & Hammer, E. (2016). Land Use Classification Workshop: LandCover6k Workshop, Paris, France, 22–23 October 2015. *PAGES Magazine, 24*, 40.

Morrison, K. D., Hammer, E., Boles, O., Madella, M., Whitehouse, N., Gaillard, M. J., … Zanon, M. (2021). Mapping past human land use using archaeological data: A new classification for global land use synthesis and data harmonization. *PloS One, 16*(4), e0246662.

Morrison, T. A., Holdo, R., & Anderson, T. M. (2016). Elephant Damage, Not Fire or Rainfall, Explains Mortality of Overstory Trees in Serengeti. *Journal of Ecology, 104*, 409–418.

Mwangi, H. M., Lariu, P., Julich, S., Patil, S. D., McDonald, M. A., & Feger, K. H. (2018). Characterizing the Intensity and Dynamics of Land-Use Change in the Mara River Basin, East Africa. *Forests, 9*(1), 8.

Myers, N. (1972). National Parks in Savannah Africa: Ecological requirements of Parks must Be Balanced Against Socioeconomic Constraints in their Environs. *Science, 178*(4067), 1255–1263.

Naidoo, R. (2004). Species Richness and Community Composition of Song-Birds in a Tropical Forest-Agricultural Landscape. *Animal Conservation, 7*, 93–105.

Naidoo, D., Roux, F. S., Dudley, A., Litvin, I., Piccirillo, B., Marrucci, L., & Forbes, A. (2016). Controlled Generation of Higher-Order Poincaré Sphere Beams from a Laser. *Nature Photonics, 10*(5), 327–332.

Neumann, K. (2005). The Romance of Farming: Plant Cultivation and Domestication in Africa. In A. B. Stahl (Ed.), *African Archaeology: A Critical Introduction*. Blackwell Publishing.

Ng'ang'a, S. K., Jalang'o, D. A., & Girvetz, E. H. (2020). Adoption of Technologies that Enhance Soil Carbon Sequestration in East Africa. What Influence Farmers' Decision? *International Soil and Water Conservation Research, 8*, 90–101.

Niang, I., Ruppel, O. C., Abdrabo, M. A., Essel, A., Lennard, C., Padgham, J., & Urquhart, T. (2014). Africa. In V. R. Barros, C. B. Field, D. J. Dokken, M. D. Mastrandrea, K. J. Mach, T. E. Bilir, M. Chatterjee, K. L. Ebi, Y. O. Estrada, R. C. Genova, B. Girma, E. S. Kissel, A. N. Levy, S. MacCracken, P. R. Mastrandrea, & L. L. White (Eds.), *Climate Change 2014: Impacts, Adaptation, and Vulnerability. Part B: Regional Aspects.*

Contribution of Working Group II to the Fifth Assessment Report of the Intergovernmental Panel on Climate Change (pp. 1199–1265). Cambridge University Press.

Nicholas, G. P. (1998). Wetlands and Hunter-Gatherers: A Global Perspective. *Current Anthropology, 39*, 720–731.

Nieves, J. J., Stevens, F. R., Gaughan, A. E., Linard, C., Sorichetta, A., Hornby, G., & Tatem, A. J. (2017). Examining the Correlates and Drivers of Human Population Distributions Across Low- and Middle-Income Countries. *Journal of the Royal Society Interface, 14*, 0401–0419.

Noss, R. F. (1990). Indicators for Monitoring Biodiversity: A Hierarchical Approach. *Conservation Biology, 4*, 355–364.

Nyumba, T. O., Sang, C. C., Olago, D. O., Marchant, R., Waruingi, L., Githiora, Y., Kago, F., Mwangi, M., Owira, G., Barasa, R., & Omangi, S. (2021). Assessing the Ecological Impacts of Transportation Infrastructure Development: A Reconnaissance Study of the Standard Gauge Railway in Kenya. *PLoS One, 16*(1), e0246248.

Öberg, H., Norström, E., Ryner, M. M., Holmgren, K., Westerberg, L. O., Risberg, J., Eddudóttir, S. D., Andersen, T. J., & Muzuka, A. (2013). Environmental Variability in Northern Tanzania from AD 1000 to 1800, as Inferred from Diatoms and Pollen in Lake Duluti. *Palaeogeography, Palaeoclimatology, Palaeoecology, 374*, 230–241.

Oldekop, J. A., Holmes, G., Harris, W. E., & Evans, K. L. (2015). A Global Assessment of the Social and Conservation Outcomes of Protected Areas. Conservation Biology [online], *30*, 133–141. https://doi.org/10.1111/cobi.12568

Opdam, P. (2013). Using Ecosystem Services in Community-Based Landscape Planning: Science is Not Ready to Deliver. In B. Fu & K. B. Jones (Eds.), *Landscape Ecology for Sustainable Environment and Culture* (pp. 77–101). Springer.

Oteros-Rozas, E., Martín-López, B., González, J. A., Plieninger, T., López, C. A., & Montes, C. (2014). Socio-cultural Valuation of Ecosystem Services in a Transhumance Social-Ecological Network. *Regional Environmental Change, 14*(4), 1269–1289.

Pearce, F. (1999). Sungbo Erdo of Nigeria, a Neglected Past. *New Scientist*, September 11, pp. 39–41.

Pfeifer, M., Burgess, N. D., Swetnam, R. D., Platts, P. J., Willcock, S., & Marchant, R. (2012). Protected Areas: Mixed Success in Conserving East Africa's Evergreen Forests. *PLoS ONE, 7*(6), e39337.

Phelps, L. N., & Kaplan, J. O. (2017). Land Use for Animal Production in Global Change Studies: Defining and Characterizing a Framework. *Global Change Biology* [online]. https://doi.org/10.1111/gcb.13732

Phelps, L. N., Chevalier, M., Shanahan, T. M., Aleman, J. C., Courtney-Mustaphi, C., Kiahtipes, C. A., Broennimann, O., Marchant, R., Shekeine, J., Quick, L. J., & Davis, B. A. (2020). Asymmetric Response of Forest and Grassy Biomes to Climate Variability Across the African Humid Period: Influenced by Anthropogenic Disturbance? *Ecography, 43*(8), 1118–1142.

Pinedo-Vasquez, M., Pasqualle, J. B., Torres, D. D. C., & Coffey, K. (2002). A Tradition of Change: The Dynamic Relationship Between Biodiversity and Society in Sector Muyuy, Peru. *Environmental Science and Policy, 5*(1), 43–53.

Platts, P. J. (2012). *Spatial modelling, phytogeography and conservation in the Eastern Arc Mountains of Tanzania and Kenya* (Doctoral dissertation).

Platts, P. J., Burgess, N. D., Gereau, R. E., Lovett, J. C., Marshall, A. R., McClean, C. J., Pellikka, P. K. E., Swetnam, R. D., & Marchant, R. (2011). Delimiting Tropical Mountain Ecoregions for Conservation. *Environmental Conservation, 38*(3), 312–324.

Platts, P. J., Gereau, R. E., Burgess, N. D., & Marchant, R. (2013). Spatial Heterogeneity of Climate Change in an Afromontane Centre of Endemism. *Ecography, 36*(4), 518–530.

Platts, P. J., Omeny, P. A., & Marchant, R. (2015). AFRICLIM: High-Resolution Climate Projections for Ecological Applications in Africa. *African Journal of Ecology, 53*(1), 103–108.

Pretty, J., Toulmin, C., & Williams, S. (2011). Sustainable Intensification in African Agriculture. *International Journal of Agricultural Sustainability, 9*, 5–24.

Prins, A. H. J. (1971). *Didemic Lamu: Social Stratification and Spatial Structure in a Muslim Maritime Town.* Instituut voor Culturele Antropologie der Rijksuniversiteit.

Prins, H. H., & van der Jeugd, H. P. (1993). Herbivore Population Crashes and Woodland Structure in East Africa. *Journal of Ecology*, pp. 305–314.

Pullan, R. A. (1988). Conservation and the Development of National Parks in the Humid Tropics of Africa. *Journal of Biogeography, 15*(1), 171–183.

Rayner, S., & Malone, E. L. (1997). Zen and the Art of Climate Maintenance. *Nature, 390*, 332–334.

Reid, R. S., Nkedianye, D., Said, M. Y., Kaelo, D., Neselle, M., Makui, O., Onetu, L., Kiruswa, S., Kamuaro, N. O., Kristjanson, P., & Ogutu, J. (2016). Evolution of Models to Support Community and Policy Action with Science: Balancing Pastoral Livelihoods and Wildlife Conservation in Savannas of East Africa. *Proceedings of the National Academy of Sciences, 113*(17), 4579–4584.

Richardson, A. D., Keenan, T. F., Migliavacca, M., Ryu, Y., Sonnentag, O., & Toomey, M. (2013). Climate Change, Phenology, and Phenological Control of Vegetation Feedbacks to the Climate System. *Agricultural and Forest Meteorology, 169*, 156–173.

Richer, S., & Gearey, B. (2017a). Rackham to REVEALS: Reflections on Palaeoecological Approaches to Woodland and Trees. *Environmental Archaeology* [online]. https://doi.org/10.1080/14614103.2017.1283765

Richer, S., & Gearey, B. (2017b). The Medicine Tree: Unsettling Palaeoecological Perceptions of Past Environments and Human Activity. *Journal of Social Archaeology, 17*(3), 239–262.

Richer, S., Stump, D., & Marchant, R. (2019). Archaeology Has No Relevance. *Internet Archaeology, 53*. https://intarch.ac.uk/journal/issue53/2/toc.html

Roberts, P., Hunt, C., Arroyo-Kalin, M., Evans, D., & Boivin, N. (2017). The Deep Human Prehistory of Global Tropical Forests and Its Relevance for Modern Conservation. *Nature Plants, 3*(8), https://doi.org/10.1038/nplants.2017.93

Robertshaw, P. T. (1986). Engaruka Revisited: Excavations of 1982. *Azania, 21*, 1–20.

Rockström, J., Steffen, W., Noone, K., Persson, Å., Chapin, F. S., Lambin, E. F., Lenton, T. M., Scheffer, M., Folke, C., Schellnhuber, H. J., & Nykvist, B. (2009). A Safe Operating Space for Humanity. *Nature, 461*(7263), 472–475.

Romijn, E., Lantican, C. B., Herold, M., Lindquist, E., Ochieng, R., Wijaya, A., & Verchot, L. (2015). Assessing Change in National Forest Monitoring Capacities of 99 Tropical Countries. *Forest Ecology and Management, 352*, 109–123.

Rowell, D. P., Booth, B. B., Nicholson, S. E., & Good, P. (2015). Reconciling Past and Future Rainfall Trends over East Africa. *Journal of Climate, 28*(24), 9768–9788.

Sánchez Goñi, M. F., Desprat, S., Daniau, A. L., Bassinot, F. C., Polanco-Martínez, J. M., Harrison, S. P., Allen, J. R. M., Anderson, R. S., Behling, H., Bonnefille, R., Burjachs, F., Carrión, J. S., Cheddadi, R., Clark, J. S., Combourieu-Nebout, N., Courtney-Mustaphi, C. J., Debusk, G. H., Dupont, L. M., Finch, J. M., … Yamamoto, M. (2017). The ACER Pollen and Charcoal Database: A Global Resource to Document Vegetation and Fire Response to Abrupt Climate Changes During the Last Glacial Period.

Earth System Science Data Discussions [online]. https://doi.org/10.5194/essd-2017-4

Sassoon, H. (1967). New Views on Engaruka Tanzania: Excavations Carried out for the Tanzania Government in 1964 and 1966. *Journal of African History, 8*, 201–217.

Schmidt-Soltau, K., & Brockington, D. (2013). *The Social Impacts of Protected Areas*. IUCN Commission on Environmental, Economic and Social Policy, Sustainable Livelihood Working Group; 2004. Accessed July 11.

Schreckenberg, K., Franks, P., Martin, A., & Lang, B. (2016). Unpacking Equity for Protected Area Conservation. *Parks, 22*(2), 11–26.

Shanahan, T. M., McKay, N. P., Hughen, K. A., Overpeck, J. T., Otto-Bliesner, B., Heil, C. W., King, J., Scholz, C. A., & Peck, J. (2015). The time-transgressive termination of the African Humid Period. *Nature Geoscience, 8*(2), 140–144.

Shaw, K., Ruyonga, G., & Nicholson, M. (2016). Enhancing Tree Conservation and Forest Restoration in East Africa. *Bgjournal, 13*(2), 28–31.

Sinclair, A. R. E., Mduma, S. A. R., & Arcese, P. (2002). Protected Areas as Biodiversity Benchmarks for Human Impact: Agriculture and the Serengeti Avifauna. *Proceedings of the Royal Society of London B, 269*, 2401–2405.

Smith N. M. (2014). Gender and Livelihood Diversification: Maasai Women's Market Activities in Northern Tanzania. *Journal of Development Studies* [online], *51*(3), 1–14. https://doi.org/10.1080/00220388.2014.957278

Sommerville, K. (2016). *Ivory: Power and Poaching in Africa*. Hurst Publishers.

Soulé, M. E., Wilcox, B. A., & Clarke, H. (1979). Benign Neglect: A Model of Faunal Collapse in the Game Reerves of East Africa. *Biological Conservation, 15*, 259–272.

Ssemmanda, I., Ryves, D. B., Bennike, O., & Appleby, P. G. (2005). Vegetation History in Western Uganda During the Last 1200 Years: A Sediment Based Reconstruction from Two Crater Lakes. *The Holocene, 15*(1), 119–132.

Struhsaker, T. T. (1981). Forest and Primate Conservation in East Africa. *African Journal of Ecology, 19*(1–2), 99–114.

Stump, D. (2006). The Development and Expansion of the Field and Irrigation System at Engaruka, Tanzania. *Azania, 41*, 69–94.

Stump, D. (2010). Ancient and Backward or Long-Lived and Sustainable? The Role of the Past in Debates Concerning Rural Livelihoods and Resource Conservation in Eastern Africa. *World Development* [online], *38*, 1251–1262. https://doi.org/10.1016/j.worlddev.2010.02.007

Sutton, J. E. G. (1978). Engaruka and Its Waters. *Azania, 13*, 37–70.

Sutton, J. E. G. (2004). Engaruka: The Success and Abandonment of an Integrated Irrigation System in an Arid part of the Rift Valley, c. 15th to 17th Centuries. In M. Widgren & J. E. G. Sutton (Eds.), *Islands of Intensive Agriculture in Eastern Africa* (pp. 114–132). James Currey.

Svenning, J. C., & Sandel, B. (2013). Disequilibrium Vegetation Dynamics under Future Climate Change. *American Journal of Botany, 100*(7), 1266–1286.

Swallow, B. M., Sang, J. K., Nyabenge, M., Bundotich, D. K., Duraiappah, A. K., & Yatich, T. B. (2009). Tradeoffs, Synergies and Traps among Ecosystem Services in the Lake Victoria Basin of East Africa. *Environmental Science and Policy, 12*(4), 504–519.

Tallis, J. H. (1991). *Plant Community History: Long-Term Changes in Plant Distribution and Diversity*. Chapman and Hall.

Terborgh, J., van Schaik, C., Davenport, L., & Rao, M. (Eds.). (2002). *Making Parks Work: Strategies for Preserving Tropical Nature*. Island Press.

Thenya, T. (2001). Challenges of Conservation of Dryland Shallow Waters, Ewaso Narok Swamp, Laikipia District, Kenya. *Hydrobiologia, 458*(1), 107–119.

Thorn, J. P. R., Aleu, R. B., Wijesinghe, A., Mdongwe, M., Marchant, R. A., & Shackleton, S. (2021). Mainstreaming nature-based solutions for climate resilient infrastructure in peri-urban sub-Saharan Africa. *Landscape and Urban Planning, 216*, 104235.

Thornton, P. K., Boone, R. B., Galvin, K. A., BurnSilver, S. B., Waithaka, M. M., Kuyiah, J., Karanja, S., González-Estrada, E., & Herrero, M. (2007). Coping Strategies in Livestock-Dependent Households in East and Southern Africa: A Synthesis of Four Case Studies. *Human Ecology, 35*(4), 461–476.

Trewartha, G. T., & Zelinsky, W. (1954). Population Patterns in Tropical Africa. *Annals of the Association of American Geographers, 44*(2), 135–162.

Tyrrell, P., Russell, S., & Western, D. (2017). Seasonal Movements of Wildlife and Livestock in a Heterogenous Pastoral Landscape: Implications for Coexistence and Community Based Conservation. *Global Ecology and Conservation, 12*, 59–72.

UNDESA. (2017) *Population Division—World Population Prospects*. Retrieved from https://www.population.un.org/wpp. Accessed 08 Apr 2018.

Vörösmarty, C. J., Green, P., Salisbury, J., & Lammers, R. B. (2000). Global Water Resources: Vulnerability from Climate Change and Population Growth. *Science, 289*, 284–288.

Warinda, E., Nyariki, D. M., Wambua, S., Muasya, R. M., & Hanjra, M. A. (2020, February). Sustainable Development in East Africa: Impact

Evaluation of Regional Agricultural Development Projects in Burundi, Kenya, Rwanda, Tanzania, and Uganda. In *Natural Resources Forum, 44*(1), pp. 3–39). Blackwell Publishing Ltd.
Wei, F., Wang, S., Fu, B., Pan, N., Feng, X., Zhao, W., & Wang, C. (2018). Vegetation Dynamic Trends and the Main Drivers Detected Using the Ensemble Empirical Mode Decomposition Method in East Africa. *Land Degradation and Development, 29*(8), 2542–2553.
Wei, F., Wang, S., Fu, B., Zhang, L., Fu, C., & Kanga, E. M. (2018). Balancing Community Livelihoods and Biodiversity Conservation of Protected Areas in East Africa. *Current Opinion in Environmental Sustainability, 33*, 26–33.
Westerberg, L. O., Holmgren, K., Börjeson, L., Håkansson, N. T., Laulumaa, V., Ryner, M., & Öberg, H. (2010). The Development of the Ancient Irrigation System at Engaruka, Northern Tanzania: Physical and Societal Factors. *The Geographical Journal, 176*, 304–318.
Western, D. (2008). *Ecotourism, Conservation and Development in East Africa: How the Philanthropic Traveller can make a Difference*. Stanford University.
Western, D., Groom, R., & Worden, J. (2009). The Impact of Subdivision and Sedentarization of Pastoral Lands on Wildlife in an African Savanna Ecosystem. *Biological Conservation* [online], *142*, 2538–2546. https://doi.org/10.1016/j.biocon.2009.05.025
Western, D., & Manzolillo Nightingale, D. L. (2003). *Environmental Change and the Vulnerability of Pastoralists to Drought: A Case Study of the Maasai in Amboseli, Kenya*. Africa Environmental Outlook Report. UNEP.
Western, D., Mose, V., Worden, J., & Maitumo, D. (2015). Predicting Extreme Droughts in Savannah Africa: A Comparison of Proxy and Direct Measures in Detecting Biomass Fluctuations, Trends and their Causes. *PLoS One* [online], 10. https://doi.org/10.1371/journal.pone.0136516
Williams, P., Gibbons, D., Margules, C., Rebelo, A., Humphries, C., & Pressey, R. (1996). A Comparison of Richness Hotspots, Rarity Hotspots, and Complementary Areas for Conserving Diversity of British Birds. *Conservation Biology, 10*, 155–174.
Willis, K. J., Gillson, L., & Brncic, T. M. (2004). How "Virgin" Is Virgin Rainforest? *Science, 304*, 402–403.
Woodcock, C. E., Loveland, T. R., Herold, M., & Bauer, M. E. (2020). Transitioning from Change Detection to Monitoring with Remote Sensing: A Paradigm Shift. *Remote Sensing of Environment, 238*, 111558.
Wynants, M., Kelly, C., Mtei, K., Munishi, L., Patrick, A., Rabinovich, A., Nasseri, M., Gilvear, D., Roberts, N., Boeckx, P., Wilson, G., Blake, W. H., & Ndakidemi, P. (2019). Drivers of Increased Soil Erosion in

East Africa's Agro-Pastoral Systems: Changing Interactions Between the Social, Economic and Natural Domains. *Regional Environmental Change, 19*, 1909–1921.

Wynne-Jones, S. (2016). *A Material Culture: Consumption and Materiality on the Coast of Precolonial East Africa*. Oxford University Press.

References

Beymer-Farris, B. A., & Bassett, T. J. (2012). The REDD Menace: Resurgent Protectionism in Tanzania's Mangrove Forests. *Global Environmental Change, 22*(2), 332–341.

Burgess, N. D., Clarke, G. P., & Rodgers, W. A. (1998). Coastal Forests of Eastern Africa: Status, Endemism Patterns and Their Potential Causes. *Biological Journal of the Linnean Society, 64*(3), 337–367.

Chrtiz, K. L., Cerling, T. E., Freeman, K. E., Hildebrand, E. A., Janzen, A., & Prendergast, M. E. (2019). Climate, Ecology and the Spreading of Herding in Eastern Africa. *Quaternary Science Reviews, 204*, 119–132.

Cuní-Sanchez, A. (in press). *Nature.*

de Menocal, P., Ortiz, J., Guilderson, T., Adkins, J., Sarnthein, M., Baker, L., & Yarusinsky, M. (2000). Abrupt Onset and Termination of the African Humid Period: Rapid Climate Responses to Gradual Insolation Forcing. *Quaternary Science Reviews, 19*, 347–361.

Eggling, W. J., & Dale, I. R. (1951). *The Indigenous Trees of the Uganda Protectorate* (2nd ed.). Government Printer.

Fairhead, J., & Leach, M. (1995). False Forest History, Complicit Social Analysis: Rethinking Some West African Environmental Narratives. *World Development, 23*(6), 1023–1035.

R. Marchant, *East Africa's Human Environment interactions*,
https://doi.org/10.1007/978-3-030-88987-6

Gaillard, M. J. (2015). LandCover6k: Global Anthropogenic Land-Cover Change and Its Role in Past Climate. *PAGES News, 23*(1), 38–39.

Karuku, G. N., Gachene, C. K. K., Karanja, N., Cornelis, W., Verplancke, H., & Kironchi, G. (2012). Soil Hydraulic Properties of a Nitisol in Kabete, Kenya. *Tropical and Subtropical Agroecosystems, 15*, 595–609.

Langdale-Brown, I., Osmaston, H. A., & Wilson, J. G. (1964). *The Vegetation of Uganda and its Bearing on Land-Use*. Govt Uganda.

Leakey, L. S. B., & Owen, W. E. (1945). *A Contribution to the Study of the Tumbian Vulture in East Africa* (Vol. 1). Coryndon Memorial Museum.

Marchant, R., Boom, A., Behling, H., Hooghiemstra, H., Melief, B., Geel, B. V., Van der Hammen, T., & Wille, M. (2004). Colombian Vegetation at the Last Glacial Maximum: A Comparison of Model-and Pollen-Based Biome Reconstructions. *Journal of Quaternary Science, 19*(7), 721–732.

Miller, J. M., & Willoughby, P. R. (2014). Radiometrically Dated Ostrich Eggshell Beads from the Middle and Later Stone Age of Magubike Rock-shelter, Southern Tanzania. *Journal of Human Evolution, 74*, 118.

Muiruri, V., Marchant, R., Rucina, S., Scott, L., & Lane, P. (2021). Late Holocene Environmental Change and Anthropogenic: Ecosystem Interaction on the Laikipia Plateau, Kenya. *Ambio*. https://doi.org/10.1007/s13280-021-01554-6

Schlebusch, C. M., Malmström, H., Günther, T., Sjödin, P., Coutinho, A., Edlund, H., Munters, A. R., Vicente, M., Steyn, M., Soodyall, H., & Lombard, M. (2017). Southern African Ancient Genomes Estimate Modern Human Divergence to 350,000 to 260,000 Years Ago. *Science, 358*(6363), 652–655.

Spear, T. (1981). *Kenya's Past: An Introduction to Historical Method in Africa*. Longman.

Thorn, J., Klein, J., Steger, C., Hopping, K., Capitani, C., Tucker, C., Nolin, A., Reid, R., Seidl, R., Chitale, V., & Marchant, R. (2021). A Systematic Review of Participatory Scenario Planning to Envision Mountain Social-Ecological Systems Futures. *Ecology and Society, 25*(3), 26. https://www.ecologyandsociety.org/vol25/iss3/art6/

Willcock, S., Phillips, O. L., Platts, P. J., Swetnam, R. D., Balmford, A., Burgess, N. D., Ahrends, A., Bayliss, J., Doggart, N., Doody, K., & Fanning, E. (2016). Land Cover Change and Carbon Emissions over 100 years in an African Biodiversity Hotspot. *Global Change Biology, 22*(8), 2787–2800.

WoldeGabriel, G., Olago, D., Dindi, E., & Owor, M. (2016). Genesis of the East African Rift System. In M. Schagerl (Ed.), *Soda Lakes of East Africa* (pp. 25–59). Cham: Springer International.

Index

A

Aberdare Mountains 11
Abolition of the slave trade 183
The Abushiri revolt 211
Acclimatisation 58
Acholi country 174
Acidification 335
Acrisols 14
Adaptation 55, 56, 58, 72, 74, 131, 220, 321, 329, 331, 333, 348, 361, 365
Adiabatic lapse rate 6, 7
Advecting clouds 6
AFR100 350
African cereal crops 124
The African pencil cedar 232
African Pollen Database 317
African socialism 245, 252, 291
African Union Development Agenda 332
Afroalpine 20, 22
Afromontane 20
Afromontane Desert 22
Agricultural expansion 82, 272, 333, 334, 349, 357, 365
Agriculturalists 25, 76, 77, 143, 163, 189
Agriculture 18, 23–25, 45, 46, 60, 74, 76, 79, 82–84, 123, 127, 129, 130, 137, 143–145, 184, 186, 188–191, 215, 219, 220, 223, 224, 228, 231, 232, 258, 261, 268–273, 296, 333–335, 337, 340, 343, 349, 352–355, 357, 364
Agriculture policy 337
Agroforestry 23, 46, 269, 280, 331
Agronomic initiatives 336
Ahakagyezi 45
Ahakagyezi swamp 47, 63

R. Marchant, *East Africa's Human Environment interactions*, https://doi.org/10.1007/978-3-030-88987-6

Aichi Targets 358
Akamba tribesmen 166
Albedo 324, 344
The albedo effect 324
Alien plants 272–274
Allah 183
Alliances 71, 77, 182, 190, 217, 251
Alluvial deposits 12
Al Masudi 146, 164
Aloe 231, 263
Aluminium 12
Ambient temperature 7
Amboseli 173, 225, 236, 271
The American War for Independence 182
Amin, Idi 252
Ammunition 170
Andosols 12
The Anglo-German agreements 170
Animal bone middens 49
Animal husbandry 78, 135, 223
Ankole 215, 217, 225
Anthrax 196
Anthropocene 60, 357
Anthropogenic destruction 17
Anthropogenic modification 45, 60
Anthropology 132, 192, 347
Apollo Kagwa 217
Apollo Milton Obote 245, 251, 252, 254
Aquaculture 269
Aquatic fauna 63
Arab sea power 146
Arabs of Oman 147
Arabuko Sokoke forest 18, 232, 266
Arabuko Sokoke Forest Adjacent Dwellers Association 263
Archaeological data 39, 124, 127, 129
Archaeology 49, 63, 313, 315
Archaeometallurgical research 83
Arenosols 14
Arid and Semi-Arid Lands projects 285
Aridification 62, 117
Aridity 14, 61, 68, 69, 75, 76, 117, 119
Arid lowlands 1, 6
Arms 5, 170, 198
Army 211, 217, 246, 251
Arrow poison 189
Arrows 189
Artefacts 49, 68–70, 77, 141
Arusha 169, 206, 277
Arusha Declaration 252, 277
Asian ivory 165
Asian rice 50, 144
Asian yam 50
Atmosphere 6, 322, 324
Avocados 337, 355

B

Bagamoyo 141, 169, 180, 288, 291, 293
The Baganda 25
Bambara groundnut 78
Bananas 127, 129
Bantu-Arab traders 25
Bantu languages 77
Barghash 183
Barley 76
Basalt 12
The BaTwa pygmies populations 228
Bayesian statistics 51, 52
Beads 56, 141, 142, 170–172, 185, 189, 190
Beans 125, 126

Beeswax 146
Before present (yr BP) 50, 51, 55, 56, 62–64, 67–84, 115, 117–129, 132, 133, 138, 143, 144, 326
The Berlin Act 170
Bhutan cypress 231
Big 5 278, 280, 356
Billiard balls 178
Biodiversity 18, 57, 60, 206, 260, 264, 272–274, 286, 287, 323, 344, 349–351, 354, 356, 358, 361, 364
Biodiversity conservation 59, 287, 290, 343, 348, 350, 356, 358, 359, 362
Biodiversity hotspot 18
Biofuel production 341
Biogenic silica record 117
Biogeochemical cycling 40, 312
Biological diversity 17
Biological hotspots 58
Biological productivity 41
Biological systems 55
Biomass 24, 80, 118, 266, 279, 351
Biome 16
Biosecurity 338
Biotic distributions 57
Bird species 278
Bismarck 209–211
Black feminism 250
Black tea 224
Black wattle 231, 274
Blood brotherhoods 190
Boma 265
Bone apatite 72
Bone points 63, 68
Boni Dodori 18
Boom and bust 183
The Borana 196
Boran bulls 137
Botanical 50, 78
Bovine pneumonia 137
Branched to Isoprenoid Tetraether (BIT) 76
Brass 171
Brass wire 172, 187
Britain 172, 210
British Anglican Church 216
The British East Africa Company 170, 197, 210, 214, 217
The British Foreign Office 197
The British South Africa Company 210
The British Treasury 214
Bronze 171
The Brussels Act 170
Brussels antislavery conference 183
Buffalo 73, 135, 196, 272, 281, 356
Buganda 121, 124, 170, 174, 208, 210, 215–218, 225, 234, 251
Bugandan royalty 251
Bukaleba 267
Bullrush millet 77
Bunyoro 124, 171, 173, 174, 225, 251
Buran 25
Burial sites 226
Burning 19, 24, 45, 80, 117, 118, 121, 123, 223, 324, 340
Burton, R.F. 119, 169, 207, 209
Bush 121, 135, 169, 194, 286
Bush fires 234, 269
Buttons 176
Bwindi-Impenetrable Forest 23, 84
Bwindi Impenetrable Forest National Park (BINP) 283

C
Calcisols 12
Cambisols 14
Camel 26, 73
Canals 131, 132
Cape Guardafui 164
Capitalism 245
Captain Johnson 171
Caravan routes 143, 164, 179, 186, 187, 190, 191, 213
Caravan trade 50, 127, 129, 133, 176, 185–189, 191, 195, 196, 205, 214, 215, 228, 235
Carbon dioxide 55, 324
Carbonised sorghum 123
Carbon offset forestry 267, 350
Carbon storage 323, 328, 347, 354
Carnivore 73, 286
Carved figures 176
Casein 176
Cash crops 217, 219, 248, 254, 274
Cashews 224
Castro, A.P. 227, 234
Catchment conditions 44
Cattle 25, 26, 71, 73, 79, 122–126, 132–135, 187–190, 196, 205, 212, 219, 221–223, 228, 286, 289, 337
Cattle fodder 7
Cattle raiding 26, 197
Cattle-rustlers 137
Cave deposits 41
Cecil Rhodes 210
Central African rebel groups 290
Ceramic 49, 63, 64, 69, 72, 74, 78, 80, 83, 121, 125, 127, 141, 144, 170
Ceramic scatters 49
Cereal grains 77
The Chagga 132
Charcoal 45, 47, 48, 51, 80, 120, 121, 283, 290
Cherangani Hills 138, 267
Chessmen 176
Chew Bahir 196
Chicken 125, 140, 143, 146
Chiefdoms 190
Chiefs 210, 215, 217, 233
China 141, 142, 147, 163, 164, 176, 214, 291–293, 296, 297, 311
The China International Telecommunication Construction Corporation 293
China's Belt and Road Initiative (BRI) 291
Christian 166, 216
Church 184, 209
Chyulu 212
Circumcision 226
City 139, 142, 288, 293, 294
Clan 125, 225, 226, 233, 234
Climate change 8, 40, 83, 206, 266–268, 286, 313, 321, 322, 326, 329, 331, 333, 337, 340, 341, 347, 351–354, 361–365
Climatic research 315
Climatic stress 81
Cloth 140, 146, 147, 170, 185, 187, 189, 190, 228
Cloud forests 352, 353
Clove cultivation 182
Clove mania 182
Clove plantations 173, 183
Cloves 173, 183, 186, 231
Coastal deposits 41
Coastal hinterland 64, 140, 147, 189

Coastal mangroves 43
Coastal rainforest 17
Coastal Swahili 24
Coastal tourism 287
Coconuts 17, 144, 173, 182, 184, 231
Coeval deforestation 82
Coffee 12, 76, 185, 213, 219, 224, 248, 254, 331, 336
Colonel Rigby 177
Colonialism 186
Colonisation 77, 210, 220, 225
Co-management regimes 357
Combs 175
Common Era (CE) 51, 118, 142, 146, 184, 205, 206, 211, 214–218, 221, 222, 224, 326
Communities 16, 18, 22, 25, 45, 55, 63, 66, 68, 70, 72–74, 77, 81, 83, 84, 121, 124–127, 133, 135, 136, 138, 141, 143, 146, 170, 187–191, 195, 206, 212, 215, 221, 226, 233, 237, 247, 249, 251, 254–259, 263, 264, 267, 268, 275, 276, 280, 282–286, 288, 296, 311, 322, 331, 336, 339–341, 343, 346, 347, 349, 351, 353–358, 362–364
Community Based Forest Management (CBFM) 264
Community Conservation Banks (COCOBA) 285
Compost 336
Condensation 7
Congo Air Boundary (CAB) 5
Congo air mass 5
Congo Basin 1, 3, 5, 209, 268, 292
Conifers 20, 232
Conservancies 281, 284, 359, 363
Conservation 18, 40, 186, 195, 225, 226, 229, 234–237, 250, 258–261, 263, 265–269, 273, 275, 277, 280, 282–286, 290, 294, 296, 312, 319, 328, 343, 344, 347, 348, 350, 351, 356–359, 361–364, 366
Contamination 221, 344
Convention on International Trade in Endangered Species (CITES) 289, 290
Convention on International Trade in Endangered Species of Wild Flora and Fauna (CITES) 264
Convergence 3, 5
Cooking pots 170
Copal 163, 171, 172, 184
Copper 144, 170, 171
Coral reef 18, 287
Corals 17, 41, 43, 143
Corruption 198, 245, 249–252, 255, 257, 258, 263, 289
Cosmopolitanism 246
COVID-19 pandemic 288, 331
Cowpea 50, 78, 140
Crater Highlands 121, 132
Crimean War 170
Crop damage 73, 284, 286
Crop growth 39
Crops 46, 52, 75, 123, 127, 130, 133, 144, 186, 191, 223, 226, 269, 274, 319, 322, 335, 338, 340, 341, 346, 355
Cross border biosafety 337
Crucifixes 175
Crude oil 294
Cuisine 147
Cultivated crops 46, 50

Cultivation 11, 46, 75, 121, 123, 124, 129, 133, 135, 137, 138, 145, 186, 223–225, 227, 228, 232, 234, 263, 269–271, 273, 274, 277, 280, 339
Cultural identity 24
Cultural traditions 50, 136, 226
Cultures 24–26, 40, 49, 72, 78, 142, 144, 146, 147, 183, 208, 228, 296, 315, 320, 346, 366
Currents 2, 40, 52, 55, 60, 66, 117, 125, 138, 141, 164, 169, 191, 197, 214, 234, 290, 296, 313, 321, 323, 324, 331, 336, 340, 346, 362–364
Customs office 184
Cutlery 166, 175
Cypress 231, 232

D

Daggers 170
Dar es Salaam vi, 164, 169, 171, 172, 288, 291, 293, 296, 344
Dating methods 51
Debarking 19, 266
Deep River 166
Deforestation 11, 13, 23, 82, 223, 258, 259, 267, 273, 280, 334, 351, 352
Deglaciated valleys and tarns 43
Deltas 63
Dendrochronology 43
Depopulation 126, 179
Deposition 9, 252
The Development through Conservation and Mountain Gorilla Project 282
Diatom 45, 62, 67, 75
Dieppe 175
Digging tools 189
Digo 189
Diseases 135, 137, 179, 187, 190, 194, 196, 205, 221, 247, 248, 286, 327, 334, 338
Dispersed subsistence agriculture 24
Displacement 228, 256, 267, 296
Disturbance regimes 16, 18, 23, 328
Diversification 68, 338, 343, 364
Domestic 26, 73, 124, 134, 140, 146, 195, 225, 233, 319
Domestication 45, 50, 71, 77, 79, 143, 185, 286
Donkey 26, 170
Dorobo 228
Downing Street 215
Dress 147
Drought 19, 24, 40, 48, 67, 76, 81, 83, 117–119, 126, 129, 133, 134, 136, 195–197, 205, 206, 212, 222, 260, 271, 275, 286, 289, 334, 337, 338, 340, 341, 353, 364, 365
Dr Stuhlmann, Franz 229
Drystone walling 126
Dung 49, 50
Dung fungal spores 48

E

Early Iron Age 52, 78, 125
Earth System models 325, 326
Earthworks 123
East Africa 1, 3–9, 11, 12, 14–18, 21–26, 39–41, 43–45, 48–51, 54, 55, 57, 61–64, 68, 69, 72, 74–77, 79, 80, 82, 84, 118, 121, 123, 130, 131, 134,

136–143, 145–147, 163–166, 169, 171, 176–180, 182, 184, 185, 189–191, 196, 198, 206–210, 214, 218, 219, 221, 224, 225, 229, 236, 245, 257–261, 263, 266–268, 276, 277, 279, 280, 287–291, 293, 294, 296, 297, 311, 313, 315, 317, 319, 321–324, 334, 337, 339–342, 344–352, 354–356, 359, 361, 364, 366
East African ivory 165, 171
East African ivory trade 164
East coast fever 137
Eastern Arc mountains (EAM) 9, 21, 45, 128, 261
Eastern Kenya and Tanzania 12
Eastern Rift Valley 5, 9, 11, 74, 126
Ecological challenges 344
Ecological driver 23
Ecological refugia 58
Ecology 73, 83, 188, 191, 192, 223, 238, 283, 313, 315, 320, 321, 337, 341, 346
Economic change 339
Economic diversification 83, 124
Economy 75, 125, 126, 132, 135, 140, 173, 186, 187, 191, 196, 208, 213, 250, 254, 260, 287, 318, 334, 340, 355, 356
Ecosystem 7, 11–13, 15–17, 20, 23, 24, 40, 41, 44, 46, 48, 53–55, 57, 58, 60, 67, 80, 117, 163, 164, 178, 179, 189–191, 194, 195, 206, 207, 228, 232, 250, 258, 259, 264, 266, 272, 283, 286, 287, 313, 322–324, 326–328, 331, 333, 335, 336, 341, 343, 347–350, 352, 354, 355, 357, 358, 361–363, 365, 366
Ecosystem and climate modeling 40, 312
Ecosystem engineers 179, 192, 238, 324, 328, 362
Ecosystem functioning 319
Ecosystem variables 41
Edaphic controls 12, 18
Edaphic savannahs 19
Education 234, 255, 258, 263, 282, 296, 317, 338, 339, 363
Elders 215
Elementeita 119
Elephant 19, 73, 164–169, 173, 174, 176–179, 186, 190–195, 197, 198, 266, 272, 281, 286, 288–290, 294, 313, 324, 327, 328, 331, 356, 362
Elephant hunting 123, 197, 198
The Elephant Trade Information System (ETIS) 290
Elephant tusks 164, 166
Eleusine 126
Elmenteitan sites 72
El Nino southern oscillation (ENSO) 8
Endemic plant and animal species 17
Engaruka 132, 133, 171, 314, 354
England 175, 177
Enslavement 124
Environment 21, 22, 40, 41, 48, 53–56, 67, 71, 79, 82, 115, 117, 129, 130, 163, 179, 186, 188, 193, 194, 197, 207, 212, 219, 227–229, 234, 250, 254, 258, 267, 268, 271, 275, 311, 313, 315, 319, 321, 325, 328,

331, 335, 341, 343, 347, 352, 353, 355, 363, 366
Environmental change 9, 39, 41, 45, 54, 58, 76, 81, 83, 84, 264, 312, 319, 321, 322, 356, 365
Environmental crime 264, 290
Environmental-cultural dynamics 40
Environmental history 41, 322, 347
Ericaceous belt 20, 22
Ericaceous scrub 17
Erosion 9, 41, 51, 62, 82, 120, 121, 130, 186, 219, 258, 260, 271, 275, 334, 340, 354, 355
Escapees 274
Ethnic 24, 25, 71, 83, 126, 137, 190, 191, 197, 205, 207, 221, 246, 249, 251–256
Ethnoarchaeological 73
Ethnohistorical information 124
Europe 142, 163, 165, 166, 171, 173, 175, 182, 183, 198, 291
Evaporation 14, 55, 76
Evapotranspiration 12
Evolution 26, 53, 54, 208, 319
Ewaso Basin 71
Exclusion 225, 231, 275, 355
Exotic species invasion 260
Exploitation 18, 64, 140, 163, 166, 167, 178, 227, 228, 232, 234, 260, 261, 266, 269, 283
Exploration 139
Export products 173
External trade 124

F

False teeth 176
Famine 119, 129, 190, 196, 197, 206, 219, 340
Far East 144, 198
Farming communities 127, 130, 315
Farming migrants 77
Fauna 54, 66, 78, 140, 187, 235, 361
Ferralsols 12–14
Fertile Nitisols 12, 14
Finger millet 47, 50, 74, 77, 79, 127, 133, 140
Fire 16, 18, 19, 23, 24, 39, 40, 45, 47, 48, 67, 73, 80, 119, 120, 194, 227, 232, 260, 327
Fire regimes 48
Fischers turaco 17
Fish 68, 78, 126, 140, 141, 287
Fisheries Act 287
Fishing 63, 64, 68, 70, 72, 78, 125, 146, 187, 287, 288, 293
Flintlock 170
Flora 22, 66, 68, 361
Fluvisols 12, 14
Fog 7, 58, 352
Food producers 60, 70, 76–79
Food security 271, 275, 336, 338, 343, 344, 347, 355
Foragers 60, 63, 64, 68, 71, 73, 78, 135, 143
Forest boundaries 16
Forest clearance 45, 48, 79, 82, 84, 184, 185, 232, 258, 349
Forest dwelling militants 84
Forest gardens 46
Forest Governance, Law Enforcement, and Trade (FGLET) 264
Forest harvesting 8
Forest policy 234
Forest refugia 58–60

Forest reserves 17, 229, 231, 233–235, 258, 259, 261, 262, 264, 361
Forests Act 264
Forest tree taxa 46
Fort Ternan 212
Forum on China-Africa Cooperation (FOCAC) 291
Fossil fungal spores 48
Fossil ostracod assemblages 75
Fragmentation 24, 58, 260, 285, 330, 344
Frankfurt Zoological Society (FZS) 285
French West Indies 182
Furniture inlays 166
Furrow based irrigation systems 132
Furrows 131–133, 337
Futures 24, 40, 53, 55, 57–60, 83, 195, 226, 229, 263, 270, 280, 286, 291, 293, 297, 311–313, 315, 319, 321, 322, 324, 326–333, 335, 336, 341, 346–349, 352, 353, 356, 357, 361, 362, 364–366

G

The Gabbra 70, 196
Game licences 197
Game regulations 174, 197, 198
Game reserves 197, 198, 235–237, 286, 361
General Climate Models (GCMs) 322
Geology 9, 12, 17
Geomorphology 43
German East Africa 178, 198, 210, 218
Giriama 189, 197
Glacial deposits 43
Glacial to interglacial oscillations 54
Glass beads 141, 144, 146, 147, 171, 172, 187
Global biodiversity hotspots 17, 261
Global warming 353
Goat 26, 125, 221, 286, 324
Gogo Falls 71, 125
Gold 140–143, 146, 147, 164, 170, 171, 179
Golden pine 231
Gorilla 261, 283, 290
Got Ramogi Hill 125
Government 171, 210, 211, 215, 217, 219, 221–223, 228, 230–234, 246–255, 257, 260, 269, 274–276, 284, 287, 293–296, 343, 352, 357, 363, 364
Grain plantations 183
Grassland 14, 16–18, 23, 62, 64, 72, 74, 80, 120, 132, 138, 179, 186, 193, 194, 233, 235, 324, 327, 328, 335, 362
Grass pollen 46, 48
Grassroots 83
Grazing optimization theory 19
Grazing regimes 23
Great Rift Valley 4
Greek 139, 141
The Greenbelt Movement 248–251
Green gram 126
Green manure 336
Green militarisation 290
Grinding stones 123, 126
Gross domestic product (GDP) 268
Guinea fowl 147
Guineo-Congolian 20

Gulf of Aden 164
Gunpowder 170
Guns 170, 217

H

Habitat 18, 19, 58, 169, 190, 192, 195, 280, 320, 321, 330, 350, 352, 357, 361, 365
Hadza 25
Handles 166, 175
Harpoons 63
Herbivores 18, 19, 48, 73, 195, 279, 327, 362
Herders 68, 70, 72, 73, 78, 187, 221, 231, 340
Herding 26, 68–70, 72, 73, 75, 78, 79, 117, 122–126, 136, 221
Hermann von Wissmann 198
Highlands 1, 3, 5, 21, 50, 64, 74–76, 78, 84, 127–129, 132, 133, 138, 173, 187, 189, 197, 212, 213, 219, 224, 268, 334, 335
High plateau 9
Hinterland 140, 142, 143, 146, 164, 170, 176
Historical research 132
The Holocene 51, 60–64, 67–69, 75, 81, 115, 130, 141, 326
Hominid evolution 53
Hominin 53–55
Honey pot 261
House construction 124
Human activity 45, 48, 56, 82, 313
Human and livestock disease 73
Human-environmental interaction 50, 320
Human impact 24, 45, 46, 82, 319, 344
Human induced erosion 186, 188
Human pressure 44
Humans 11, 15, 23, 24, 40, 41, 44–50, 52–56, 63, 66, 73, 75, 79, 80, 82, 84, 117, 119, 121, 126, 127, 129, 132, 138, 183, 188, 205–207, 229, 235, 252, 258, 260, 265, 266, 271, 274, 280, 283, 286, 290, 311, 312, 316, 321–325, 329–331, 333, 336, 337, 343, 349, 354, 358, 359, 361, 366
Human wildlife conflict 280
Humidity 57, 117
Hunter-gatherer 66, 71, 79
Hunting 65, 70, 78, 126, 140, 167, 179, 188, 190, 225, 228, 235, 236, 289, 327, 349
Hunting parties 236
Hunting reserves 356
Hurricane 183
Hydroclimatic changes 42
Hydro-electric dams 260
Hydrological budget 42, 62
Hydrological change 130

I

Ibo 182
Ice Age 48
Ice sheet melting 40
Illegal slave trade 183
Iloikop Wars 137
Immigration 77, 146
Imported goods 144, 170
Incense 164
Increasing global temperatures 40

Index 76, 324
India 141, 142, 147, 163, 164, 171, 172, 176, 184, 211, 214, 311
Indian Ocean 3–5, 7, 8, 11, 50, 52, 60, 73, 79, 127, 138, 139, 143, 144, 147, 170, 180, 218
Indian Ocean dipole (IOD) 8
Indian Ocean marine sediments 41
Industrial Revolution 129, 166, 180
Infrastructure 131, 191, 217, 258, 288, 291, 293, 294, 296, 333, 344, 353, 366
Insolation 6
Interactions 3, 15, 17, 18, 39–41, 45, 48–50, 53, 55–57, 71, 80, 138, 189, 194, 197, 210, 216, 219, 223, 260, 289, 311–313, 315, 319, 321, 322, 326, 333, 343, 345, 361, 366
Inter-community raiding 187
Interglacial periods 52, 54, 55
Interior 132, 142–144, 166, 167, 169, 170, 172, 173, 178, 179, 183, 185, 186, 189, 190, 210, 291
Internal conflicts 216, 217, 254
International scramble for control 147
Internet Data Centre 293
Inter-Tropical Convergence Zone (ITCZ) 4, 5, 8
Invasive alien plant species 273
IPCC 40, 322, 324, 340
Iron 12, 77, 82, 83, 121, 124, 126–128, 140, 146, 185, 189, 266
Iron Age package 77
Iron smelting 82, 83, 128
Iron technology 77
Irrigation 39, 127, 128, 130, 132–134, 137, 138, 171, 187, 259, 339, 353, 354, 364
Irrigation farming 126
Irrigation schemes 11, 134, 276
Islam 142, 146
Isolated massifs 6
Isotopic character 67
Ivory 57, 134, 139–141, 143, 146, 147, 163–180, 183, 184, 186, 188–190, 192, 196–198, 216, 288–290, 313, 328, 362
Ivory carving industry 166
Ivory fans 175
Ivory-handled umbrellas 175
Ivory objects 123
Ivory snuffboxes 175
Ivoryton 166, 177

J

Jacaranda 231
Jeevanjee Gardens 250
Johnson, Henry 215, 217
Johnston, H.H. 125, 126, 217
Jointed jumping dolls 170

K

The Kabaka 208, 216, 217, 245, 251
Kafu River 174
The Kalenjin 26, 136
Kamba 189, 190, 246
Kamurasi, King of Bunyoro 173
The Kanga 147
Kansyore pottery 64
Kaolinite mining 124
The Karamojong 26

Karamojong pastoralists 196, 289
Kenya 1, 2, 5, 11, 17, 18, 24–26, 45, 48, 50, 52, 54, 64, 67, 70–72
Kenya African National Union (KANU) 248
Kenya Africa Union 246
Kenya Forests Act 250
Kenya kamba ancestors 189
Kenya Land Freedom Army 246–249
Kenya Rail 295
Kenya SGR 292, 294, 296
Kenyatta, Jomo 245, 247–249, 277, 296
Kericho 212, 224
Khedive Ismail Pasha 215
The Kikuyu 25, 166, 206, 223, 226, 248, 271
Kikuyuland 166
Kikuyu Red Clays 12
Kilimanjaro 1, 3, 8, 9, 22, 23, 67, 76, 81, 127, 128, 132, 189, 261, 269, 277, 336, 353
Kilombero Growth Corridor 293
Kilwa 140, 142, 144, 146, 169, 170, 180–182
Kilwa Kisiwani 182
Kinships 189, 190
Kipepeo (Butterfly) project 262
Kipini 171
Kisima ceramic 126
The Kisite–Mpunguti MNP 287
Kismayu 146, 171
Kiswahili 2, 147
Kitchenware 140
Kiu 212
Kruger, Eugen 229
Kuwasenkoko Swamp 63

L

Lacustrine 41, 54
Lagoons 18
Laikipia plateau 48, 50
Lake Albert 122
Lake Baringo 118, 126, 133, 186
Lake Bogoria 67, 117
Lake Challa 51, 76, 118
Lake Duluti 118, 121
Lake Emakat 115, 120, 121, 132
Lake Eyasi 25, 64, 71
Lake Kivu 67, 68
Lake Manyara National Park 279, 289
Lake Masoko 45, 82, 115
Lake Naivasha 50, 75, 76, 115, 118, 129
Lake Nyasa 169, 185, 209
Lake Rukwa 67, 119
Lake Simbi 117–120
Lake Tanganyika 11, 45, 62, 67, 75, 76, 115, 117, 119, 169, 170, 185, 188, 209
Lake Turkana 12, 62, 68–71, 75, 81, 83, 115, 137, 173, 189
Lake Victoria 1, 4, 5, 11, 13, 14, 21, 25, 52, 62, 64, 71, 73, 78, 80, 115, 119, 121, 124, 169, 210, 213, 214, 224, 333
Lamu 18, 63, 140, 142, 146, 171, 288, 291, 294
Land Act No. 4 264
Land clearance 82, 133, 138, 224
Land-cover change 39, 44, 146, 311, 315, 324, 326, 332, 350, 353
Land degradation 219, 223, 254, 271, 274, 275, 334, 337, 346, 355
Landsat 332

Landscape management 23
Land security 272
Land tenure 218, 219, 224, 254, 259, 264, 334
Land use systems 2
Langata barracks 247
Last glacial maximum (LGM) 43, 56, 57, 60, 62
Late-Holocene 42
Late Stone Age (LSA) 63–65, 69
League of Nations 218, 230
Leishmaniasis 205
Leopard orchid 356
Liberalisation of latent heat 7
Lieutenant-Colonel Martyr 174
Lindblom, G. 186, 189
Linguistic 49, 53, 71, 78, 121, 124, 127
Lira 174
Lithology 13, 14
Livelihoods 3, 24–26, 126, 186, 225, 229, 233, 246, 250, 258, 259, 264, 265, 268, 271, 273, 275, 276, 280, 282, 285, 287, 288, 296, 318, 319, 326, 330, 334, 338, 340, 341, 343, 344, 347, 352, 353, 355–359, 361, 363, 365
Livestock 26, 49, 68, 70–73, 75, 117, 123, 125–127, 134, 135, 137, 189, 196, 219, 221, 222, 269, 270, 275, 277, 284, 286, 338–340, 353, 358, 364
Livingstone, David 20, 67, 177, 185, 186, 190, 207, 209, 217
Lixisols 14
Logging 8, 232, 261, 263
Loita Hills 268
Loitokitok-Sultan Hamud pipeline 271
The London agreement 210
Long rains, Masika 5
Looking glasses 170
Lord's Resistance Army (LRA) 290
Love potions 176
Lowlands 1, 9, 11, 21, 57, 64, 126, 127, 138, 197, 212, 215, 219, 223, 258, 268, 340, 341, 353, 361
LPJ-Guess 328
Luguard, Frederick 212
Lukenya Hill 64
Lumber 235
The Luo 25, 125, 126
Luo pottery 125

M

Maa 25
The Maasai 26, 127, 134, 136–138, 187, 189, 190, 196, 205, 206, 212, 213, 220–223, 228, 236, 246, 276, 279, 282
Maasai Mara 236, 279
Maa speaking pastoralists 220
Maathai, Wangarii 249
Machakos 189
Macro and mesoscale drivers 8
Madaraka Day 296
Mahoganies 232
Maize 47, 50, 125, 143, 184, 185, 224, 227, 270, 274, 337, 338, 355
The Maji Maji rebellion 211
Major Delme Radcliffe 174
The Makinnion Sclater road 214

Malignant catarrhal fever (MCF) 135
Malindi 142, 171, 184
Man and Biosphere Reserve 279
Man-eating lions of Tsavo 214
Mangrove 17, 18
Mangrove forests 17, 63, 146, 164, 228
Mangrove poles 140, 146, 163, 173
Manuring systems 187
Mara River 11, 352
Marco Polo 164
Marginal farms 271
Marine 78, 140, 286
Marine sediments 43
Maritime Silk Road (MSR) 291, 293, 294
Maritime trade 138, 139, 311
Markets 142, 165, 169, 171–174, 176–178, 183, 184, 187, 188, 191, 196, 208, 221, 222, 225, 227, 228, 232, 255, 258, 276, 290, 292, 296, 331, 336–338, 347, 351, 362
Market towns 146
Marriage alliances 135
Mass migration 132, 196
Masters 172
Mathematical instruments 175
Matoke (plantains) 185, 224
Mau Forest 11
Mau Highlands 11, 224, 247
Mau Mau 246–248
Mauritius 180
Mbogo Noho 217
Medicines 176, 206, 228, 346
Megalithic cemeteries 70
Menouthias Island 164
Merchant classes 182
The Meru 25
Mesic highlands 6
Mesozoic 9
Metallurgy 77, 79, 80
Metal technologies 68, 77
Metamorphic rock 14
Microfaunal communities 49, 73
Microfossils 50
Microphyllous 22
The Middle East 138, 163
Middle Stone Age (MSA) 55
Migori County 125, 126
Migration 5, 25, 49, 53, 58, 77, 119, 129, 143, 174, 222, 294, 311
Migration corridors 294, 357
Militarization of conservation 264, 290
Military expansion 124
Millet 47, 125, 134, 173, 224, 337, 355
Mines 231
Miombo woodland 224, 225, 259
Mires 41, 43, 82
Missionary 166, 170, 209
The Mkomazi Game Reserve 280
Mobile herding 68
Modern farmers 259
Modern humans – homo sapiens 53
Mohammed, M.U. 48, 115, 183
Moi, Daniel Arap 249, 250
Moist air mass 5
Mombasa 119, 140, 142, 144, 146, 147, 169, 171, 173, 184, 189, 213–215, 217, 231, 287, 288, 291–293
Monitoring the Illegal Killing of Elephants (MIKE) vi, 290
Monocrops 335

Monsieur Morice 181
Monsoonal rainfall 5
Monsoon winds 138, 141, 164
Montane forest 17, 20–23, 57, 67, 261
Montane vegetation 20, 23
Monterey cypress 231
Mortality 123, 177, 205, 286
Mortuary traditions 72
Mosaics 24, 50, 56, 60, 77, 121, 143, 186, 324, 331, 340
Moshi 169
Mosques 142, 144
Mosquito 234
Mountain base sites 77
Mountain chains 6, 9
Mountain ridge sites 77
Mountains 1, 3, 6, 7, 9, 14, 22, 25, 82, 84, 127, 128, 132, 186, 207, 224, 229, 260, 261, 279, 340, 352
Mount Elgon 21–23, 67, 267
Mount Kenya 1, 4, 9, 11, 21–23, 67, 75, 118, 189, 190, 247, 259
Mount Marsabit 4, 7, 83, 266, 340
Mount Meru 127
Mount Shengena 45
Movement of people 146
Mozambique 17, 138, 181, 182
Muchoya swamp 45, 47, 63
Mulanje cedar 231
Mulching 269, 336
Multi-disciplinary research teams 320
Multi-economic 77
Multi-ethnic 77
Multiple Use zones 262
Munsa 47, 48, 50, 118, 123
Murchison Falls 174
Musical instruments 166, 176
Muslims 216
Musoma Bay 11
Mvule 232
The Mwakisenge famine 196
Mwanga II 216, 217
Myrrh 164

N

Nairobi vi, 213, 215, 224, 247, 250, 292–294, 296
Nairobi National Park 236, 294, 295
Namwasa 267
The Nandi people 214
Nanuki 255, 344
Napoleonic Wars 182
National Forest Policy 233, 264
National parks 23, 235–237, 261, 266, 271, 279, 280, 282–284, 287, 289, 356, 357, 360
National Parks Ordinance 235
Native Authority Forest Reserves 231
Native councils 215
Native Lands Trust Ordinance 228
Natural resources 44, 227, 229, 230, 250, 257, 258, 263, 266, 267, 269, 275, 277, 286, 323, 336, 343, 344, 346, 347, 350, 356, 359, 364–366
Naval and cavalry sabres 170
Navy 217
Nderit ceramics 70
Neem tree 274
New World crops 143
New York 172, 208
Ngorogoro Conservation Area (NCA) 279

The Ngorongoro crater 278
The Nile 25, 137, 209, 216, 217
Nile River 1
Niolotes 24
Nival zone 22
Non-deposition or post-depositional erosion 46
Non-governmental organisations 266, 317, 363
Non-precipitating mists 6
Non-timber forest products (NTFP) 264, 283
North America 166, 171, 172, 183, 185, 311
Northeast trade winds 5
North Rangeland Trust 285
Norwegian International Climate and Forestry Initiative (NICFI) 267
Nyagah, Jeremiah 250
Nyerere, Julius Mwalimu 245, 252, 254, 277

O

Obsidian artefacts 83
Occult precipitation 7, 8, 22, 353
Ogiek 351
Oil-palm 78
Omani Arabs 182
Operation Anvil 247
Origins 9, 26, 50, 52–54, 68, 77, 129, 132, 147, 224, 344, 356, 357
Oromo-speaking Warra Daaya 71
Out of Africa migration 53, 56

P

Pacific Ocean 8
Palaeoecological archives 46
Palaeoecology 313, 352
Paleoenvironmental insights 50
Paleolakes Drilling Project 54
Palm wine 189
Palustrine 41
Pangani 169, 173
Pangani Valley 119, 140, 143, 187
Panga Ya Saidi (PYS) 56
Paper-cutters 175
Parasols 170
The Pare 25, 50, 82
Park boundaries 280, 281, 288, 344, 358
Parliament 215, 218, 257
Participatory forest management (PFM) 263, 264, 266, 268
Participatory scenarios 329–331
Pastoral Iron Age (PIA) 74, 77
Pastoralism 24, 68, 70–74, 81, 122, 135, 138, 187, 221, 223, 339, 343
Pastoral Neolithic (PN) 52, 68, 70, 72–74, 83, 125
Pastoral societies 137, 190, 212, 255, 318, 364
Paul Emil von Letto-Vorbeck 218
Payment for Ecosystem Services (PES) 347–349
Pearl divers 184
Pearl millet 50, 77, 79, 132, 140
Peas 126
Peat deposits 43
Peat geoarchives 41
Peatlands 41, 48
Pemba 51, 140, 144, 164, 173, 184
Perfumes 140

Persian Gulf 139, 140, 147, 173
Personal and social identity 77
Pests 269, 270, 274, 334
Peters, Carl 210
Phaeozems 14
Phosphate-fixation capacity 12
Photosynthesis 328
Physical systems 55
Physiological drought 22, 58
Pianos 177
Picture books 170
Pinaceae 274
Plains oasis sites 77
Planosols 13
Plantains 185, 224
Plant-animal interactions 40
Plantations 123, 124, 129, 173, 180, 182, 183, 185–187, 209, 219, 224, 231, 233, 235, 259, 349
Plantation slavery 180
Plant cultivation 26
Plant genera 17
Pleistocene 42, 56–58, 63
Pleuropneumonia 196
Plio-Pleistocene 53, 54
Poaching 193, 198, 279, 280, 283, 285, 286, 289, 290, 362
The Pokot 133, 138
Polar ice caps 55
Polio 187
Political 2, 18, 24, 74, 82, 123, 129, 147, 163, 167, 184, 188, 189, 191, 210, 214, 234, 245, 249, 250, 255, 256, 258, 268, 276, 287, 288, 346, 347, 355, 357, 362, 365
Political complexity 117, 123, 124
Political figures 144
Political unrest 129
Pollen 44–48, 62, 67, 82, 115, 117, 123, 315, 318, 320
Pollution 260, 335, 352
Population growth 146, 223, 271, 280, 331, 333, 343, 344, 346
Porters 167, 170, 173, 174, 183, 185, 188, 190, 209
Port of Malindi 190
Portugal 311
Post colonial 225, 235, 245, 252, 255, 259, 269, 277, 287, 350, 353, 355
Potatoes 185, 224
Pots to peoples 49
Pottery 49, 52, 70, 71, 77, 78, 125–127, 138, 141
Power 5, 39, 123, 172, 188, 189, 207, 209, 210, 214, 216–219, 233, 245, 246, 248, 252, 255
Precambrian crystalline rocks 9
Precipitation 5, 8, 15, 18, 23, 24, 41, 46, 54, 55, 57, 58, 62, 67, 348
Predators 18, 75, 279
Privatisation 234, 275, 276, 355
Professional poison makers 190
Protected areas 18, 23, 179, 229, 236, 261, 272, 279, 280, 282–284, 286, 291, 333, 344, 349, 356–360, 363, 364
Protestants 209, 216, 217
Pumpkin 126

Q

Quaternary 41, 58
Quaternary glaciations 9
Quaternary period 54
Queen Elizabeth Park 193

Queen Victoria 217

R

Radiation balance 7
Radiocarbon 51, 63, 81, 128
Radiocarbon dating 51, 81
Radiocarbon determinations 51, 132, 317
Railway 169, 210, 212–215, 217, 231, 235, 292–294, 333, 363
Rainfall 3–5, 7–9, 12, 22, 46, 57, 58, 67, 72, 76, 118, 119, 130, 275, 334, 340, 352
Rainfall pattern 8
Rainy season 5, 8
Ras Hafun 164
Reaping knives 123
The Red Sea 164
Reducing Emissions from Deforestation and forest Degradation (REDD+) 264, 266, 267, 268, 347, 350, 351
Regionalism 246
Regosols 12
Religions 77, 209
Religious ceremonies 226
Religious figures 144
Reunion 180, 224
Rhapta 164
Rhinoceros horn 146
Rifting process 11, 54
Rift System 1, 9, 11
Rift Valley area 12
Rift valley fever (RVF) 135
Rift valleys 1, 3, 5, 11–13, 52, 64, 71, 72, 74, 75, 117, 119, 133, 137, 220, 249
Rinderpest epidemic 196
Rings 170
The Rio Earth summit 280
Rising levels of atmospheric carbon dioxide 40
Rites of passage 226
Ritual 70, 123
Rock crystals 146
Rock shelter refugia 64
Roman 139, 216
Roman empire 141, 163
Root systems 6, 355
Rotational grazing 275
Ruaha River wetlands 14
Rubber 219
Rufiji river 164
Rufiji river floodplains 14
Rural 141, 145, 254, 257, 259, 334, 344, 348, 359
Ruwenzori 9, 22, 23
Rwanda 63, 67, 78, 84, 282

S

The Sabaki River 190
Sacred groves 226, 234
Sacred Lake 45, 67, 75, 118
Sacrifices 226
Safari tourism 236
Sahara 69, 82, 133
Said ibn Sultan 172, 183
Salem 171, 172
Saline groundwater 14
Saline soils 12
Salinization 335
Salt production 122
The Samburu 26, 196, 236
Savanna grassland 135, 193
Savannas 1
Sawmills 232, 235

Scenarios 326, 352, 365
Scientific and Forestry Department 232
Scissors 170
Sea level 9, 54, 55, 63, 76, 121, 142
Sea level change 63
Seasonal cycle 9
Sea surface temperatures (SST) 8
Sedentarisation 254, 324, 339, 355
Sedimentary deposits 14
Sedimentary rock 14
Sedimentary sequences 44, 45, 62, 347
Sedimentation 62, 120
Sedimentation rates 82
Sediment cores 75, 313
Sediment formation 316
Sediment stratigraphy 41
Semi arid flood plains 11
Senecio forest 22
Senegal 182
Sengwer 351
Serengeti 13, 14, 19, 236, 237, 277, 285, 286, 363
Serengeti National Park 236, 278
Serviette rings 175
Sesame 126, 182
Settlement hierarchy 124, 144
Settlement patterns 72, 124, 133
Settlements 48, 50, 52, 63, 64, 73, 74, 122–126, 128, 132, 134, 137, 138, 141, 146, 166, 184, 186, 188, 191, 197, 212, 223, 229, 235, 253, 254, 361
The Seven Years War 182
Seychelles 180
Seyyid Said 173
Shallow crater lakes 118
Sheep 26, 125, 221, 286, 324
Shellfish 78
Shells 171
She-oak 231
Shimba Hills 17, 179, 266
Shimba hills national reserve 18
Short rains, Vuli 5
Silk Road Economic Belt (SREB) 291
Silted dams 223
Silver 141, 164
Sir Elliot, Charles 215
Sirikwa holes/hollows 74
Sisal plantations 213, 219, 223
Skins 26, 146
Slash and burn agriculture 273
Slaughter 123
Slave raiding 186, 190
Slaves 140, 143, 146, 147, 163, 165, 173, 179–189, 197, 207, 216, 246
Sleeping sickness 190, 194
Sludge 336
Smallholders 219, 224, 225, 259, 268, 270, 274, 275, 337
Smallpox 205
Snuff bottles 189
Social-ecological history 346
Social fragmentation 206
The Society for German Colonisation 210
Socio-economic change 24
Sodic Solonetz 12
Softwood plantations 259
Soil 12, 17, 73, 130, 186, 194, 215, 219, 229, 249, 250, 254, 258, 268, 269, 271, 272, 320, 324, 326, 335, 336, 338, 339, 341, 343, 347, 349, 354–356

Soil erosion 44, 83, 127, 128, 132, 223, 234, 258, 259, 261, 271, 346, 349, 354–356
Soil exhaustion 258, 355
Soil types 12
Solonchacks 12
Solonetz 12, 14
Somalia 17, 138, 249, 255
Sorghum 47, 50, 77–79, 125–127, 132–134, 140, 173, 185, 270, 274, 337, 338, 355
South Arabia 173
Southeast trade winds 5
South Pare Mountains 187
Spatial disconnection 80
Spatial variability 324
Special Economic Zone 293
Species composition 18, 19, 192, 194, 232
Species distribution 23, 58, 326
Speckled maize 126
Speke, John Hanning 169, 183, 207–209
Speleothems 43
Spices 140, 184
Spiny shrubs 274
Spiritual activities 228
Stable isotope analyses 49
Standard Gauge Railway (SGR) 214, 292–294, 296
Stanislaus Mugwanya 217
Stanley, Henry M. 208
State Forestry and Beekeeping Division 261
Statuettes 175
Steamer 171, 210, 231
Stem borers 270, 274
Stone rings 127
Stone terracing 126
Stone tools 49, 52, 56, 69
Stone towns 143, 144, 184
Storage pits 123
Striga weed infestation 270
Subalpine arborescent Ericaceous 22
Sub-Saharan Africa 70, 163, 318, 350
Subsistence 26, 50, 63–66, 68, 70–72, 78, 79, 125, 129, 133, 137, 144, 187, 190, 196, 197, 219, 221, 225, 264, 340
Subtropical high-pressure cells 5
Sugar 186
The Sukuma 25
The Sultan 173
Sustainability 276, 336, 345, 346, 366
Sustainable development agenda 357
Sustainable Development Goals 332, 366
Swahili 25, 138, 140–144, 146, 147, 170, 182, 189, 228, 315, 330
Swamp forests 130
Swamps 41, 48, 50, 62, 117, 118, 234, 353, 354
Sweet potatoes 126, 185
Syphilis 187

T

Tabora 169, 187
Tana River 11
Tanga 169, 173
Tanga Coral Gardens Marine Reserve 287
Tanganyika African National Union (TANU) 254
Tanganyika Basin 81
Tanganyika Territory 218

Tanzania v, vi, 1, 2, 11, 13, 14, 21, 24, 25, 45, 50, 64, 67, 70–74, 78, 82, 84, 118–121, 129, 132, 133, 137, 140, 141, 143, 169, 173, 189, 190, 197, 206, 211, 218, 223–225, 227–229, 232, 236, 237, 245, 246, 251, 252, 254, 257, 259–261, 264, 267–270, 276, 277, 279, 280, 286, 287, 290, 293, 294, 314, 354
Tanzanian Craton 12
Taro 50, 143
Taxation 173, 217
Tea v, 12, 213, 219, 224
Teak 231
Technological development 129
Technologies 65, 77, 78, 80, 83, 141, 146, 163, 179, 191, 206, 258, 290, 311, 336–338, 346, 356
Tectonic activity 9, 11
Tectonically modified topography 12
Terrace 130, 132, 133, 269
Terrestrial depositional systems 41
Terrestrial ecosystem 81, 326
Thermodynamics 7, 322
The Third World Parks Congress 280, 358
Thorn brush fences 265
Threatened species 17
Ticks 24
Timber 140, 143, 163, 173, 228, 231–235, 258, 259, 261, 263, 266, 269, 290, 347, 349
Tippu Tip 173
Tobacco 126, 143, 184, 185, 224
Tomatoes 143
Tools 24, 45, 48, 80, 124, 195, 265, 279, 284, 319, 320, 322, 328, 330, 348, 350, 351, 353, 358, 359
Topography 3, 6, 7, 9, 12, 14, 15, 17
Topsoil 12, 14
Toro 215, 217, 225
Tortoise shell 146
Tourism 232, 235, 261, 277, 282–284, 287, 288, 291, 294, 295, 356, 359, 364
Trade 5, 121, 127, 138–144, 146, 147, 163–174, 176, 178–180, 182–192, 196–198, 206, 207, 209, 216, 221, 222, 235, 236, 245, 266, 288–291, 313, 315, 328, 337, 338, 350, 358, 362, 364
Trade barriers 338
Trading hubs 188
Trading posts 184
Trading routes 143, 188, 210
Trading ships 185
Traditional knowledge 337, 346
Traditional migration routes 265
Traditions 24, 25, 43, 63, 74, 129, 138, 176, 189, 277
Transhumance 212, 219–221, 225, 338
Transoceanic biological transfers 143
Transport 130, 164, 166, 183, 185, 190, 214, 217, 263, 291, 294, 296, 344
Treaty 171, 172, 181, 183
Tree rings 41
Tribal chieftans 164
Tribe 24, 26, 136, 207, 220
Tropical Africa 81

Tropical cyclones 4
Trypanosomiasis 190
Tsavo 19, 146, 169, 190, 196, 236, 271
Tsavo National Park 294
Tsetse fly 73, 179, 190, 194
Turkana Basin 63, 64, 69–71
Turkana Boy (homo ergaster) 54
Turkana channel jet 7
Turkana jet stream 9
The Twa 25

U

Udzungwa 23
Udzungwa National Park 286
Uganda v, 1, 2, 5, 12, 24–26, 45, 47, 48, 62–64, 76, 78, 82, 84, 117, 118, 120, 122, 123, 133, 136, 138, 169, 171, 174, 185, 193, 196, 198, 210, 212, 214, 215, 217, 220, 224, 225, 228, 232–234, 237, 245, 246, 251, 252, 254, 259, 267, 269, 270, 276, 277, 283, 294, 315, 336, 340
Uganda National Congress 251
The Uganda Protectorate 217, 218
Ugandan Constitutional Conference 251
Ugogo 185, 186
Uhuru Park 248, 250
Ujamaa 252, 254, 259
Ujiji 169, 187, 188
Uluguru violet backed sunbird 17
Undulating topography 14
UNESCO World Heritage Site 278
Ungulates 19, 279, 286
Ungwana Bay 11
United National Framework Convention on Climate Change (UNFCCC) 264, 266, 322, 350
United Nations (UN) 218, 250
Unpalatable grass 24
Unyamwezi 170
Unyanyembe 169
Uplifted geology 9
Uprisings 211, 217, 246, 248
Urambo scheme 254
Urban 18, 143, 288, 344
Urban centres 44, 143, 147, 342
Urban communities 185
Urbanization 233, 287, 343, 344
Urewe ceramics 60, 77
Usambara forest gecko 17
Usambara short horned chameleon 17

V

Vasco da Gama 147, 184
Vegetable ivory 176
Vegetable oil 182
Vegetation 6, 16, 17, 19–23, 41, 44, 45, 47, 50, 57, 58, 66, 67, 73, 74, 80–82, 117, 121, 132, 186, 193, 286, 294, 324, 340, 344, 354, 362
Vegetation change 58, 80, 81, 324
Vegetation distribution 15
Vegetation type 46
Victorian 175
Victoriana 166
Victorian British explorers 207
Village Land Act 264
Village nucleus 254
Villagisation 247, 248, 253

Violence 187, 254, 256, 257
Voi River Super Bridge 294
Volcanoes 12
Volcanoes National Park 282

W

Wage labour 338
Walker circulation 8
Warehouse 184
Warfare 26, 179, 247
War for biodiversity 351
Warrior 134, 137
WaTaita 190
Watamu 142, 287
Water harvesting 269
Water pollution 273, 355
Water source 126, 354
Water towers 6, 250, 341, 353, 361
Weapons 164
The Wehehe rebellion 211
West Atlantic slave trade 181
Western Indian Ocean 14, 184, 245, 246
Western Nilotic language 125
Wetlands 117, 344, 353, 354
White Highlands Policy 211
The White Nile 174
Wildebeest 135
Wildebeest migration 278
Wild game 126
Wild herbivores 272, 281
Wildlife 73, 74, 178, 190, 191, 205, 207, 237, 238, 264–266, 276, 277, 279–284, 286, 291, 294–296, 339, 351, 353, 358, 359, 363, 364, 366
Wildlife Conservation and Management bill 264
Wildlife population monitoring 279
Wildlife reserves 223, 236, 290
Winam Gulf 125, 126
Wine 140
Wooded savannah 17, 120
Wood harvesting 39
Woodland 19, 20, 121, 179, 186, 190, 193, 194, 228, 235, 327, 328, 336
Woody plant growth 18
Woody plants 120
World Bank's Forest Carbon Partnership Facility (FCPF) 267
World War One 218
World War Two 235
Wound beads 141, 142

Y

Yala Swamp 125
Yams 78
Yoweri Museveni 252, 254

Z

Zakaria Kizito 217
Zanzibar 51, 56, 63, 138, 140, 144, 146, 164, 171–173, 176–178, 180–184, 186, 198, 208–210, 218, 245, 246, 248, 288
Zebu cows 137
Zigua communities 187
Zinj 164
Zoonotic diseases 68

The manufacturer's authorised representative in the EU is Springer Nature Customer Service Centre GmbH, Europaplatz 3, 69115 Heidelberg, Germany. If you have any concerns regarding our products, please contact ProductSafety@springernature.com

Printed and bound by CPI Group (UK) Ltd, Croydon, CR0 4YY
30/09/2026
02225456-0013